Probability and Finance Theory

Probability and Finance Theory

Kian Guan Lim

Singapore Management University, Singapore

NEW JERSEY • LONDON • SINGAPORE • BEIJING • SHANGHAI • HONG KONG • TAIPEI • CHENNAI

Published by

World Scientific Publishing Co. Pte. Ltd.
5 Toh Tuck Link, Singapore 596224
USA office: 27 Warren Street, Suite 401-402, Hackensack, NJ 07601
UK office: 57 Shelton Street, Covent Garden, London WC2H 9HE

Library of Congress Cataloging-in-Publication Data
Lim, Kian Guan.
Probability and finance theory / by Kian Guan Lim.
p. cm.
Includes index.
ISBN-13: 978-981-4307-93-2
ISBN-10: 981-4307-93-9
1. Finance--Mathematical models. 2. Probabilities. I. Title.
HG106.L558 2011
332.01'5192--dc22
2011015845

British Library Cataloguing-in-Publication Data
A catalogue record for this book is available from the British Library.

Typeset by Stallion Press
Email: enquiries@stallionpress.com

Printed in Singapore.

To *my parents Teck Hock and Lay Chop*

PREFACE

I like to quote from two brilliant mathematicians:

> "For since the fabric of the universe is most perfect and the work of a most wise Creator, nothing at all takes place in the universe in which some rule of maximum or minimum does not appear."
>
> – Leonhard Euler, 1707–1783

To me, this is the foundation of mathematical analysis, upon which even the fragilities of the financial marketplace cannot escape analytic scrutiny. One key analytical tool is probability theory.

> "We see that the theory of probability is at the bottom only common sense reduced to calculation; it makes us appreciate with exactitude what reasonable minds feel by a sort of instinct, often without being able to account for it... It is remarkable that this science, which originated in the consideration of games of chance, should have become the most important object of human knowledge... The most important questions of life are, for the most part, really problems of probability."
>
> – Pierre Simon Lapace, 1749–1827

This book is an introduction to the mathematical analysis of probability theory and provides some understanding of how probability is used to model random phenomena of uncertainty, specifically in the context of finance theory and applications. Measure-theoretic probability, distribution theory, the limit laws, and continuous-time stochastic processes are covered in basic terms. Applications are made in modeling state securities under market equilibrium. The martingale is studied, leading to a consideration of equivalent martingale measures, the fundamental theorems of asset pricing, the change of numéraire and discounting, risk-adjusted and forward-neutral measures, and hedging using contingent claims, Markovian models, and the existence of martingale measures preserving the Markov property. Stochastic calculus and multi-period financial asset pricing models leading to no-arbitrage pricing of contingent claims are also studied. The theory of risk aversion and utility is examined, leading to ideas of risk premia. We also look at the theory of Markov chains and its appropriate applications in credit risk modeling. Other application topics include optimal consumption and investment problems and interest rate modeling.

The integrated coverage of both basic probability theory and finance theory makes this book useful reading for advanced undergraduate students or for first-year postgraduate students in a quantitative finance course. The reader will hopefully be equipped to understand key probability results and how probability can be effectively used to model random financial phenomena. (Modern) finance theory since the time of Markowitz's portfolio theory and Modigliani-Miller's de facto no-arbitrage theorem of irrelevance in financing alternatives half a century ago has become a mature discipline in its own right though it is intellectually indebted to neo-classical economics. For that matter, it is sometimes apt to refer to the finance discipline as financial economics. In this book, one will also find that many finance-theoretic concepts are connected to economic ones.

The book, supplemented with the referenced articles or other resource materials listed in footnotes, can be used for either a one- or two-semester course depending on the speed of the student's progress. The exposition in the book is mathematical in style for the most part, although formalism is kept at a lower level compared to that of a book assigned in a mathematics degree course. Therefore, theorems will have attached proofs only if they are necessary for understanding the material within the intended scope of this book. Application examples are provided throughout the chapters to illustrate the relevance to finance. At the end of each chapter is a little problem set for dessert. Suggested answers are provided in the Appendix. Any future updates or any post-publication errata will be available at http://www.mysmu.edu/faculty/kglim.

Back in 1996 through 2001, my growing interest in quantitative finance and in constructive (rather than destructive) financial engineering was motivated by the enormous sophistication and accelerating development of financial markets in the 1990s, and the need to organize concerted and research-based advanced training in this field in Singapore. I have been privileged to receive encouragement and assistance from various distinguished international scholars, including Professors Andrew Chen of Southern Methodist University, Bruce Grundy of Melbourne Business School, Hang Chang Chieh of NUS (who was also our NUS Deputy Vice-Chancellor at that time), Krishna Ramaswamy of Wharton Business School, and Suresh Sundaresan of Columbia Business School, among many other colleagues. Supervising and interacting with the pioneer batch of Masters in Financial Engineering at NUS was a most enjoyable and memorable experience for the great quality of minds and spirited camaraderie.

In my view, the global financial crisis that erupted in 2008 and the main street disdain of complex derivatives that ensued could have been abated if finance professionals and regulators worldwide had been more adequately trained in understanding the risks involved, as well as the financial economics behind it. I mentioned "constructive" versus "destructive" financial engineering. An analogy might be the use of rockets launching navigation satellites versus rockets carrying warheads. When financial ideas were engineered constructively as in producing farm commodity futures, they brought about improved farming productivity and welfare for both the farming communities and consumers in society. There have been many other successful examples such as the wide utilization of currency options by exporters and importers for minimizing exposures to currency risks, and the use of interest rate swaps to reduce a firm's funding costs. However, unnecessary overexposure to these same instruments constitute speculations that will entail severe losses that are as probable as huge gains. A major problem is the asymmetric outcome of this gamble. For investment institutions that are deemed too big to fail, or those listed investment companies whose shareholders do not get to vote effectively on executive payouts and corporate governance, it is heads their corporate leaders or senior executives win (and win big) and tails the public taxpayers or the small shareholders lose (and lose big time).

When market avarice got out of hand, the financial markets turned into a universal casino gambling on receiving incredibly high yields from collaterized debt obligations (CDOs) by some, and colossal downside bets by others. We are witnesses to the 2008 global financial meltdown that followed such mindless excesses. Without collaterization and transparency, financial engineering can produce such derivatives that "... are financial weapons of mass destruction," as Warren Buffett is famously quoted.

Today, the need for more education, advanced training, and appropriate applications in this quantitative finance field remains. Similarly, more effective — not simply more — market regulations and governance for public interest and positive externalities, more transparencies and information disclosures for effective shareholder votes, and a more responsible and risk-minded global banking structure must be implemented, and soon.

Finally, but not least, I wish to thank Agnes Ng and Gan Jhia Huei for their diligent editorial assistance in the publication process.

Kian Guan Lim
Singapore, October 2010

ABOUT THE AUTHOR

Kian Guan Lim received his doctorate from Stanford University in 1986 and works in the field of risk management and financial asset pricing. He is Professor of Quantitative Finance in the Business School at Singapore Management University (SMU) and adjunct Professor in the Mathematics Department of the National University of Singapore (NUS). Prior to joining SMU, Kian Guan was at NUS and founded the University Center for Financial Engineering. He also started the Master of Science program in Financial Engineering. He has been a consultant for several banks in risk validation and valuation. He was a reservist captain in the Singapore Armed Forces and had served in administrative positions at SMU and NUS over many years.

CONTENTS

Chapter 1

PROBABILITY DISTRIBUTIONS

Probability is everywhere and is an important part of our lives. Most people make decisions based on the likelihood or chance, *de facto* probability, of one outcome versus others, that best enhances their well-being. For a more exact analytical modeling of financial problems and issues, a rigorous framework is required for studying this phenomenon of chance.

1.1. Basic Probability Concepts

A sample space Ω is the set of all possible simple outcomes of an experiment where each simple outcome or sample point ω_j is uniquely different from another, and each simple outcome is not a set containing other simple outcomes, i.e., not $\{\omega_1, \omega_2\}$, for example. An experiment could be anything happening with uncertainty in the outcomes, such as throwing of a dice in which case the sample space is $\Omega = \{1, 2, 3, 4, 5, 6\}$, and $\omega_i = j$ for $j = 1$ or 2 or 3 or 4 or 5 or 6, or a more complicated case of investing in a portfolio of N stocks, in which case the sample space could be $\Omega = \{(R_1^1, \ldots, R_1^N), (R_2^1, \ldots, R_2^N), \ldots\ldots\}$ where R_j^k denotes the return rate of the kth stock under the jth outcome. Another possible sample space could be $\Omega = \{R_1^P, R_2^P, R_3^P, \ldots\ldots\}$, where $R_j^P = \frac{1}{N}\sum_{k=1}^{N} R_j^k$ denotes the return rate of the equal-weighted portfolio P under the jth outcome.

Each simple outcome or sample point ω_j is also called an "atom" or "elementary event". A more complicated outcome involving more than a sample point, such as $\{2, 4, 6\}$ or "even numbers in a dice throw" which is a subset of Ω is called an event. Technically, a sample point $\{2\}$ is also an event. Therefore, we shall use "events" as descriptions of outcomes, which may include the cases of sample points as events themselves.

As another example, in a simultaneous throw of two dices, the sample space consists of 36 sample points in the form $(i, j) \in \Omega$ where each $i, j \in \{1, 2, 3, 4, 5, 6\}$. An event could describe an outcome whereby the sum of the two numbers on the dices is larger than 8, in which case the event is said to happen if any of the following sample points or simple outcomes happen, (3, 6), (4, 5), (4, 6), (5, 4), (5, 5), (5, 6), (6, 3), (6, 4), (6, 5), and (6, 6). This event is described by the set $\{(3,6),(4,5),(4,6),(5,4),(5,5),(5,6),(6,3),(6,4),(6,5),(6,6)\} \subset \Omega$. Another event could be an outcome whereby the sum of the two numbers on the dices is smaller than 4, in which case the event is described by $\{(1,1),(1,2),(2,1)\} \subset \Omega$. Another event could be "either the sum is larger than 10 or smaller than 4" or represented by $\{(5,6),(6,5),(6,6),(1,1),(1,2),(2,1)\} \subset \Omega$. Thus, an event is a subset of the sample space.

An event E is often a relevant and important description of what happens in an experiment, over and above sample points. In the November 2008 US presidential election, US citizens in each of the 50 states plus the District of Columbia voted by majority in each state for either the Democratic Party or the Republican Party. Each of the states plus DC is allocated a fixed number of electoral votes based on the population proportion. (Each electoral vote is rested on an elector, but electors in all but 2 states, Maine and Nebraska, cast their electoral college vote on a winner-takes-all basis from the result of the popular votes by the citizens of each state plus DC.) To become president, a candidate needs 270 electoral votes out of 538.

In this historic experiment, assuming all electors follow the popular vote carried for the state, the sample space is the set of simple outcomes where $\omega_i = ([d_{1i}, r_{1i}], [d_{2i}, r_{2i}], \ldots, [d_{ji}, r_{ji}], \ldots, [d_{51i}, r_{51i}])$, j is the state identity, and d_{ji}, r_{ji} are the electoral votes carried for state j under the Democrats and the Republicans, respectively in simple outcome i. Since each $[d_{ji}, r_{ji}]$ is either $[N_j, 0]$ or $[0, N_j]$ where N_j is the total number of eligible electoral votes for state j, there are altogether 2^{51} combinations or simple outcomes in Ω.

In the election outcome, presidential candidates Barack Obama and John McCain each won 365 and 173 electoral votes, respectively. If an event is described as "Who won?" and the outcome is "Obama won", this event E is as meaningful a description of experimental outcomes as simple ones. How would you describe this event in terms of the simple outcomes? Assuming all electors follow the popular vote carried for the state, this would be $E = \{\omega_a, \omega_b, \omega_c, \ldots\}$ such that the sum of the Democrats' electoral votes in each $\omega_i \in E$, or $\sum_{j=1}^{51} d_{ji}$, adds up to 270 or more. Thus, E is a subset of Ω.

Collection of Events

Suppose there is a sample space $\Omega = \{a_1, a_2, a_3, \ldots, a_6\}$. We may form events E_i as follows

$$E_1 = \{a_1, a_2\}$$
$$E_2 = \{a_3, a_4, a_5, a_6\}$$
$$E_3 = \{a_1, a_2, a_3\}$$
$$E_4 = \{a_4, a_5, a_6\}$$

$\phi = \{\,\}$, or the empty set, and Ω itself are also events. All these events are subsets of Ω.

The set of events is also called a collection of events or a family of events. (It is really a set of subsets of Ω.) We denote the collection by

$$\mathcal{F} = \{\phi, \Omega, E_1, E_2, E_3, E_4\}.$$

If a collection $\mathcal{F}$ satisfies the following 3 properties, it is called an algebra or a field:

(1a) $\Omega \in \mathcal{F}$
(1b) If $E_j \in \mathcal{F}$, then $E_j^c \in \mathcal{F}$
(1c) If $E_i \in \mathcal{F}$ and $E_j \in \mathcal{F}$, then $E_i \bigcup E_j \in \mathcal{F}$

If there is an infinite sequence of $E_n \in \mathcal{F}$, and if $\bigcup_{n=1}^{\infty} E_n \in \mathcal{F}$, then $\mathcal{F}$ is called a σ-algebra or σ-field in Ω.

For association between sets, we need to formalize the idea of mapping. A map is a function. A function $f(x)$ assigns to an element $x \in D(f)$, where $D(f)$ is called the domain set of f, a unique value $y = f(x) \in R(f)$, where $R(f)$ is called the range set of f. It is written $f : D(f) \longrightarrow R(f)$ (set to set) or equivalently, $f : x \mapsto y$ (element to element).

A function $f(x) = y$ is injective if and only if it also has an inverse function $f^{-1}(y) = x$ (unique) for every $x \in D(f)$. A function is surjective (onto) if for every $y \in R(f)$, there exists $x = f^{-1}(y)$. A surjective function needs not be injective as there could be $x \neq x'$ such that $f(x) = f(x') = y$. An injective function needs not be surjective as there may exist $y' \in R(f)$ where $f(x) \neq y'$ for all $x \in D(f)$. A function that is both injective and surjective is called bijective (or one-to-one correspondence).

A probability measure P is a function mapping $\mathcal{F}$ into the unit interval $[0,1]$, $P : \mathcal{F} \longrightarrow [0,1]$, or equivalently $P : E \in \mathcal{F} \mapsto x \in [0,1]$, such that

(2a) $0 \leq P(E) \leq 1$ for $E \in \mathcal{F}$
(2b) $P(\Omega) = 1$
(2c) For any sequence of disjoint events E_n of $\mathcal{F}$

$$P\left(\bigcup_{n=1}^{\infty} E_n\right) = \sum_{n=1}^{\infty} P(E_n).$$

The triple $(\Omega, \mathcal{F}, P)$ is called a probability space.

For finite sample space, the probability measure P poses no problem as each of the finite number of outcomes or sample points can carry a strictly positive probability number or P-measure. For example, if there are N equally likely occurrences in an experiment, each sample point has probability $\frac{1}{N} > 0$, no matter how large N is.

However, in a continuous sample space, such as occurrences on a real line $[A, B] \in \mathcal{R}$, suppose any outcome x which is a point, $A \leq x \leq B$, has probability $q > 0$. We know that a rational number is countably infinite or denumerable, which means the set of rational numbers can be put in a one-to-one correspondence with natural numbers. Real numbers in $[A, B]$ are a lot more in quantity than natural numbers $(1, 2, 3, \ldots)$; they are uncountable or nondenumerable. The total of all probabilities of the outcomes of real numbers is then $\geq \sum_i^{\infty} q = \infty$ which is certainly not finite. Therefore, $P(\Omega) = 1$ is not satisfied.

In the above, we define the probability for all possible subsets of Ω, including a sample point of Ω being given a probability mass of $q > 0$. In other words, we had included all possible subsets, including any sample point, as in $\omega \in \mathcal{F}$. However, this is shown not to be feasible as it violates condition (2b). The way to resolve this is to define probability over a smaller sets that can be constructed as events in the field $\mathcal{F}$ but are points. The latter will include only events that are "measurable", and exclude others such as an arbitrary single sample point which is "not measurable". Technically, it is possible to assign a probability of zero to some "not measurable" sets that can be constructed as events in the field $\mathcal{F}$ but are points in $[A, B]$. If we are able to assign zero measure to all possible events, including sample points $\omega_j \in \mathcal{F}$, that are subsets of an event with zero probability, then the probability space is said to be complete.

Thus, for continuous sample space, the probability measure has to be defined on a collection of measurable subsets of Ω, and not on the entire collection of all subsets of Ω. The latter collection is sometimes expressed as a power set 2^{Ω}. One way of construction is to define the probability measure on half-open sets $(A, B]$, or $A < x \leq B$. In this case, $A < B$. Thus, points are excluded as measurable sets. Moreover, each event as in $(A, B]$ has a non-zero length $B - A > 0$ which can be used as a definition of the measure, no matter how infinitesimally small the measure is. This is also called a Lebesgue measure. As an immediate application of the Lebesgue measure, if an integral is taken over an indicator function f on support $[0, 1]$, where

$$\begin{aligned} f(x) &= 1 \quad if\ x \in [0,1] \text{ is a rational number} \\ &= 0 \quad \text{otherwise} \end{aligned}$$

the Riemann integral cannot be found. However, using the idea that this integral is basically the sum of $1\times$ the total lengths of the rational numbers on $[0, 1]$, since Lebesgue measure of all these total lengths is zero, the Lebesgue integral of the above is zero. Another famous case of measure-zero set is the Cantor set where the elements are infinitely many, but they all add up to zero length. Half-open sets $\omega \subset \mathcal{R}$ and events equal to their intersections or unions that form elements of $\mathcal{F}$ are also called Borel sets.

Random Variables

In a horse race involving 6 horses, simply called A, B, C, D, E, and F, a simple outcome of the race is a 6-tuple or 6-element vector viz. ω_B = (B,D,F,A,C,E) denoting horse B coming in first, D second, and so on. Assuming there is no chance of a tie, and no horse drops off, there are 6! permutations or $6 \times 5 \times 4 \times 3 \times 2 \times 1 = 720$ possible outcomes. However, for most people going to the Sunday derbies to wager, they are more interested in some function of the simple outcomes rather than the simple outcome itself. For example, if they had wagered on horse B for a payout of 3 to 1 for the winner, then the variables relevant to them are the returns to their wager.

There are $5! = 120$ permutations with B as winner or a return rate of 200%. There are $5 \times 5! = 600$ permutations where B is not a winner or a return rate of -100%. The function

$f : \omega_B \mapsto 200\%$, if ω_B is a 6-tuple (B, ...), otherwise -100%, is called a random variable.

Formally, let $(\Omega, \mathcal{F}, P)$ be an arbitrary probability space where $\mathcal{F}$ is a σ-algebra or collection of measurable subsets of Ω, or collection of events. Let X be a real-valued function on Ω; in other words, $X : \Omega \longrightarrow \mathcal{R}$ or $X : \omega \in \Omega \mapsto x \in \mathcal{R}$.

X is a random variable (RV) if it is a measurable function from Ω to $\mathcal{R}$. It is a measurable function if for any Borel set $A \subset \mathcal{R}$, its inverse

$$X^{-1}(A) = \{\omega : X(\omega) \in A\} \in \mathcal{F}.$$

Hence, $X^{-1}(A)$ is seen to be an element of $\mathcal{F}$ or a subset of Ω that is measurable or that can be assigned a suitable probability. (Note that $X^{-1}(A)$ can be $\phi \in \mathcal{F}$.) If not, the RV is not well-defined.

Sometimes, when a RV is defined, it may not be necessary or convenient to refer to the more fundamental sample space Ω from which the RV is derived. In the example above, each horse race permutation may be the simple event $\omega_k \in \Omega$. However, the relevant RV outcome is the return rate $\tilde{r}_k \in \{-100\%, K\%\}$ if the bet is on horse K. Or we can simply take the sample space as $\Omega = \{-100\%, K\%\}$, and directly define (as a short-hand) probability on the RV. In the same way that investment problems are studied by considering the stock i's return $\tilde{r}_i$ as a RV taking values in $[-1, +\infty)$, we can regard the sample space as $\Omega = [-1, +\infty) \subset \mathcal{R}$.

Now, X is a simple RV if it has a finite number of values of $X(\omega) = x$, for each real $x \in \mathcal{R}$ that is finite, and if $\{\omega : X(\omega) = x\} \in \mathcal{F}$. Graphically, this can be depicted in Fig. 1.1 as follows.

An example of a commonly used simple RV is the indicator variable $1_{\omega \in A}$ where this RV takes the value 1 when ω is in set A, and 0 otherwise. The indicator variable is a useful tool in developing analytical solutions to some probability problems. For example, $E(1_{\omega \in A}) = P(A)$. This allows the concepts of expectation and probability distribution to be interconnected. A more general simple RV is $X = \sum_{i=1}^{N} x_i 1_{A_i}$ where X takes value x_i in the event $A_i = \{\omega : X(\omega) = x_i\} \in \mathcal{F}$, and $A_i \cap A_j = \phi$, for $i \neq j$. Simple RVs are easily measurable.

It is usually very easy to prove (and intuitively understand) complicated interchange of limits and expectations (integrals) using simple RVs. For example, suppose $X_n(\omega)$ is a sequence of simple RVs that are uniformly

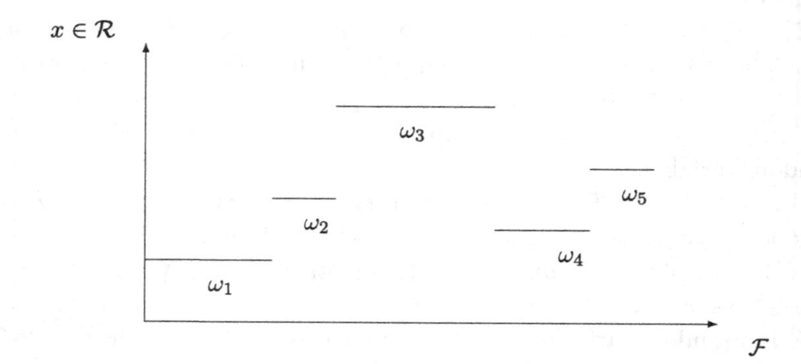

Fig. 1.1: A Simple RV Function

bounded, i.e., for all n and all ω, $|X_n(\omega)| \leq K$ for a finite constant K. Suppose $\lim_{n\to\infty} X_n = X$. $X_n(\omega \in A) = K - 1/n$, $X_n(\omega \in A^c) = K - 1 - 2/n$; $P(A) + P(A^c) = p_1 + p_2 = P(A \cup A^c) = 1$. Then, $X_n \to X$ such that $X(\omega \in A) = K$ and $X(\omega \in A^c) = K - 1$. $E(X_n) = p_1(K - 1/n) + (1 - p_1)(K - 1 - 2/n) = K + p_1 - 1 + (p_1 - 2)/n$. $E(X) = p_1 K + (1 - p_1)(K - 1) = K + p_1 - 1$. Thus, $\lim E(X_n) = K + p_1 - 1 = E(\lim X_n)$.

To continue with the general exposition, since $\mathrm{E}_b = (-\infty, b]$ is a Borel set in $\mathcal{R}$, a measurable RV X can be defined as a mapping $X : \Omega \longrightarrow \mathcal{R}$ such that for any event $E_b \in \Omega$, $X(E_b) = b \in \mathcal{R}$. We shall see how this provides for X as a proper RV.

Distribution Function

The distribution function, also called probability distribution function or cumulative distribution function (CDF) of a random variable X is a mapping from $\mathcal{R}$ to $[0, 1]$ defined by $F(b) = P(X \leq b)$.

A distribution function has the following properties:

(3a) F is a nondecreasing (or monotone increasing) function, i.e., if $a \leq b$, then $F(a) \leq F(b)$

(3b) $\lim_{b\to-\infty} F(b) = 0$ and $\lim_{b\to+\infty} F(b) = 1$

(3c) F is right continuous, i.e., $\forall b \in \mathcal{R}$, $\lim_{h\downarrow 0} F(b + h) = F(b)$

Right continuity of $F(x)$ ensures that even if F is discontinuous at x, the lumpy probability of a single point x for all $x \in \mathcal{R}$ can be evaluated as $F(x) - F(x-)$. Ambiguity in the value of this probability at x could arise if instead we had used other notions of continuity at such a jump point, e.g., $F(x-) < F(x) < F(x+)$ or left continuity.

When a RV X is discrete, X takes only finitely many or countably many values, a_i, $i = 1, 2, \ldots$. The probability of event $\{X \leq a_i\}$ occurring is $F(a_i) = \sum_{x \leq a_i} p(x)$. The probability of a_i occurring is $p(a_i) = F(a_i) - F(a_i-)$. In the above, $p(x)$ is called the probability mass function (PMF) for a discrete RV X.

For a continuous RV X, there exists a non-negative function $f(x)$ such that

$$P(a < X \leq b) = \int_a^b f(x)dx, \quad \text{for } -\infty < a < b < +\infty.$$

The function $f(x)$ is called a probability density function (PDF) of RV X.

Some Moments

For a RV X, its nth moment is $E(X^n)$. The mean of X is its first moment when $n = 1$.

For a continuous RV X with PDF $f(x)$

$$\mu = E(X) = \int_{-\infty}^{+\infty} xf(x)dx.$$

In the discrete case, it is given by

$$\mu = E(X) = \sum_x xp(x),$$

where $p(x)$ is the PMF. The variance of X is

$$\sigma^2(X) = \text{var(X)} = E(X - \mu)^2$$

The moment-generating function (MGF) of RV X is

$$M(\theta) = E(e^{\theta X}) = \int_{-\infty}^{+\infty} e^{\theta x} dF(x).$$

Differentiating M w.r.t. its argument θ

$$M'(\theta) = \frac{d}{d\theta}E(e^{\theta X}) = E\left[\frac{d}{d\theta}(e^{\theta X})\right] = E\big[Xe^{\theta X}\big].$$

Likewise, the second derivative is

$$M''(\theta) = \frac{d}{d\theta}M'(\theta) = \frac{d}{d\theta}E\big(Xe^{\theta X}\big) = E\left[\frac{d}{d\theta}\big(Xe^{\theta X}\big)\right] = E\big[X^2e^{\theta X}\big].$$

Similarly, we can show $\frac{d^n}{d\theta^n}M(\theta) = E\big[X^n e^{\theta X}\big]$. Thus, the nth non-central moment of X can be recovered by putting $\theta = 0$ in the nth derivative of the MGF $M(\theta)$. $E(X^n) = E\big[X^n e^{\theta X}\big]|_{\theta=0}$.

The covariance of two jointly distributed (both realizations occurring simultaneously) RVs X and Y is

$$\sigma_{XY} = \text{cov(X,Y)} = E(X - \mu_X)(Y - \mu_Y).$$

Continuous jointly distributed RVs X and Y have a joint PDF $f(x, y)$. The marginal distribution of any one of the joint RVs is obtainable by integration, viz.

$$f_X(x) = \int_y f(x, y)dy.$$

Likewise for discrete RVs, the marginal PMF of RV X is

$$P_X(x) = \sum_y P(x, y).$$

The correlation coefficient of two RVs X and Y is $(\sigma_{XY})/(\sigma_X \sigma_Y)$. Note that for notations, we employ X to denote a RV, and small letter x to denote its sample value.

Independence

For discrete RVs X and Y, they are independent if and only if (iff) $P(x, y) = P_X(x) \times P_Y(y)$. For continuous RVs X and Y, they are independent iff $f(x, y) = f_X(x) \times f_Y(y)$.

When X and Y are independent, their covariance is obtained as

$$\begin{aligned}
&\int_y \int_x (x - \mu_x)(y - \mu_Y) f(x, y) dx dy \\
&\quad = \int_y \int_x (x - \mu_X)(y - \mu_Y) f_X(x) f_Y(y) dx dy \\
&\quad = \int_y (y - \mu_Y) \left(\int_x (x - \mu_X) f_X(x) dx \right) f_Y(y) dy \\
&\quad = 0.
\end{aligned}$$

Therefore, independent RVs have zero covariance. However, the converse is generally not true. For the case of normal RVs, zero correlation does imply independence.

More generally, a sequence of RVs $X_1, X_2, \ldots, X_n$ are independent if and only if (iff)

$$\begin{aligned}
&P(X_1 \in A_1, X_2 \in A_2, \ldots, X_n \in A_n) \\
&\quad = P(X_1 \in A_1) \times P(X_2 \in A_2) \times \cdots\cdots \times P(X_n \in A_n).
\end{aligned}$$

1.2. Binomial Distribution

Suppose we have an experiment or trial in which there are only two outcomes. Without loss of generality, we call the two outcomes "head (H)" or "tail (T)" as in a coin toss. This type of trial is also called a Bernoulli trial. The probability of H is p, while the probability of T is therefore $1 - p$.

Suppose the experiment is repeated and we perform N independent trials. Let X equal the number of observed H's in the N trials. X can take the values (outcomes) $0, 1, 2, \ldots, N$. The RV X is a binomial RV $B(N, p)$ with parameters N and p, and the discrete probability distribution of X is a binomial distribution.

For $0 \leq k \leq N$,

$$P(X = k) = \binom{N}{k} p^k (1-p)^{N-k}.$$

$$\begin{aligned}
E(X) &= \sum_{j=0}^{N} j \binom{N}{j} p^j (1-p)^{N-j} \\
&= \sum_{j=1}^{N} j \frac{N!}{j!(N-j)!} p^j (1-p)^{N-j} \\
&= \sum_{j=1}^{N} N \left(\frac{(N-1)!}{(j-1)!(N-j)!} \right) p^j (1-p)^{N-j} \\
&= Np \sum_{j=1}^{N} \left(\frac{(N-1)!}{(j-1)!(N-j)!} \right) p^{j-1} (1-p)^{N-j} \\
&= Np \sum_{i=0}^{N-1} \left(\frac{(N-1)!}{i!(N-i-1)!} \right) p^i (1-p)^{N-1-i} \\
&= Np \sum_{i=0}^{N-1} \binom{N-1}{i} p^i (1-p)^{N-1-i} \\
&= Np.
\end{aligned}$$

In the same way, we can show $\sigma_X^2 = Np(1-p)$.

The moment-generating function of $B(N, p)$ is

$$\begin{aligned}
M(\theta) &= E\left[e^{\theta X}\right] \\
&= \sum_{x=0}^{N} e^{\theta x} \binom{N}{x} p^x (1-p)^{N-x} \\
&= \sum_{x=0}^{N} \binom{N}{x} (pe^{\theta})^x (1-p)^{N-x} \\
&= \left[pe^{\theta} + (1-p)\right]^N.
\end{aligned}$$

From the MGF of X,

$$\mu_X = \left.\frac{dM}{d\theta}\right|_{\theta=0} = M'(0) = N\big[pe^\theta + (1-p)\big]^{N-1}(pe^\theta)|_{\theta=0} = Np.$$

$$\begin{aligned} E(X^2) &= M''(0) \\ &= \left.\Big\{N(N-1)\big[pe^\theta + (1-p)\big]^{N-2}(pe^\theta)^2 + M'(\theta)\Big\}\right|_{\theta=0} \\ &= N(N-1)p^2 + Np \\ &= (Np)^2 + Np(1-p) = \mu_X^2 + Np(1-p). \end{aligned}$$

Hence, $\sigma^2(X) = E(X^2) - E^2(X) = Np(1-p)$.

The Pascal arithmetic triangle, Fig. 1.2, is closely associated with the development of the binomial distribution. Each interior number on any line in the Pascal triangle is the sum of the two adjoining numbers immediately above on the previous line.

Pascal (1623–1662) showed that numbers on the triangle can be interpreted as combinatorial numbers as in Fig. 1.3. Pascal also showed that ${}^{n+1}C_k = {}^nC_{k-1} + {}^nC_k$. What is interesting is that each nth line on the Pascal triangle contains the (binomial) coefficients in the expansion of $(a+b)^n$. The binomial distribution probabilities are a special case of the binomial expansion where $a = p$ and $b = 1 - p$.

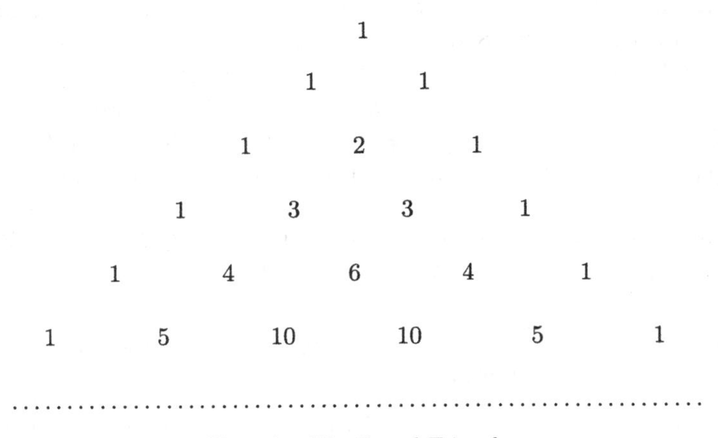

Fig. 1.2: The Pascal Triangle

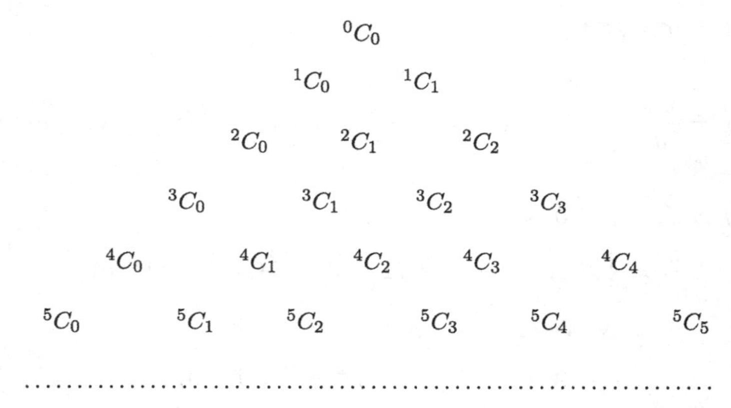

Fig. 1.3: The Pascal Triangle, in Combinatorial Numbers

Applications

In a casino[1] roulette game, there are usually 38 (sometimes 37) numbers on the wheel, $00, 0, 1, 2, \ldots, 36$. Even numbers are red and odd numbers are black. "00" and zero are green. There are 18 reds and 18 blacks to bet on. If green shows up, the banker wins. A \$1 bet on a red (black) returns \$2, including the bet amount, if red (black) turns up, or else \$0 with the loss of the bet amount. For each color-bet, there are 20 losing numbers to 18 winning ones. John decides to try his luck on red numbers. He will put \$1 bets 19 times on red. What is John's expected gain or loss after 19 games?

Assuming the wheel is fair, the probability of a win for John is 18/38. The probability of a loss is 20/38. His dollar win in each game is \$1. His dollar loss in each game is also \$1. Thus, John's outcome in each game is +\$1 with probability 18/38 and −\$1 with probability 20/38.

Let Z be the number of wins after 19 games. Z can take any of the values $0, 1, 2, \ldots, 19$. Let X be John's dollar gain/loss after 19 games. $X = Z \times 1 + (19 - Z) \times (-1) = 2Z - 19$. Therefore, X takes values $-19, -17, -15, \ldots, -1, 1, \ldots, 15, 17, 19$, with a total of 20 possible outcomes from $B(19, \frac{18}{38})$.

After 19 games, his expected number of wins is $E(Z) = 19 \times \frac{18}{38} = 9$. His expected \$ gain/loss is $E(X) = 2E(Z) - 19 = -1$.

---

[1]The application examples of casino games should not be construed as an encouragement of casino gambling. Table-betting games have been around for thousands of years and the odds are quite objective if there is no cheating. However, one should also realize that the casino business usually makes lots of money because playing against the house is similar to retail customers speculating against big hedge funds that have deep pockets in the capital markets.

In the casino game of craps, two dice are rolled. The rules are that if you roll a total of 7 or 11 on the first roll, you win. If you roll a total of 2, 3, or 12 on the first roll, you lose. But if you roll a total of 4, 5, 6, 8, 9, or 10 on your first roll, the game is not ended and continues with more rolls. The total number in the first roll becomes your "point". This "point" is fixed for the game. If in subsequent rolls, you hit your "point" again before you hit a total of 7, then you win. If you roll a total of 7 before your hit your "point", then you lose. The rolling would continue until your "point" is hit or else 7 is hit. The game payoff is $1 for a $1 bet. You either win a dollar or you lose a dollar in each game. One wins or loses against the casino or banker.

First, we list all the possible outcomes in Table 1.1. The outcomes represent the total of the numbers on the two dice.

Table 1.1: Total of Numbers on Two Dice

Total	1	2	3	4	5	6
1	2	3	4	5	6	7
2	3	4	5	6	7	8
3	4	5	6	7	8	9
4	5	6	7	8	9	10
5	6	7	8	9	10	11
6	7	8	9	10	11	12

There are 8 outcomes with "7" or "11" out of 36. Thus, the probability of a win out of a first roll is 8/36. There are 4 outcomes with a "2", "3", or "12". Thus, the probability of a loss out of a first roll is 4/36.

If the first roll is "4", the game continues. The probability of the next roll being "4" and thus a win is 3/36. The probability of the next roll being "7" and thus a loss is 6/36. The game continues to a third and subsequent roll if the total is not "4" or "7". Thus, the probability of an eventual win if the first roll is "4" is $\frac{3}{36} + (1 - \frac{3}{36} - \frac{6}{36}) \times \frac{3}{36} + (1 - \frac{3}{36} - \frac{6}{36})^2 \times \frac{3}{36} + (1 - \frac{3}{36} - \frac{6}{36})^3 \times \frac{3}{36} + \cdots$, or $\frac{3}{36}(1 + \frac{27}{36} + [\frac{27}{36}]^2 + [\frac{27}{36}]^3 + \cdots) = \frac{3}{36} \times \frac{1}{1-\frac{27}{36}} = \frac{1}{3}$. Notice that the idea of independence of the outcome in each roll is assumed in the probability computations.

The contingency table of the probabilities of the various outcomes is shown in Table 1.2. In Table 1.2, the last column shows that the probability of winning = probability of winning given first roll × probability of first roll.

Thus, the probability of winning in a game of craps by rolling is $\frac{8}{36} + 2 \times (\frac{1}{36} + \frac{4}{90} + \frac{25}{396})$. This is 49.2929%. The probability of loss is 50.7071%. Thus, the casino has a 1.4142% advantage over your bet.

Table 1.2: Outcome Probabilities in Craps

First roll	Probability of first roll	Probability of winning given first roll	Probability of winning
4	3/36	1/3	1/36
5	4/36	4/10	4/90
6	5/36	5/11	25/396
8	5/36	5/11	25/396
9	4/36	4/10	4/90
10	3/36	1/3	1/36

From a historical perspective, the Franciscan monk Friar Luca Pacioli (1445–1517) posed the following question in 1494. In those early days there were no computers, no video games, and no gadgets for people to play with, except perhaps coins. Two players would spend their leisure time tossing a coin in a match. If a head came up in a toss, player 1 won the game, otherwise player 2 won the game. The player who was the first to win a total of 6 games won the match and collected the entire prize pool that both had contributed to. This sounds simple enough. However, Pacioli's question was: How would the prize pool be distributed if, for some reason, the games had to stop before the final winner was determined? This problem posed in the late 1400s came to be known as "the problem of points", and it remained unsolved for nearly 200 years until Fermat and Pascal came along. The issue was to decide how the stakes of a game of chance should be divided if that game was not completed for whatever reason. It appears that Pacioli had proposed that if player 1 was up by 5 game wins to player 2's 3 wins, then they could divide the stakes 5/8 to player 1 and 3/8 to player 2. However, there were fierce objections to this proposed solution.

Blaise Pascal (1623–1662) and Pierre de Fermat (1601–1665) began a series of letters around 1654 that led to the solution of the problem of points and expansion of the foundation for classical probability. Pascal's solution can be represented by a binomial tree shown in Fig. 1.4. The full dot indicates the current state of 5 wins and 3 losses for player 1 (or equivalently 5 losses and 3 wins for player 2). The empty dots with number j indicate the completion of the match with j as winner.

Assuming that player 1 and 2 each has a probability 1/2 of winning in each game, player 1 will end up winning in 3 possibilities: {6 wins, 3 losses} with probability 1/2, {6 wins, 4 losses} with probability 1/4, {6 wins, 5 losses} with probability 1/8, and player 2 will win in only one possibility {6 wins, 5 losses} with probability 1/8. These 4 possibilities total up to a probability of 1. Hence, after 5 wins and 3 losses, player 1 has a total

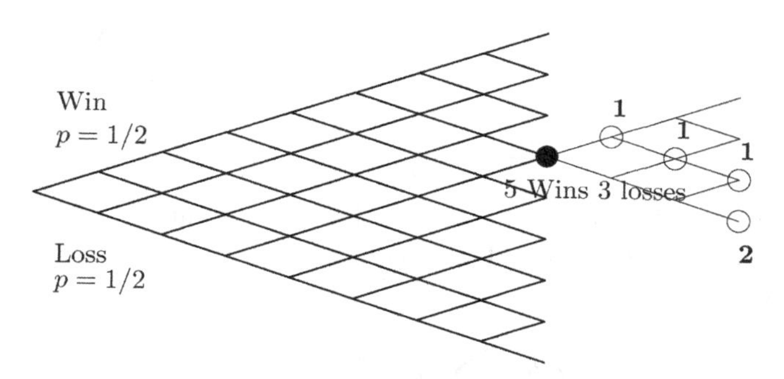

Fig. 1.4: Binomial Tree for the Solution to the Problem of Points

probability of winning the match of $1/2 + 1/4 + 1/8 = 7/8$ while player 2 has a probability of winning at $1/8$. A more reasonable way of dividing the stakes is therefore $7/8$ to player 1 and $1/8$ to player 2.

The concept is akin to looking forward in expected risk-return relationships instead of looking backward to sunk costs as in historical accounting.

1.3. Poisson Distribution

X is a Poisson RV with parameter λ if X takes values $0, 1, 2, \ldots$ with probabilities

$$P(X = k) = \frac{\lambda^k}{k!} e^{-\lambda}.$$

If we sum all the probabilities

$$\sum_{k=0}^{\infty} P(X = k) = e^{-\lambda} \left(1 + \frac{\lambda}{1!} + \frac{\lambda^2}{2!} + \frac{\lambda^3}{3!} + \cdots\right) = e^{-\lambda} e^{\lambda} = 1$$

where we have used the Taylor series expansion of the exponential function e.

For a binomial RV $B(N, p)$ where N is large and p small such that the mean of X, $Np = \lambda$ is $O(1)$, then

$$\begin{aligned} P(X = k) &= \frac{N!}{(N-k)!k!} p^k (1-p)^{N-k} \\ &= \frac{N!}{(N-k)!k!} \left(\frac{\lambda}{N}\right)^k \left(1 - \frac{\lambda}{N}\right)^{N-k} \end{aligned}$$

$$= \frac{N(N-1)\cdots(N-k+1)}{N^k}\frac{\lambda^k}{k!}\frac{(1-\lambda/N)^N}{(1-\lambda/N)^k}$$

$$\approx 1 \times \frac{\lambda^k}{k!}\frac{e^{-\lambda}}{1}$$

when N is large. Parameter λ denotes the average number of occurrences over the time period whereby RV X is measured.

The MGF of a Poisson $X(\lambda)$ is

$$\begin{aligned} M(\theta) &= \sum_{x=0}^{\infty} e^{\theta x}\frac{\lambda^x e^{-\lambda}}{x!} \\ &= e^{-\lambda}\sum_{x=0}^{\infty}\frac{(\lambda e^{\theta})^x}{x!} \\ &= e^{-\lambda}e^{\lambda e^{\theta}} = e^{\lambda(e^{\theta}-1)}, \end{aligned}$$

for all real values of θ.

Thus, the mean and variance of Poisson $X(\lambda)$ are found respectively as

$$\mu_X = M'(0) = \lambda e^{\theta}e^{\lambda(e^{\theta}-1)}|_{\theta=0} = \lambda$$

and

$$\begin{aligned} \sigma_X^2 &= M''(0) - \mu_X^2 \\ &= \lambda[\lambda e^{2\theta}e^{\lambda(e^{\theta}-1)} + e^{\theta}e^{\lambda(e^{\theta}-1)}]|_{\theta=0} - \mu_X^2 \\ &= \lambda(\lambda+1) - \lambda^2 = \lambda. \end{aligned}$$

The Poisson process can be motivated and derived from a few reasonable axioms characterizing the type of phenomenon to be modeled.

A counting process $N(t)$, $t \geq 0$, taking values $0, 1, 2, \ldots$ is called a Poisson process if it has an occurrence rate λ per unit time, where at each occurrence the count increases by 1, and

(4a) $N(0) = 0$
(4b) $N(t+h) - N(t)$ and $N(t'+h) - N(t')$ are stationary regardless of t or t', and are independent for $t' \geq t+h$.
(4c) $P[N(h) = 1] = \lambda h + o(h)$ for small time interval h
(4d) $P[N(h) \geq 2] = o(h)$.

Properties (4c) and (4d) indicate that the probability of a single occurrence of event is approximately proportional to the time that elapsed without an event. During a short period of time h, the probability of more than one event occurring is negligible. Clearly, (4b) indicates that the increments are stationary and independent, and thus memoryless. In other words, what happened in a previous period will not affect what will happen next.

Now

$$\begin{aligned} P[N(t+h)=0] &= P[N(t)=0, N(t+h)-N(t)=0] \\ &= P[N(t)=0]P[N(t+h)-N(t)=0] \\ &= P[N(t)=0][1-\lambda h+o(h)]. \end{aligned}$$

Then

$$\frac{P[N(t+h)=0]-P[N(t)=0]}{h} = -\lambda P[N(t)=0] + \frac{o(h)}{h}.$$

By definition, $\lim_{h\mapsto 0}\frac{o(h)}{h}=0$.

Let

$$P_0(t) = P[N(t)=0].$$

As $h \downarrow 0$,

$$\frac{dP_0(t)}{dt} = -\lambda P_0(t).$$

Thus, $d\ln P_0(t) = -\lambda dt$, hence

$$P_0(t) = \exp(-\lambda t). \tag{1.1}$$

More generally

$$\begin{aligned} P_n(t+h) &= P_n(t)P_0(h) + P_{n-1}(t)\,P_1(h) + P_{n-2}(t)\,P_2(h) \\ &\quad + \cdots + P_0(t)P_n(h) \\ &= (1-\lambda h)P_n(t) + \lambda h P_{n-1}(t) + o(h). \end{aligned}$$

Then

$$\frac{P_n(t+h)-P_n(t)}{h} = -\lambda P_n(t) + \lambda P_{n-1}(t) + \frac{o(h)}{h}$$

As $h \downarrow 0$

$$\frac{dP_n(t)}{dt} = -\lambda P_n(t) + \lambda P_{n-1}(t).$$

Thus

$$e^{\lambda t}[P_n'(t) + \lambda P_n(t)] = \lambda e^{\lambda t} P_{n-1}(t).$$

So

$$\frac{d}{dt}[e^{\lambda t}P_n(t)] = \lambda e^{\lambda t}P_{n-1}(t).$$

Suppose the solution is

$$e^{\lambda t}P_n(t) = \frac{(\lambda t)^n}{n!}, \quad \text{for any } t. \tag{1.2}$$

To prove by mathematical induction, by first putting $n = 0$ in Eq. (1.2), $P_0(t) = \exp(-\lambda t)$ which is true as in Eq. (1.1). Second, for $n = 1$

$$\frac{d}{dt}[e^{\lambda t}P_1(t)] = \lambda e^{\lambda t}P_0(t) = \lambda.$$

Integrating w.r.t. t, $e^{\lambda t}P_1(t) = \lambda t + c$. At $t = 0$, probability of $n = 1$ arrival is zero, so $c = 0$.

Therefore, $P_1(t) = \lambda t \exp(-\lambda t)$, which follows Eq. (1.2).

Next, suppose Eq. (1.2) applies for $t = n - 1$. So

$$\frac{d}{dt}[e^{\lambda t}P_n(t)] = \lambda e^{\lambda t}P_{n-1}(t) = \lambda e^{\lambda t}e^{-\lambda t}\frac{(\lambda t)^{n-1}}{(n-1)!} = \frac{\lambda^n}{(n-1)!}t^{n-1}.$$

Integrating w.r.t. t,

$$e^{\lambda t}P_n(t) = \frac{(\lambda t)^n}{n!} + c$$

Since $P_n(0) = 0$, $c = 0$. Thus, Eq. (1.2) is also satisfied for $t = n$. By mathematical induction, since $t = 1$ in Eq. (1.2) is true, then $t = 2$ in Eq. (1.2) is also true, and so on, for all $t \geq 0$.

Poisson events occur with a discrete time interval in-between that we call the interarrival or waiting time between events. Let T_n, for $n = 1, 2, \ldots$, be the interarrival time between the $(n-1)$th and the nth Poisson events.

$P[T_1 > t] = P[N(t) = 0|N(0) = 0] = \exp(-\lambda t)$, from Eq. (1.1). Then, T_1 has an exponential distribution with rate λ. The CDF of RV T_1 taking values in $[0, t]$ is $F(t) = 1 - \exp(-\lambda t)$ for time $t \geq 0$. Differentiating, its PDF is $\lambda \exp(-\lambda t)$ for time $t \geq 0$.

Its mean is

$$\int_0^\infty t\lambda \exp(-\lambda t)dt = -\int_0^\infty t\, d(e^{-\lambda t}) = -\left[e^{-\lambda t}\left(\frac{1}{\lambda} + t\right)\right]_0^\infty = \frac{1}{\lambda}.$$

The variance is $(\frac{1}{\lambda})^2$.

From the above, it is seen that when the interarrival time of events or occurrences is exponentially distributed, the number of occurrences or events in a given time interval has a Poisson distribution, and vice-versa. A Poisson process can also be defined as an activity whose interarrival time has an exponential distribution. The exponential distribution is the only continuous distribution that has the memoryless property in (4b). Here

$$P[T_1 > t, T_1 > t + h] = P[T_1 > t + h | T_1 > t] P[T_1 > t]$$

or

$$P[T_1 > t + h] = P[T_1 - t > h | T_1 > t] P[T_1 > t].$$

Since the exponential distribution of the waiting time till an event happens, $P[T_1 \leq t] = 1 - \exp(-\lambda t)$, is only a function of the waiting time t from the time of the last event regardless of the past, therefore we can also write $P[T_1 - t > h | T_1 > t] = P[T_1' > h] = \exp(-\lambda h)$ regardless of the time just past, t, and defining T_1' as the new waiting time starting at t. Then

$$P[T_1 > t + h] = P[T_1 > h] P[T_1 > t].$$

Application

In major tennis tournaments, a match winner is the first one to win three sets. A set consists of games, and games, in turn, consist of points. Typically, a player wins a set by being the first to win a total of six games. If there is a tie at a score of 5–5, then the winner is the one who would win 2 consecutive games subsequently. However, a tiebreaker rule allows a set winner with a score of 7–6. Only in the final sets of matches at the Australian Open, the French Open, Wimbledon, the Olympic Games, Davis Cup, and Fed Cup are tiebreakers not played. In these cases, sets are played indefinitely until one player has a two game lead. (Caveat: even in major sports, game rules do change from time to time.)

Suppose you are the convenor of an important match between two renowned players on tiebreaker rule. The match will start in 2 hours' time, and weather forecast warns of a large-scale storm hitting in 5 hours. You have a choice now to announce a postponement of the match, but you figure you will do that only if the chance of the storm hitting while the match is still progressing is more than 20%.

Looking at the historical records of similar tournaments, the average number of games completed per hour is 20. Therefore, assuming the maximum of 5 sets and 13 games per set, or a total of 65 games, the probability of completing up to 65 games in 3 hours is 76.5%. Hence the chance that the match is not completed yet within 3 hours is 23.5%. Since this is larger than 20%, you postpone the game.

In past economic crises since 1970, suppose the number of days within a month in which stock price index changes from day to day by more than 5% averages 3. In the month of October 2008 in the recent global financial crisis, the number of days where stock price index changed by more than 5% a day was 10. What is this probability given $\lambda = 3$? It is $e^{-3}\frac{3^{10}}{10!} = 0.00081$. The probability is so small that it casts doubt on the suitability of using the memoryless Poisson process for stock price movements in October 2008. The more acceptable scenario was that the market carried memories of what had happened over the days and weeks, and reacted accordingly.

1.4. Normal Distribution

Consider $\int_{-\infty}^{+\infty} e^{1-|y|}dy = \int_{-\infty}^{0} e^{1+y}dy + \int_{0}^{+\infty} e^{1-y}dy = [e^{1+y}]_{-\infty}^{0} - [e^{1-y}]_{0}^{+\infty} = 2e$. For $y > 0$, $y^2 - 2|y| + 2 \equiv y^2 - 2y + 2 = (y-1)^2 + 1 > 0$. For $y < 0$, $y^2 - 2|y| + 2 \equiv y^2 + 2y + 2 = (y+1)^2 + 1 > 0$.

Then, $y^2 - 2|y| + 2 > 0$ for all y. Or, $-\frac{y^2}{2} < 1 - |y|$. Hence, $J = \int_{-\infty}^{+\infty} e^{-y^2/2}dy < 2e$, and is bounded.

We first find $I^2 = \int_{-\infty}^{\infty}\int_{-\infty}^{\infty} \exp\left(-\frac{x^2+y^2}{2}\right) dxdy$. Changing to polar coordinates by putting $x = r\cos\theta$ and $y = r\sin\theta$

$$I^2 = \int_0^{2\pi}\int_0^{\infty} e^{-r^2/2}\,|J|\,dr\,d\theta$$

where J is the Jacobian matrix and $|J|$ is its Jacobian determinant or simply "Jacobian":

$$|J| = \begin{vmatrix} \frac{dx}{dr} & \frac{dx}{d\theta} \\ \frac{dy}{dr} & \frac{dy}{d\theta} \end{vmatrix} = \begin{vmatrix} \cos\theta & -r\sin\theta \\ \sin\theta & r\cos\theta \end{vmatrix} = r\cos^2\theta + r\sin^2\theta = r.$$

Therefore

$$I^2 = \int_0^{2\pi}\int_0^{\infty} e^{-r^2/2}\,r\,dr\,d\theta = 2\pi.$$

Hence,

$$2\pi = \int_{-\infty}^{\infty}\int_{-\infty}^{\infty} \exp\left(-\frac{x^2+y^2}{2}\right) dx\,dy = \left(\int_{-\infty}^{\infty} \exp\left(-\frac{y^2}{2}\right) dy\right)^2.$$

Thus, $\int_{-\infty}^{\infty} \frac{1}{\sqrt{2\pi}} e^{-\frac{y^2}{2}} dy = 1$. Applying a change of variable $y = \frac{x-\mu}{\sigma}$, we obtain

$$\int_{-\infty}^{\infty} \frac{1}{\sigma\sqrt{2\pi}} \exp\left[-\frac{(x-\mu)^2}{2\sigma^2}\right] dx = 1,$$

so that the normal PDF is $f(x) = \frac{1}{\sigma\sqrt{2\pi}} \exp\left[-\frac{(x-\mu)^2}{2\sigma^2}\right]$.

We shall see that the normal distribution is indeed a fascinating, if not one of the most celebrated results in mathematical statistics, via the central limit theorem. It comes naturally from common phenomena such as aggregation and averaging. Its distribution is also found to describe well the frequencies of occurrences in natural processes such as heights of people, IQs of students, spatial densities in plant growth, and so on. Not surprisingly, it is also used in describing the distributions of stock returns. While this is reasonable in normal times, it would appear from the many incidences of market turbulence in history, that unquestioned application could be greatly flawed. Even during normal times, stock returns generally display some skewness and a larger kurtosis or fatter tails than those of a normal distribution. Some more complicated functions of the normal RV are often used to describe stock returns. In any case, normal distributions are excellent basic pillars to more complicated constructions.

The MGF of a standardized (or "unit") normal RV $X \sim N(0,1)$ is

$$\begin{aligned}
M(\theta) &= \int_{-\infty}^{\infty} e^{\theta x} \frac{1}{\sqrt{2\pi}} e^{-\frac{1}{2}x^2} dx \\
&= \frac{1}{\sqrt{2\pi}} \int_{-\infty}^{\infty} \exp\left\{-\frac{(x^2 - 2\theta x)}{2}\right\} dx \\
&= \frac{1}{\sqrt{2\pi}} \int_{-\infty}^{\infty} \exp\left\{-\frac{(x-\theta)^2}{2} + \frac{\theta^2}{2}\right\} dx \\
&= e^{\frac{1}{2}\theta^2}.
\end{aligned}$$

A related distribution is the lognormal distribution. A RV X has a lognormal distribution when $\log(X) = Y$ is normally distributed:

$$Y \overset{d}{\sim} N(\mu, \sigma^2).$$

Then

$$E(X) = E(e^Y) = \int_{-\infty}^{\infty} e^y \frac{1}{\sigma\sqrt{2\pi}} e^{-\frac{1}{2\sigma^2}(y-\mu)^2} dy$$

$$= \int_{-\infty}^{\infty} \frac{1}{\sigma\sqrt{2\pi}} e^{-\frac{1}{2\sigma^2}[(y-\mu)^2 - 2\sigma^2 y]} dy$$

$$= \int_{-\infty}^{\infty} \frac{1}{\sigma\sqrt{2\pi}} e^{-\frac{1}{2\sigma^2}[(y-(\mu+\sigma^2))^2 - \sigma^4 - 2\mu\sigma^2]} dy$$

$$= e^{\mu + 1/2\sigma^2}.$$

Similarly, it can be shown that $\text{var}(X) = e^{2\mu}[e^{2\sigma^2} - e^{\sigma^2}]$.

1.5. Uniform Distribution

A continuous RV X has a uniform distribution over interval (a, b) if its PDF is given by

$$f(x) = \begin{cases} \frac{1}{(b-a)} & a < x < b \\ 0 & \text{otherwise} \end{cases}$$

$$E(X) = \int_a^b \frac{x}{b-a} dx = \frac{b^2 - a^2}{2(b-a)} = \frac{a+b}{2}$$

$$E(X^2) = \int_a^b \frac{x^2}{b-a} dx = \frac{b^3 - a^3}{3(b-a)} = \frac{a^2 + ab + b^2}{3}.$$

Thus

$$\text{var}(X) = E(X^2) - E^2(X) = \frac{a^2 + ab + b^2}{3} - \frac{(a+b)^2}{4} = \frac{(b-a)^2}{12}.$$

There is a simple but useful theorem when the uniform distribution appears naturally.

Theorem 1.1 *Given any RV X with a distribution function $F(X) = U, U$ is a RV with uniform distribution.*

We provide proof for a more specific case where X is a continuous RV and $F(x)$ is continuous strictly increasing. The more general case can also be proved. ■

Proof. $P(F(x) \leq y) = P(x \leq F^{-1}(y))$. Note that to each y, a unique $F^{-1}(y)$ exists, since $F(y)$ is continuous strictly increasing. But $P(x \leq F^{-1}(y))$ by definition is $F(F^{-1}(y)) = y$.

If $F(X)$ is a RV denoted as U, then the above shows $P(U \leq y) = y$. Note that $0 \leq y \leq 1$. But this is the characterization of a uniform $U(0,1)$ distribution. Hence, $F(X) = U$ is $U(0,1)$. ■

A very useful application of this theorem is to enable the generation of random values of X with any given CDF $F(X)$ where its inverse could be computed. Generate a random u from $U(0,1)$, then compute $x = F^{-1}(u)$ to obtain the random value of X.

1.6. Problem Set 1

1. From properties (1a), (1b), and (1c) of a field, show how an equivalent version of (1c) could lead to the following. For any sequence of disjoint events E_n of $\mathcal{F}$, for $n = 1, 2, \ldots, N$, $P(\bigcup_{n=1}^{N} E_n) = \sum_{n=1}^{N} P(E_n)$. .
2. Show that $\sigma^2(X) = E(X^2) - \mu^2$.
3. Show that for jointly distributed stationary RVs X_i, $i = 1, 2, \ldots, K$, each with mean μ,

$$\mathrm{var}\left(\sum_{i=1}^{K} X_i\right) = \sum_{i=1}^{K} \mathrm{var}(X_i) + 2 \sum_{1 \le i < j \le K} \mathrm{cov}(X_i, X_j).$$

4. Suppose the cumulative probability distribution of X has a discontinuity at point x_1 but is continuous at point x_2. Is $P(x_1) > P(x_2)$?
5. If RVs X and Y are independent, and RVs Y and Z are independent, are RVs X and Z necessarily independent?
6. Show $|\sigma_{XY}| \le |\sigma_X \sigma_Y|$ and hence that correlation coefficient has magnitude ≤ 1.
7. For simple RVs X and Y where $X = \sum_i x_i 1_{A_i}$ and $Y = \sum_j y_j 1_{B_j}$ on $\mathcal{R}$, show (from first principles) how $E(aX + bY) = aE(X) + bE(Y)$ where a and b are constants.
8. A fair die is thrown and the number that shows is noted. What is the probability that 5 or fewer throws are needed before an even number appears the third time? What is the expected number of throws before hitting an even number three times? (This is a negative binomial distribution.)
9. Using John's one-dollar a bet strategy at roulette, how many games do you expect John to be able to play before he loses all his initial $19?
10. In the game of craps, it is possible for you to bet with the casino, and against the shooter. To do this, you put your money on the Don't Pass Line. However, if you place a Don't Pass Bet, and the shooter initially rolls a 12, the casino wins, but you don't. You don't lose your bet, but you don't win anything — it is a tie. If you place a Don't Pass Bet, what is your probability of winning?

11. In May 1997, IBM's Deep Blue Supercomputer played a set of 6 chess games with the reigning world chess champion, Garry Kasparov. Suppose there is a fictitious rematch and assuming no draw in any game, Kasparov is up by one game at 2 wins and 1 loss out of a total of 6 games. If the game ends with 3 wins and 3 losses, the pot is split even. Suppose the computer crashes, and the umpire decides to split the pot of 1 million dollars. How much should Kasparov expect to get if the prior probability is that both are equally matched?
12. Show that the sum of independent Poission RVs is Poisson.
13. Show that the sum of any two independent standard normal RVs is $N(0, 2)$. (Hint: Let $Z = X + Y$, where X, Y are independent $N(0, 1)$. Then, evaluate CDF of Z,

$$F(z) = \int_{-\infty}^{\infty}\left(\int_{-\infty}^{z-x}\frac{1}{2\pi}\exp\left[-1/2(x^2+y^2)\right]dy\right)dx.$$

Then, find $F'(z)$.
14. Find the MGF of the normal RV $X(\mu, \sigma^2)$. Use this to find the skewness $E\left[\frac{(X-\mu)^3}{\sigma^3}\right]$ and kurtosis $E\left[\frac{(X-\mu)^4}{\sigma^4}\right]$ of X.
15. Show that the variance of the exponential distribution is $(\frac{1}{\lambda})^2$.
16. A bank officer is given a project to study how many ATM machines to install at a central location. He studies the traffic at a competitor ATM machine in a similar locality and notes that on average 2 customers arrive per minute to use the machine. He wants to keep the probability of potential customers queueing to less than $1/3$. You may assume that each customer takes on average 1 minute at the machine. How many ATM machines should be installed?

Chapter 2

CONDITIONAL PROBABILITY

The idea of conditional probability is central to any probabilistic investigation. Climbing up the mountain trail more likely leads to vistas of the distant hills than descending the valleys would to seeing meandering brooks and meadows. Imagine a totally flat landscape on earth in which nowhere one goes or whichever direction one turns would produce any difference in scenery. If conditional probability were useless, then it amounts to such a landscape where any random event is independent of whatever else has happened or is happening. Fortunately, it is not so, as famous Scottish-born American naturalist and explorer John Muir wrote, "God never made an ugly landscape. All that the sun shines on is beautiful, so long as it is wild."

2.1. Set Operations

Consider the sample space Ω and events A, B, C represented by sets in the Venn diagram, Fig. 2.1, shown below.

Suppose simple sample points or elements of Ω are shown as $e_1, e_2, e_3, e_4, e_5, e_6$. By definition, only one of the sample points can occur in any one experimental outcome. Events may contain one or more sample points, and as discussed in Chap. 1, are subsets of the universal set Ω. When e_1 occurs, event A is also said to occur. Likewise, when e_2 occurs, event B is said to have occurred. When e_4 occurs, both events A and B are said to have occurred, and we can say that the intersection event $A \cap B$ has occurred.

De Morgan's law in set theory states that

$$(A \cup B)^c = A^c \cap B^c, \quad \text{and}$$
$$(A \cap B)^c = A^c \cup B^c.$$

Event $\Omega \backslash A$ is the same as A^c. $A \backslash B$ is the same as $A \cap B^c$.

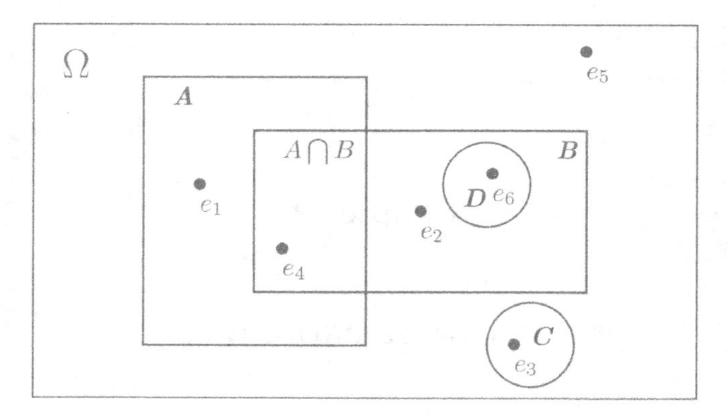

Fig. 2.1: Venn Diagram

When e_5 occurs, event $\Omega\backslash(A \cup B \cup C)$ has occurred. Using the above relationship

$$\begin{aligned}\Omega\backslash(A \cup B \cup C) &= (A \cup B \cup C)^c \\ &= A^c \cap B^c \cap C^c.\end{aligned}$$

Hence, $e_5 \in A^c \cap B^c \cap C^c$.

From the diagram, clearly $A \cap C = B \cap C = \phi$, where ϕ is the empty or null set. In addition, $D \subset B$, which means that if event D occurs, then B is also said to have occurred.

Suppose there are $n_{A\backslash B}$, $n_{B\backslash A}$, $n_{A\cap B}$, and n_Ω sample points in events $A\backslash B$, $B\backslash A$, $A \cap B$, and Ω, respectively. Assume each sample point can occur with equal probability. Using the frequentist notion of probability (taking "long-run" relative frequency as probability), the probability of A happening is $\frac{(n_{A\backslash B}+n_{A\cap B})}{n_\Omega}$. That of B happening is $\frac{(n_{B\backslash A}+n_{A\cap B})}{n_\Omega}$. That of event $A \cap B$ happening is $\frac{(n_{A\cap B})}{n_\Omega}$. Since $A \cup B = A\backslash B + A \cap B + B\backslash A$, where "+" denotes union of disjoint sets, then probability of event $A \cup B$ is $\frac{(n_{A\backslash B}+n_{A\cap B}+n_{B\backslash A})}{n_\Omega}$.

Suppose we are given the information that event B has happened. Conditional on (or given) this information, what is the probability that another event, say A, has happened?

The conditional probability of A given B is

$$P(A|B) = \frac{P(A \cap B)}{P(B)} \tag{2.1}$$

which is $\frac{n_{A\cap B}}{n_B}$.

Intuitively this is correct since $P(A \cap B) = P(A|B) \times P(B) = \frac{n_{A\cap B}}{n_B} \times \frac{n_B}{n_\Omega} = \frac{n_{A\cap B}}{n_\Omega}$. For the case of D, $P(B|D) = \frac{P(B\cap D)}{P(D)} = P(D)/P(D) = 1$.

Suppose there are disjoint sets G_i, $i = 1, 2, 3, \ldots, M$, such that $B \subseteq \cup_{i=1}^{M} G_i$. Then, Bayes' formula is obtained as follows.

Bayes' Formula

$$\begin{aligned}
P(A|B) &= \frac{P(A \cap B)}{P(B)} \\
&= \frac{P(B \cap A)}{P(B \cap G_1) + P(B \cap G_2) + \cdots + P(B \cap G_M)} \\
&= \frac{P(B|A)P(A)}{P(B|G_1)P(G_1) + P(B|G_2)P(G_2) + \cdots + P(B|G_M)P(G_M)} \\
&= \frac{P(B|A)P(A)}{\sum_{i=1}^{M} P(B|G_i)P(G_i)}. \qquad (2.2)
\end{aligned}$$

The denominator on the right-hand side (RHS) of Eq. (2.2) exists if $B \subseteq \cup_{i=1}^{M} G_i$. A stronger sufficient condition is that there is a partition of Ω by sets G_i, $i = 1, 2, 3, \ldots, M$, i.e., disjoint sets G_i such that $\cup_{i=1}^{M} G_i = \Omega$, and hence, necessarily $B \subseteq \cup_{i=1}^{M} G_i$. The probability of an event B as a union of subevents $(B \cap G_i)$, resulting in $P(B) = \sum_{i=1}^{M} P(B|G_i)P(G_i)$, is sometimes called the Law of Total Probability.

Application

As an example in a casino setting, suppose on a table of blackjack with 4 players and a dealer, the following hands are dealt in a shoe-game (open cards).

Assume a simple version of the game where there are no splits, no doubling up or down, and it is your turn. The dealer has one card that is not disclosed. The other players have settled at the shown hands. You can fold and retire or you can hit, which means you get another card from the stack. Assume for simplicity that the stack is just what is left over from a normal 52-card stack. After your decision, the dealer will open his card, and continue to hit until his hand is at least a total of 17. If your final total is larger than the dealer's, provided total is less than 22 — you win. Otherwise you lose. If the dealer gets over 21 and you are still under 22, you win. If you get over 21 and the dealer stays under, you lose. If the dealer

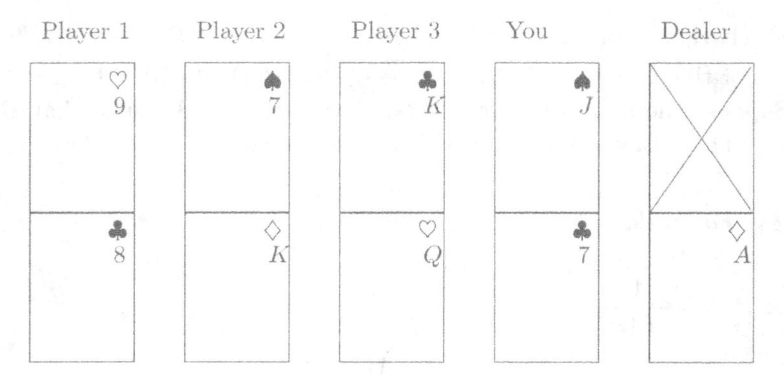

Fig. 2.2: Cards Dealt in a Blackjack Game

hits black jack, i.e., "a king, queen, jack, or a ten" and an "ace" to make 21 points out of 2 cards, you lose even if you hit 21. If both tie at the same score or both hit black jack, there is a draw. You play only with the dealer and do not worry about other players except to note or "count" their cards. (In black jack, an ace doubles as 1 or as 11.)

Suppose the game does not allow you to see all the cards as shown in Fig. 2.2, and you only see your own cards, what is your probability of obtaining a total of "21"? To get "21", you will need a "4". $P(\text{"4"}) = P(\text{"4"}|$ all four "4"s dealt)P(all four "4"s dealt) $+$ $P(\text{"4"}|$ only three "4"s dealt) P(only three "4"s dealt) $+$ $P(\text{"4"}|$ only two "4"s dealt)P(only two "4"s dealt) $+$ $P(\text{"4"}|$ only one "4" dealt)P(only one "4" dealt) $+$ $P(\text{"4"}|$no "4" dealt)P(no "4" dealt) $= 0 + \frac{1}{42} \times \frac{{}^{46}C_5 \times {}^4C_3}{{}^{50}C_8} + \frac{2}{42} \times \frac{{}^{46}C_6 \times {}^4C_2}{{}^{50}C_8} + \frac{3}{42} \times \frac{{}^{46}C_7 \times {}^4C_1}{{}^{50}C_8} + \frac{4}{42} \times \frac{{}^{46}C_8 \times {}^4C_0}{{}^{50}C_8}$.

This equals $0 + \frac{1}{42} \times 0.0102128 + \frac{2}{42} \times 0.1046809 + \frac{3}{42} \times 0.3987842 + \frac{4}{42} \times 0.4860182 = 0.08$. Thus, there is only an 8% chance of making "21". However, conditional on the shown cards, only one "4" is dealt at most. Hence, the conditional probability of drawing a "4" is $P(\text{"4"}|$shown hands$) = P(\text{"4"}|$dealer's closed card is "4", shown hands$) \times P($dealer's closed card is "4"$|$shown hands$) + P(\text{"4"}|$dealer's closed card is not "4", shown hands$) \times P($dealer's closed card is not "4"$|$shown hands$) = \frac{3}{42} \times \frac{4}{43} + \frac{4}{42} \times \frac{39}{43} = 0.0066 + 0.0864 = 0.0930$. Thus, with the information of the shown hands, the chance of drawing a "4" and making "21" is now 9.3%. The chance of the dealer hitting black jack, assuming you decide not to hit, is $12/43 = 27.9\%$. In this game so far, it appears the dealer is loaded to win.

2.2. First Look into Investments

Before the 1990s, the Chicago School of full rational expectations modeling dominated the thinking behind asset pricing and investments. As an alternative, the competing behavioral finance school argues that many financial phenomena can plausibly be understood using models in which at least some if not many agents are not fully rational. There are other behavioral schools of thought originating from the field of psychology that suggest bounded rationality, ecological rationality, and so on.

One major implication of behavioral finance is that risky arbitrage opportunities and abnormal profit opportunities arise in the market because investor behavior is governed by psychology, which prompts deviations from full rationality that we might otherwise expect.

There is a huge variety of psychological effects modeled to explain aberrations from empirical validations of rational asset pricing models. Some examples are regret theory, anchoring behavior, prospect theory by Kahneman and Tversky (1979),[1] mental accounting by Thaler (1980),[2] and so on.

For example, regret theory is a theory that says people expect to regret if they make a wrong choice, and the regret will cause aversion especially to the type of decisions that in the past had produced regrets. It could run both ways, implying more risk aversion if in the past the decision-maker had taken risk and suffered heavy losses, or less risk aversion if in the past the decision-maker had been conservative and regretted missing multiplying his or her wealth during the boom.

Anchoring behavior is the use of irrelevant information as a reference for estimating or expecting some unknown quantities. For example, in assessing the fair price of a small firm's stock, the investor could be using the price of another small firm's stock for comparison, even though the latter information is irrelevant because the two stocks are in different industries and at different levels of risks.

Prospect theory postulates that preferences will depend on how a problem is framed. Preference is a function of decision weights on outcomes, and the weights do not correspond exactly to the outcome probabilities.

---

[1]Kahneman, D and A Tversky (1979). Prospect theory: An analysis of decision under risk. *Econometrica*, 47(2), 263–292.

[2]See Thaler, RH (1980). Toward a positive theory of consumer choice. *Journal of Economic Behavior and Organization*, 1, 39–60, and also Thaler, RH (1985). Mental accounting and consumer choice. *Marketing Science*, 4, 199–214.

Specifically, prospect theory predicts that most decision-makers tend to overweigh small probabilities on huge losses and underweigh moderate and high probabilities on moderate gains or returns. Hence, prospect theory is better able to explain phenomenon such as loss aversion, as in selling stocks after a major drop for fear of further drop.

Mental accounting theorists argue that people behave as if their assets are compartmentalized into a number of non-fungible (non-interchangeable) mental accounts such as current income or current wealth. The marginal propensities to consume out of the different accounts are all different, and thus an investor with a larger mental account in current income may indeed invest more, while a similarly wealthy investor with a larger mental account in current wealth may consume more and invest less.

Market Efficiency

The concept of (informational) market efficiency was investigated by Fama (1970)[3] and many others. Fama surveyed the idea of an informationally efficient capital market, and made the following famous definition: "A market in which prices always 'fully reflect' available information is called 'efficient'". Three forms of the efficient market hypothesis (EMH) are often cited. The weak-form asserts that all past market prices or their history are fully reflected in securities prices. An immediate implication of this version of the EMH is that charting and technical analyses are of no use in making abnormal profit. Technical analysis and charting rely on the belief that past stock prices show enough patterns and trends for profitable forecasting. This possibility is opposed to the notion of stock prices "following" random walks. When prices adjust instantaneously, past returns are entirely useless for predicting future returns. The semi-strong form asserts that all publicly available information, including historical prices, is fully reflected in securities prices. The implication is that fundamental analyses such as analyses of a company's balance sheet, income statement, and corporate news and development, are of no use in making abnormal profit. Finally, the strong-form asserts that all available information including public and private information is fully reflected in securities prices. If true, the implication is that even insider information is of no use in making abnormal profit.

[3]Fama, E (1970). Efficient capital markets: A review of theory and empirical work. *Journal of Finance*, 25(2), 383–417.

We shall consider semi-strong form market efficiency in more detail. The market is represented by the collective body of investors at time t who make the best use of whatever available information (thus rational investors) to predict future period stock price at time $t+1$, i.e., price S_{t+1}. Suppose Φ_t is all relevant publicly available information available at t. Think of information Φ_t as a conditioning RV that is jointly distributed with S_{t+1}. In addition, Φ_M is the information actually used by the market, and is at most all of Φ_t.

The true conditional probability of next period price is $P(S_{t+1}|\Phi_t)$ while the market's conditional probability is $P(S_{t+1}|\Phi_M)$. We assume the market knows the true joint distribution of S_{t+1} and Φ_t.

The market is semi-strong form informationally efficient if and only if

$$P(S_{t+1}|\Phi_M) \stackrel{d}{=} P(S_{t+1}|\Phi_t),$$

i.e., distributionally the same. One implication is that the market forecast $E(S_{t+1}|\Phi_M) = E(S_{t+1}|\Phi_t)$.

Suppose not all of the available information Φ_t is used by the market, and $E(S_{t+1}|\Phi_M) \neq E(S_{t+1}|\Phi_t)$, then the market is informationally inefficient. In this case, all available information is not instantaneously incorporated into price at t, S_t. In the next instance, when more of the information gets absorbed by the market, the price S_t will adjust toward equilibrium. Hence, an informationally inefficient market will see price adjustments over a discrete time interval, and not instantaneously, to any substantive news.

Many tests of asset pricing in the literature employ the framework of rational investors and (informational) market efficiency. A framework such as the Sharpe-Lintner capital asset pricing model typically employs additional assumptions such as exogenous price processes and an explicit or implicit homogeneous preference function (typically a standard strictly monotone concave utility function) for all investors making up the market. In addition, explicitly or implicitly, the von Neumann–Morgenstern expected utility hypothesis is usually employed. Aberrations or non-validation in the test results were pointed out as evidence of market inefficiencies or sometimes coined as market anomalies (meaning something yet to be explained). Behavioral finance arose in this context to help explain the anomalies. It may agree with informational efficiency, but not with the rationality framework mostly to do with the standard preference assumption and the expected utility hypothesis. Some examples of early anomalies were the size effect, day-of-the-week, month-of-the-year effects, value versus growth premium, and so on. Later anomalies against the implication of random walk or approximate random walk by rational asset pricing models include contrarian strategies and momentum trading profits.[4] We provide an illustration of market efficiency as follows.

Suppose at time $t = 0$, there was information about whether the December 2008 GM, Ford, Chrysler bailout plan of $25 billion would pass through Senate. A particular stock price at $t = 0$ was $3. If the bailout were successful, the stock price would increase to either $5 or $4 at $t = 1$. The latter variation is due to other risk factors and uncertainties. If the bailout were unsuccessful, the stock price would drop to either $2 or $1 at $t = 1$. All probabilities of the Bernoulli outcomes were 50%. Assume that the risk-adjusted discount rate from $t = 0$ to $t = 1$ was 0. This is depicted in Fig. 2.3.

If the market did not know about the outcome (i.e., did not have the information even when it was known), then at $t = 0$, its expected stock price at $t = 1$ was $\frac{1}{4}(\$5 + \$4 + \$2 + \$1) = \$3$. However, if the market was informationally efficient, then at $t = 0_+$, its conditional (upon the bailout outcome information) expectation of price at $t = 1$ was either $4.50 if the outcome were successful or $1.50 if the outcome were unsuccessful. Looking at this simple setup, it is easy to see that if at $t = 0_+$, the stock price did not

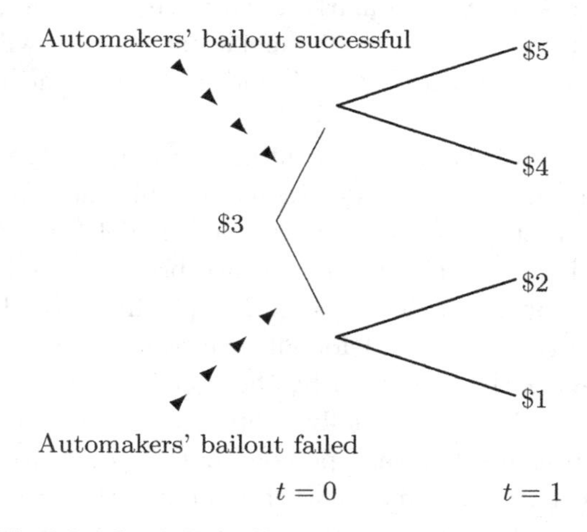

Fig. 2.3: Stock Price Changes Contingent on Bailout News

[4]A useful undergraduate investment textbook for reference in these background readings is Bodie, Kane and Marcus's *Investments*, 8th edition, McGraw Hill, 2009.

quickly move away from \$3, then the market was informationally inefficient as it did not capture the information immediately.

2.3. Conditional Expectation

A conditional probability is constituted by a joint probability and the marginal probabilities. Let's analyze this from a building-blocks perspective. The automakers' story is depicted as follows in a Venn diagram (Fig. 2.4).

Here, one and only one sample point can occur, represented by the bullets. E is the event that the automakers' bailout is successful. We can also represent this by a RV that is an indicator function, X (or denoted using indicator notation 1_E) where $x = 1$ if the bailout is successful, i.e., event E, and $x = 0$ if the bailout is not successful, i.e., event E^C. Let S be the event that a stock takes strictly positive prices, and the RV Y that corresponds to elements of S takes dollar values $y \in \{5, 4, 2, 1\}$. Let each sample point be a joint outcome of the automakers' bailout and the stock price, (x, y). Note that we may allow an event S^c to denote the firm's bankruptcy and hence a stock price of zero. However, in our probability model here, $P(S^c) = 0$.

$\Omega = \{(1, 5), (1, 4), (0, 2), (0, 1)\}$. Each point occurs with equal probability 1/4. Therefore, conditional probability

$$P(y \in S|x = 1) = \frac{P(S \cap E)}{P(E)} = \frac{P(1,5) + P(1,4)}{1/2} = 1.$$

This may be a trivial exercise in verifying the framework of the Bayesian formula since the Venn diagram clearly implies that $P(S) = 1$ independent of whatever the outcome of X. We had seen that $P(S^c) = 0$ earlier.

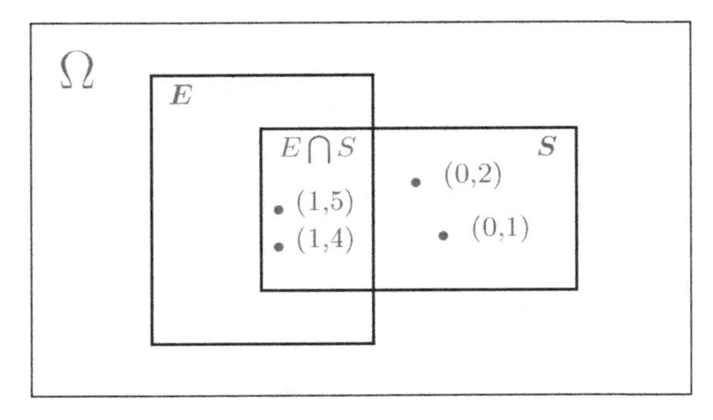

Fig. 2.4: Venn Diagram on Automakers' Events

Conditional expectation

$$
\begin{aligned}
E(\tilde{Y}|x=1) &= \sum_{y \in S} y P(y|x=1) \\
&= \sum_i y_i P(x=1, y_i)/P(x=1) \\
&= 5 \times P(1,5)/0.5 + 4 \times P(1,4)/0.5 \\
&= 5 \times \frac{1}{2} + 4 \times \frac{1}{2} = 4.50.
\end{aligned}
$$

Similarly, we can show $E(\tilde{Y}|x=0) = 1.50$.

In a more general setting involving two random variables $X \in \mathcal{R}$ and $Y \in \mathcal{R}$, the outcome probabilities may be represented by Table 2.1 and also the Venn diagram in Fig. 2.5. The table shows $P(X = x,\ Y = y)$.

Table 2.1: Joint Probabilities of Events (x, y)

	$x = 1$	$x = 2$	$x = 3$	$x = 4$	$\ldots$
$y = 1$	0.02	0.03	0.01	0.05	$\ldots$
$y = 2$	0.03	0.03	0.02	0.01	$\ldots$
$y = 3$	0.04	0.03	0.01	0.02	$\ldots$
$y = 4$	0.05	0.02	0.01	0.05	$\ldots$
$\vdots$	$\vdots$	$\vdots$	$\vdots$	$\vdots$	$\ddots$

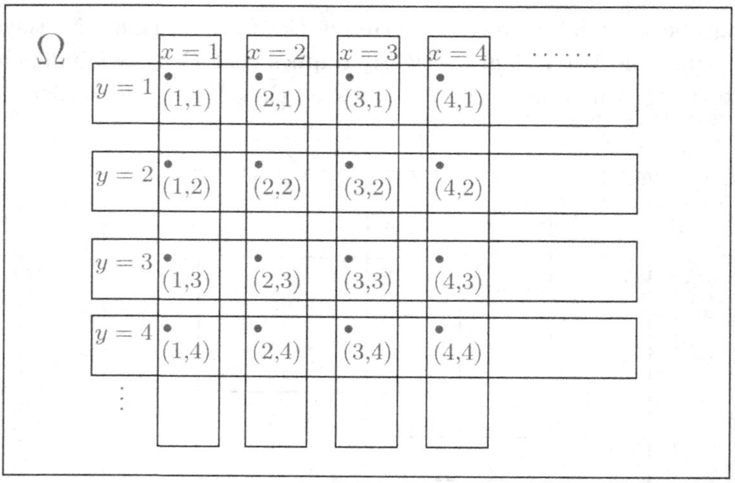

Fig. 2.5: Venn Diagram of Joint Probabilities

Sample points in the Venn diagram are (x, y). The set $\{x = k\}$ denotes $\{(k,1),(k,2),\ldots\}$, and the set $\{y = j\}$ denotes $\{(1,j),(2,j),\ldots\}$. Thus in a multivariable sample space, using the Venn diagram, it is natural to use marginal variable values $x = 1$, $x = 2$, etc. as defining sets, i.e., all elements with $x = 1$, or with $x = 2$, etc. It is easily seen that $P(Y = j) = P(1,j) + P(2,j) + \cdots$ etc., and $P(X = k) = P(k,1) + P(k,2) + \cdots$ etc. $P(Y = j|X = k) = P(k,j)/P(X = k)$.

When the probability distribution is continuous so that PDF $f(x, y)$ is used, it is often more convenient to employ analytical methods than use Venn diagrams which can become quite clumsy.

For continuous $x \in R(x)$ and $y \in R(y)$, where $R(x)$, $R(y)$ denote the support sets over which the integrations take place,

$$\begin{aligned}
E^{X,Y}(Y) &= \int_{R(x)} \int_{R(y)} y\, f(x,y)\, dy\, dx \\
&= \int_{R(y)} y \left(\int_{R(x)} f(x,y)\, dx \right) dy \\
&= \int_{R(y)} y f_Y(y)\, dy = E^Y(y),
\end{aligned}$$

where $f_Y(y)$ indicates the marginal PDF of RV Y integrated out from the joint PDF $f(x, y)$. For clarity, we have put superscripts on the expectation operator $E(\cdot)$ denoting the joint distribution (X, Y) or marginal distribution Y underlying the integration.

In addition,

$$\begin{aligned}
E^{X,Y}(Y) &= \int_{R(x)} \int_{R(y)} y\, f(x,y)\, dy\, dx \\
&= \int_{R(x)} \int_{R(y)} y\, \frac{f(x,y)}{f_X(x)} f_X(x)\, dy\, dx \\
&= \int_{R(x)} \left(\int_{R(y)} y f(y|x)\, dy \right) f_X(x)\, dx \\
&= \int_{R(x)} E^{Y|X}(Y|X) f_X(x)\, dx \\
&= E^X\, E^{Y|X}(Y|X).
\end{aligned}$$

Conditional probability is just a special case of conditional expectation if we put $Y = 1_{y \le k}$ taking values 1 or 0. In this case

$$E^{Y|X}(Y|X) = \int_{R(y)} 1_{y \le k} f(y|x)\, dy = P(Y \le k|x).$$

It should also be noted that $E^{Y|X}(Y|X)$ is a RV varying with X, and can be expressed as a certain function $g(X)$ of X.

Application: Value-at-Risk

Suppose RV X is distributed as $N(\mu, \sigma^2)$ and X is the change in capital of a financial institution as a result of market forces on its investments. If $X > (<)\ 0$, there is a gain (loss). At $(1-q)$ level of confidence, the worst loss is v whereby $\int_{-\infty}^{v} f(x)dx = q$ where the left-hand side (LHS) is the area under the curve of the normally distributed X from $-\infty$ to v. $f(x)$ is the PDF of the normal X. v (or usually its absolute value $|v|$, v being understood to be negative) is called the investment's (absolute) value-at-risk (VaR) at the $(1-q)$ level of confidence.

Suppose $|v|$ is established as the VaR(q) at qth percentile, or equivalently VaR at the $(1-q)$ level of confidence. The expected loss conditional on hitting the VaR is

$$E(X|X < v) = \frac{\int_{-\infty}^{v} x f(x)dx}{\int_{-\infty}^{v} f(x)dx}.$$

In risk language, this quantity is also called the conditional value-at-risk or expected shortfall or conditional loss or expected tail loss. We shall find this quantity as follows, remembering that $X \sim N(\mu, \sigma^2)$.

First, let $z = \frac{x-\mu}{\sigma}$, so $z \in$ RV $Z \sim N(0,1)$. Now,

$$\begin{aligned} E(X|X < v) &= E(\mu + \sigma Z|\mu + \sigma Z < v) \\ &= \mu + \sigma E\left(Z|Z < \frac{v-\mu}{\sigma}\right). \end{aligned}$$

The second step is to find $E(Z|Z < v')$ where $v' = \frac{v-\mu}{\sigma}$.
Let $c = \frac{1}{\sqrt{2\pi}}$. Since

$$d\left(ce^{-\frac{1}{2}z^2}\right) = -z\left(ce^{-\frac{1}{2}z^2}\right) dz,$$

integrating over $(-\infty, v']$, we have

$$\left[ce^{-\frac{1}{2}z^2}\right]_{-\infty}^{v'} = -\int_{-\infty}^{v'} zce^{-\frac{1}{2}z^2} dz + k.$$

Therefore

$$\phi(v') - 0 = -\int_{-\infty}^{v'} zce^{-\frac{1}{2}z^2} dz + k.$$

As $v' \uparrow +\infty$, the LHS approaches zero. First term on the RHS approaches the mean of the standard normal RV, which is zero. Therefore, constant of integration $k = 0$. Hence

$$\int_{-\infty}^{v'} zce^{-\frac{1}{2}z^2} dz = -\phi(v').$$

Then,

$$\begin{aligned} E(Z|Z < v') &= \frac{\int_{-\infty}^{v'} zf(z)dz}{\int_{-\infty}^{v'} f(z)dz} \\ &= -\frac{\phi(v')}{\Phi(v')}. \end{aligned}$$

Thus,

$$E(X|X < v) = \mu - \sigma\frac{\phi(v')}{\Phi(v')} < v.$$

2.4. Moving Across Time

So far, we have dealt with joint probability distribution $f^{X,Y}(x,y)$ and conditional probability distribution $f^{Y|X}(y|x)$ where each of these yields a certain number in $\mathcal{R}^1$, and are $\in [0,1]$. Random variables X and Y occur at one point in time.

When we are in a single time period or instance, conditioning is quite simple using the conditional probability rule in Eq. (2.1) or the extended Bayes' formula in (2.2). It all happens within the same time–space on the same Venn diagram.

However, when events unfold across time, more apparatus is needed.[5] To generate some more serious results involving conditional probability

[5] *Introduction to Mathematical Finance*: Discrete Time Models by SR Pliska (1997), Blackwell Publishers, has a detailed description of some of the concepts here.

distributions (over and above the conditional probabilities of events shown earlier), and which leads to martingale theory at a deeper end of probability theories and applications, we need a more formal structure and architecture with regard to the probability space and σ-fields.

First, it is needful to explain some ideas about information structure and information sets. Suppose the sample space is $\Omega = \{\omega_1, \omega_2, \omega_3\}$. The largest field or algebra is the set

$$\mathcal{F}_b = \{\phi, \Omega, \{\omega_1\}, \{\omega_2\}, \{\omega_3\}, \{\omega_1, \omega_2\}, \{\omega_1, \omega_3\}, \{\omega_2, \omega_3\}\}$$

consisting of $2^3 = 8$ events $E_i \in \mathcal{F}$. It is also called a field generated by $\{\{\omega_1\}, \{\omega_2\}, \{\omega_3\}\}$ which means the smallest field containing $\{\omega_1\}$, $\{\omega_2\}$, and $\{\omega_3\}$. Thus we can always find a field by picking a subset of Ω and using it to generate a field satisfying conditions (1a), (1b), and (1c) of Chap. 1. Obviously, the bigger the subset, the bigger the generated field. Fields will always include ϕ and Ω. The largest field associated with sample space Ω, i.e., generated by Ω, will contain all elements of Ω and all possible unions of these elements, as well as ϕ. In a continuous state space, $\mathcal{F}$ will be a σ-algebra. Note that in the above case, $\mathcal{F}_b = \mathcal{F}$, the algebra generated by Ω. The smallest field is $\mathcal{F}_0 = \{\phi, \Omega\}$.

A smaller field than $\mathcal{F}_b$ could be $\mathcal{F}_a = \{\phi, \Omega, \{\omega_3\}, \{\omega_1, \omega_2\}\}$, being generated by $\{\{\omega_3\}, \{\omega_1, \omega_2\}\}$. The algebras or fields are collections of subsets of Ω or more conveniently termed as collections of events.

For $\mathcal{F}_b$, let the events E_i at $t = 2$ be defined as

$$E_0 \equiv e_0^b = \phi \quad \text{with} \quad P(e_0^b) = P(\phi) = 0$$

$$E_\Omega \equiv e_\Omega^b = \Omega \quad \text{with} \quad P(e_\Omega^b) = P(\Omega) = 1$$

$$E_1 \equiv e_1^b = \{\omega_1\} \quad \text{with} \quad P(e_1^b) = P(\{\omega_1\}) = p_1$$

$$E_2 \equiv e_2^b = \{\omega_2\} \quad \text{with} \quad P(e_2^b) = P(\{\omega_2\}) = p_2$$

$$E_3 \equiv e_3^b = \{\omega_3\} \quad \text{with} \quad P(e_3^b) = P(\{\omega_3\}) = p_3$$

$$E_4 \equiv e_4^b = \{\omega_1, \omega_2\} \quad \text{with} \quad P(e_4^b) = P(\{\omega_1, \omega_2\}) = p_1 + p_2$$

$$E_5 \equiv e_5^b = \{\omega_1, \omega_3\} \quad \text{with} \quad P(e_5^b) = P(\{\omega_1, \omega_3\}) = p_1 + p_3$$

$$E_6 \equiv e_6^b = \{\omega_2, \omega_3\} \quad \text{with} \quad P(e_6^b) = P(\{\omega_2, \omega_3\}) = p_2 + p_3.$$

For $\mathcal{F}_a$, let the events E_i at $t = 1$ be defined as

$$E_0 \equiv e_0^a = \phi \quad \text{with} \quad P(e_0^a) = P(\phi) = 0$$

$$E_\Omega \equiv e_\Omega^a = \Omega \quad \text{with} \quad P(e_\Omega^a) = P(\Omega) = 1$$

$$E_1 \equiv e_1^a = \{\omega_1, \omega_2\} \quad \text{with} \quad P(e_1^a) = P(\{\omega_1, \omega_2\}) = p_1 + p_2$$

$$E_2 \equiv e_2^a = \{\omega_3\} \quad \text{with} \quad P(e_2^a) = P(\{\omega_3\}) = p_3.$$

Now, over time in the period $[0, T]$, the information structure in the market is a time-sequence of increasing algebras such that each future algebra is a superset of past algebras (or past algebras are subsets of future algebras). An information structure can be conveniently represented by an evolving tree in Fig. 2.6 as follows. The events corresponding to nodes on the information tree show increasing partitioning of existing sets as time moves forward into the future. Suppose there are 2 discrete periods so that there are time points $t = 0$, $t = 1$, and $t = T = 2$.

Clearly, $\mathcal{F}_0 \subset \mathcal{F}_a \subset \mathcal{F}_b$. The information structure, which is assumed to be known by all investors at the start $t = 0$, says that at $t = 0$, there is trivial information: either null event ϕ or that all future events are possible, i.e., Ω, with probability 1. At $t = 1$, events are either e_1^a with probability $p_1 + p_2$ or e_2^a with probability p_3. Then, at $t = 2$, it is either e_1^b or e_2^b if e_1^a at $t = 1$, or e_3^b if e_2^a at $t = 1$. Thus, at $t = 1$, the information set is $\mathcal{F}_a$ whereby investors can tell which event in $\mathcal{F}_a$ has occurred. At $t = 1$, investors cannot tell which of e_1^b or e_2^b has occurred since these events do not belong to $\mathcal{F}_a$. However, at $t = 2$, there is greater resolution of uncertainty, and the information set enlarges (becomes finer) to become field $\mathcal{F}_b$ so that the investors then know which of e_1^b, e_2^b, or e_3^b has occurred.

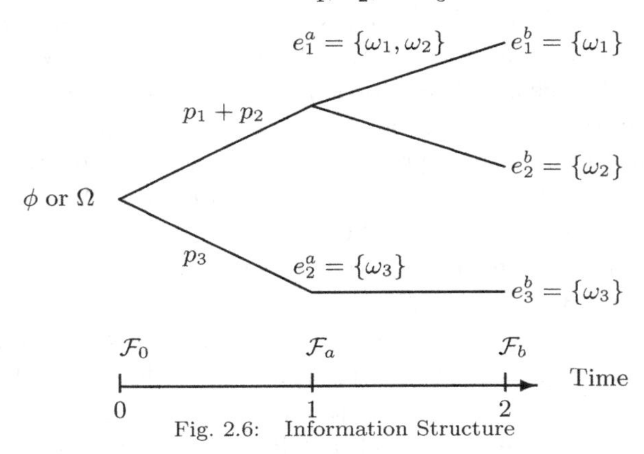

Fig. 2.6: Information Structure

For the algebra on the LHS in Fig. 2.7 below, we can find a RV Y as a mapping

$$Y : E_i \mapsto Y(E_i), \text{ for } i = 1, 2, 3, \ldots$$

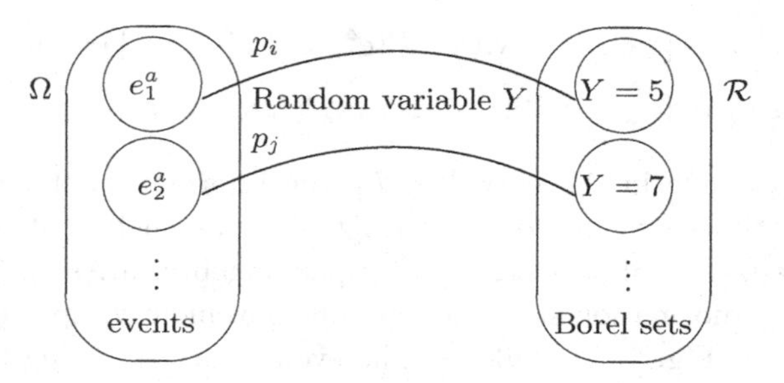

Fig. 2.7: Random Variable as a Function $Y(t = 1)$

Clearly, the RV is a function from events in $\mathcal{F}_a$ to $\mathcal{R}$. Sometimes, more than one sample point maps to the same value that the RV takes. But when we collect all those sample points that map to the same value on the RHS in $\mathcal{R}$, we can group those points as belonging to the same event and redefine the RV if necessary. Then, there is a one-to-one correspondence or bijection between disjoint events and unique values in $\mathcal{R}$ when the range set is appropriately defined. In the earlier example, a RV can be defined so that event $E_1 \equiv e_1^a$ or $E_1 = \{\omega_1, \omega_2\}$ is equivalent to RV $\tilde{Y}$ with its value $y = 5$. Similarly, event $E_2 \equiv e_2^a$ or $E_2 = \{\omega_3\}$ is equivalent to RV $\tilde{Y}$ with its value $y = 7$.

Thus, $P(Y = 5) = P(E_1)$, and $P(Y = 7) = P(E_2)$. Since we say algebra or field $\mathcal{F}_a$ is generated by E_1 and E_2, we can equivalently say $\mathcal{F}_a$ is generated by $Y = 5$ and $Y = 7$. Thus, in general, we can say an algebra at $t = j$ is generated by events at $t = j$ or generated by the RV Y at $t = j$, i.e., $\mathcal{F}_t = \sigma(\tilde{Y}_t)$. In this case, when we say a RV Y is measurable on field $\mathcal{F}$, we may also say that it is measurable on $\sigma(Y)$, the field generated by Y.

For continuous RVs, a properly defined RV may be constructed as a bijection from a σ-field to Borel sets in $\mathcal{R}$ (for discrete RV, it is mapped onto at most countably infinite numbers in $\mathcal{R}$), so any probability measure p on the events in the σ-field is equivalent to the probability measure p on the corresponding Borel sets in $\mathcal{R}$.

We have shown how an information set at time t, Φ_t, corresponds to a particular field at t, $\mathcal{F}_t$. Consider information set $\Phi_b \equiv \mathcal{F}_b$ where $\mathcal{F}_b = \{\phi, \Omega, \{\omega_1\}, \{\omega_2\}, \{\omega_3\}, \{\omega_1, \omega_2\}, \{\omega_1, \omega_3\}, \text{ or } \{\omega_2, \omega_3\}\}$. With a properly defined probability space, we can always find the conditional probability $P(E_i|\Phi_t)$, or equivalently $P(E_i|\mathcal{F}_t)$, on the measurable events $\{E_i\}$.

The probability Table 2.2 for RV X at $t = 0$ can be shown as follows. The probabilities are in the distribution $P(X|\mathcal{F}_0)$ or $P(X|\phi, \Omega)$. There is no information about which event E_i has occurred, so the unconditional probabilities are p_i's.

Table 2.2: Unconditional Distribution

State ω	RV $X(\omega)$	Probability
ω_1	$X_1 = X(\omega_1)$	p_1
ω_2	$X_2 = X(\omega_2)$	p_2
ω_3	$X_3 = X(\omega_3)$	p_3

When we condition on an information set or algebra $\mathcal{F}_t$, the defined meaning is that we are conditioning on the event which is known to have occurred at t. However, for analytical purpose on an *ex ante* basis, the conditioning is done on all events of $\mathcal{F}_t$ so that we know if one event occurs, what the conditional probability of future events given that is.

Based on the last statement, we are looking at a two-sided table as follows for $\mathcal{F}_a$. The events in $\mathcal{F}_a$ are $\{\omega_1, \omega_2\}$, and ω_3. We omit the trivial events of null and Ω. Each column in the following Table 2.3 shows conditional probability of a sample point (or possible future event) given the various events in $\mathcal{F}_a$.

Note that when event $\{\omega_1, \omega_2\}$ occurred, it is not possible to distinguish which state ω_1 or ω_2 had occurred.

Earlier we see how events and RV values may be put in one-to-one correspondence. Conditioning on the events of algebra $\mathcal{F}_a$ can also be written as conditioning on RV $\tilde{Y}$. In particular, $P(\{\omega_j\}|\{\omega_1, \omega_2\}) \equiv P(\{\omega_j\}|Y = 5)$. In addition, $P(\{\omega_j\}|\{\omega_3\}) \equiv P(\{\omega_j\}|Y = 7)$.

For $\mathcal{F}_b$, the conditional probabilities are shown in Table 2.4.

Table 2.3: Conditional Probabilities Given $\mathcal{F}_a$

Event E_i:	$P(\{\omega_1\}|E_i)$	$P(\{\omega_2\}|E_i)$	$P(\{\omega_3\}|E_i)$
$\{\omega_1, \omega_2\}$	$\dfrac{p_1}{p_1 + p_2}$	$\dfrac{p_2}{p_1 + p_2}$	0
$\{\omega_3\}$	0	0	1

Table 2.4: Conditional Probabilities Given $\mathcal{F}_b$

Event E_i:	$P(\{\omega_1\}|E_i)$	$P(\{\omega_2\}|E_i)$	$P(\{\omega_3\}|E_i)$
$\{\omega_1\}$	1	0	0
$\{\omega_2\}$	0	1	0
$\{\omega_3\}$	0	0	1

We may define $P(X|\mathcal{F}_t)$ as a $M \times N$ matrix or table in which the ijth element is $P(\{\omega_j\}|E_i)$, (recall in Table 2.2, $\exists$ (there exists) RV $X \ni$ (such that) $\omega_j \mapsto x \in X$) and there are N simple sample points $\{\omega_j\}$ in Ω. M is the number of events, $E_1, E_2, \ldots, E_M$, excluding ϕ, Ω, in $\mathcal{F}_t$. We can also write $P(X|\mathcal{F}_t)$ as $P(X|\Phi_t)$. When $\exists$ RV Y with bijection $g : E_i \mapsto y \in Y$, then we can also write $P(X|\mathcal{F}_t) \equiv P(X|Y)$.

Now, we define $E(X|\mathcal{F}_t) \equiv E(X|\Phi_t) \equiv E(X|Y)$ as a $M \times 1$ vector

$$\begin{pmatrix} E(X|E_1) \\ E(X|E_2) \\ \vdots \\ E(X|E_M) \end{pmatrix}.$$

Back to the example of $\Omega = \{\omega_1, \omega_2, \omega_3\}$ and information $\mathcal{F}_0$, $\mathcal{F}_a$, and $\mathcal{F}_b$, conditional on event $E_1 = \{\omega_1, \omega_2\}$ at $t = 1$,

$$\begin{aligned} E(X|\{\omega_1, \omega_2\} \in \mathcal{F}_a) &\equiv Y_a(\{\omega_1, \omega_2\}) \\ &= \sum_{i=1}^{3} X(\omega_i) \times P(\omega_i|E_1) \\ &= X_1 \frac{p_1}{p_1 + p_2} + X_2 \frac{p_2}{p_1 + p_2} + X_3 \times 0 \\ &= \frac{X_1 p_1 + X_2 p_2}{p_1 + p_2}. \end{aligned}$$

Conditional on event $E_2 = \{\omega_3\}$,

$$\begin{aligned} E(X|\{\omega_3\} \in \mathcal{F}_a) \equiv Y_a(\{\omega_3\}) &= \sum_{i=1}^{3} X(\omega_i) \times P(\omega_i|E_2) \\ &= X_1 \times 0 + X_2 \times 0 + X_3 \times 1 \\ &= X_3. \end{aligned}$$

Hence

$$\begin{aligned} E(X|\mathcal{F}_a) &= \begin{pmatrix} Y_a(E_1) \\ Y_a(E_2) \end{pmatrix} \\ &= \begin{pmatrix} \frac{X_1 p_1 + X_2 p_2}{p_1 + p_2} \\ X_3 \end{pmatrix}. \end{aligned}$$

If we take the unconditional expectation, the scalar

$$\begin{aligned}E[E(X|\mathcal{F}_a)] &= P(\{\omega_1,\omega_2\}) \times Y_a(\{\omega_1,\omega_2\}) + P(\{\omega_3\}) \times Y_a(\{\omega_3\}) \\ &= (p_1+p_2)\frac{X_1p_1+X_2p_2}{p_1+p_2} + p_3X_3 \\ &= p_1X_1+p_2X_2+p_3X_3\end{aligned}$$

which is equal to $E[X]$.

Under information set $\mathcal{F}_b$, $E(X|\mathcal{F}_b)$ is a vector $[E(X|E_i \in \mathcal{F}_b)]$.

For event $\{\omega_1\}$,

$$\begin{aligned}E(X|\{\omega_1\} \in \mathcal{F}_b) \equiv Y_b(\{\omega_1\}) &= \sum_{i=1}^{3} X(\omega_i) \times P(\omega_i|\omega_1) \\ &= X_1 \times 1 + 0 + 0 \\ &= X_1.\end{aligned}$$

For event $\{\omega_2\}$,

$$\begin{aligned}E(X|\{\omega_2\} \in \mathcal{F}_b) \equiv Y_b(\{\omega_2\}) &= \sum_{i=1}^{3} X(\omega_i) \times P(\omega_i|\omega_2) \\ &= 0 + X_2 \times 1 + 0 \\ &= X_2.\end{aligned}$$

For event $\{\omega_3\}$,

$$\begin{aligned}E(X|\{\omega_3\} \in \mathcal{F}_b) \equiv Y_b(\{\omega_3\}) &= \sum_{i=1}^{3} X(\omega_i) \times P(\omega_i|\omega_3) \\ &= 0 + 0 + X_3 \times 1 \\ &= X_3.\end{aligned}$$

Hence, if X is measurable w.r.t. $\mathcal{F} \equiv \sigma(X)$, $E(X|\mathcal{F}) = X$.

If we take the unconditional expectation

$$E[E(X|\mathcal{F}_b)] = p_1X_1 + p_2X_2 + p_3X_3,$$

which is equal to $E[X]$.

How about $E[E(X|\mathcal{F}_b)|\mathcal{F}_a]$? We saw how conditioning on $\mathcal{F}_a \neq \mathcal{F}_0$ produces a RV.

Notationally, $E[E(X|\mathcal{F}_b)|\mathcal{F}_a] = E[Y_b|\mathcal{F}_a]$. Conditional on event $E_1 = \{\omega_1, \omega_2\} \in \mathcal{F}_a$,

$$\begin{aligned} E(Y_b|E_1) &= \sum_{i=1}^{3} Y_b(\omega_i) \times P(\omega_i|E_1) \\ &= X_1 \frac{p_1}{p_1 + p_2} + X_2 \frac{p_2}{p_1 + p_2} + X_3 \times 0 \\ &= \frac{X_1 p_1 + X_2 p_2}{p_1 + p_2}. \end{aligned}$$

Conditional on event $E_2 = \{\omega_3\} \in \mathcal{F}_a$,

$$\begin{aligned} E(Y_b|E_2) &= \sum_{i=1}^{3} Y_b(\omega_i) \times P(\omega_i|E_2) \\ &= X_1 \times 0 + X_2 \times 0 + X_3 \times 1 \\ &= X_3. \end{aligned}$$

The probability table for $E[E(X|\mathcal{F}_b)|\mathcal{F}_a]$ can be shown below.

Event E_i:	RV $E[E(X\|\mathcal{F}_b)\|E_i]$
$\{\omega_1, \omega_2\}$	$\frac{X_1 p_1 + X_2 p_2}{p_1 + p_2}$
$\{\omega_3\}$	X_3

This is identical with the RV $E(X|\mathcal{F}_a)$ a.s. It is important to recapitulate the usage of information sets. When we say an investor has an information set Φ_t equivalent to the algebra $\mathcal{F}_t$, we mean that the investor would have known which particular event had occurred in that algebra. The vector $P(X(\omega_i)|\Phi_t) \equiv P(X(\omega_i)|\mathcal{F}_t)$ provides the conditional probability on an *ex ante* basis of each $\{\omega_i\}$ given each event in $\Phi_t \equiv \mathcal{F}_t$.

2.5. Law of Iterated Expectations

From the previous sub-section, we have the law of iterated expectations. To be more precise, consider the law as expressed in the following lemma.

Lemma 2.1 Suppose $\mathcal{G} \subset \mathcal{F}$, then

$$E(E(X|\mathcal{F})|\mathcal{G}) = E(X|\mathcal{G}) = E(E(X|\mathcal{G})|\mathcal{F}), \text{ a.s.}$$

where $\mathcal{G}$ is a sub-algebra of $\mathcal{F}$. ■

We note that for any σ-field $\mathcal{G}$, $E(E(X|\mathcal{G})|\mathcal{G}) = E(X|\mathcal{G})$. We also note that $E(X|\mathcal{F}_0) = E(X)$, and hence as a special case of the lemma, $E[E(X|\mathcal{G})|\mathcal{F}_0] = E(X|\mathcal{F}_0) = E(X)$. Or for a suitable RV Z, $E[(EX|Z)] = E(X)$. This is sometimes called a smoothing lemma.

Proof. Let $\mathcal{F} = \sigma(W)$ and $\mathcal{G} = \sigma(Z)$ where W, Z are RVs. Since $E(X|W)$ is a RV in W,

$$\begin{aligned} E(E(X|W)|Z) &= \int_{R(w)} \left(\int_{R(x)} x f_{X|W}(x|w) dx \right) f_{W|Z}(w|z)\, dw \\ &= \int_{R(w)} \left(\int_{R(x)} x f_{X|W,Z}(x|w,z)\, dx \right) \frac{f_{W,Z}(w,z)}{f_Z(z)} dw \end{aligned}$$

where we use the fact $\mathcal{G} \subset \mathcal{F} \Leftrightarrow f(x|w,z) = f(x|w)$ since $\{w's\} \subset \{z's\}$, and if event w occurs, z occurs w.p.1, so z yields no additional information over w. Then

$$\begin{aligned} E(E(X|W)|Z) &= \int_{R(w)} \left(\int_{R(x)} x \frac{f_{X,W,Z}(x,w,z)}{f_{W,Z}(w,z)}\, dx \right) \frac{f_{W,Z}(w,z)}{f_Z(z)} dw \\ &= \int_{R(x)} x \left(\int_{R(w)} \frac{f_{X,W,Z}(x,w,z)}{f_Z(z)}\, dw \right) dx \\ &= \int_{R(x)} x \frac{1}{f_Z(z)} \left(\int_{R(w)} f_{X,W,Z}(x,w,z)\, dw \right) dx \\ &= \int_{R(x)} x \frac{1}{f_Z(z)} f_{X,Z}(x,z)\, dx \\ &= \int_{R(x)} x f_{X|Z}(x|z)\, dx = E(X|Z). \end{aligned}$$

■

Consider another useful result:

$$\begin{aligned} E(E[Y|X]g(X)) &= \int_{R(x)} \left(\int_{R(y)} y f(y|x) dy \right) g(x) f_X(x) dx \\ &= \int_{R(x)} \left(\int_{R(y)} y\, g(x) f(y|x) f_X(x)\, dy \right) dx \end{aligned}$$

$$= \int_{R(x)} \int_{R(y)} y\, g(x)\, f(x,y)\, dy\, dx$$

$$= E^{XY}(Y g(X)). \quad (2.3)$$

As a corollary from Eq. (2.3), we put $g(X) \equiv 1$.

Corollary 2.1

$$E(E[Y|X]) = E^{XY}(Y) = E(Y) \triangleq E(Y|\mathcal{F}_0). \qquad \blacksquare$$

In the earlier discussion, there is a time dimension, and information set $\mathcal{F}_0 = \{\phi, \Omega\}$ occurs at time $t = 0$, $\mathcal{F}_a \equiv \mathcal{F}_1$ at time $t = 1$, and $\mathcal{F}_b \equiv \mathcal{F}_2$ at time $t = 2$. We can see that the information sets become finer and richer as time progresses: $\mathcal{F}_0 \subseteq \mathcal{F}_1 \subseteq \mathcal{F}_2 \ldots$. This fits with intuition about how rational agents would know more as time progresses (assuming no loss of memory)! In such a time setup, the information set stochastic process $\mathcal{F}_t$ is called a filtration. Sometimes the probability space is enhanced to show a filtration, i.e., a filtered probability space $(\Omega, \mathcal{F}, \{\mathcal{F}_t\}, \mathcal{P})$.

If there is a sequence of RVs Y_t that are measurable with respect to each $\mathcal{F}_t$, then we say the sequence $\{Y_t\}$ is adapted w.r.t. the filtration $\{\mathcal{F}_t\}$. $\{\omega_1, \omega_2\}$ and $\{\omega_3\}$ are adapted to $\mathcal{F}_a$, so are $\{\omega_1\}$, $\{\omega_2\}$, $\{\omega_3\}$, $\{\omega_1, \omega_2\}$, $\{\omega_1, \omega_3\}$, and $\{\omega_2, \omega_3\}$ adapted to $\mathcal{F}_b$. Similarly, Y_a is adapted to $\mathcal{F}_a$, and Y_b is adapted to $\mathcal{F}_b$.

Applications

The Law of Iterated Expectations in application to a filtration says that if $\mathcal{F}_t \subseteq \mathcal{F}_{t+1} \subseteq \mathcal{F}_{t+2} \subseteq \ldots$, then

$$E(E(X|\mathcal{F}_{t+1})|\mathcal{F}_t) = E(X|\mathcal{F}_t).$$

This result is used commonly in testing asset pricing models in the empirical finance and economics literature. In finance theory under rational expectations, a traded asset price is determined by

$$p_t = E(m_{t+1} p_{t+1} | \Phi_t)$$

where $m_{t+1} = \left(\beta \frac{U'(C_{t+1})}{U'(C_t)}\right)$ is the marginal rate of substitution or pricing kernel of an economy agent consuming C_t with von Neumann–Morgenstern utility function $U(\cdot), p_t$ is the asset price and Φ_t is the agent's information set. However, in trying to test this model, an econometrician cannot observe the agent's information Φ_t. At time t, p_t is known, so it can be treated as a constant. Then, we can take iterated expectations on the asset pricing formula to obtain the unconditional version $p_t = E(m_{t+1}p_{t+1})$ consistent with the conditional one, and be able to test it using econometric methods such as the generalized method of moments.

The results above can also be used to show that the best predictor (in the sense of minimum mean square error) of a random variable Y that is correlated with random variable X is given by the conditional mean of Y given $X = x$, or $E(Y|X = x)$. Y and X are jointly distributed.

Suppose function $\pi(X)$ is the best predictor of Y. The mean square error of prediction is

$$\begin{aligned}
&E[Y - \pi(X)]^2 \\
&\quad = E\{[Y - E(Y|X)] + [E(Y|X) - \pi(X)]\}^2 \\
&\quad = E[Y - E(Y|X)]^2 + E[E(Y|X) - \pi(X)]^2 \\
&\qquad + 2E\{[Y - E(Y|X)][E(Y|X) - \pi(X)]\} \\
&\quad = E[Y - E(Y|X)]^2 + E[E(Y|X) - \pi(X)]^2 \\
&\quad \geq E[Y - E(Y|X)]^2.
\end{aligned}$$

Therefore, the best predictor in the sense of minimum possible mean square error is $\pi(X) = E(Y|X)$ which is a RV varying with X. For now, let $E(Y|X) = g(X)$. In the above derivation, the middle term becomes zero by using Eq. (2.3), as seen below:

$$\begin{aligned}
&E\{[Y - E(Y|X)][E(Y|X) - \pi(X)]\} \\
&\quad = E\{Y\ E(Y|X)\} - E\{Y\ \pi(X)\} \\
&\qquad - E\{E(Y|X)\ E(Y|X)\} + E\{E(Y|X)\ \pi(X)\} \\
&\quad = E\{Y\ g(X)\} - E\{Y\ \pi(X)\} \\
&\qquad - E\{E(Y\ g(X)|X)\} + E\{E(Y\ \pi(X)|X)\} \\
&\quad = E\{Y\ g(X)\} - E\{Y\ \pi(X)\} - E\{Y\ g(X)\} + E\{Y\ \pi(X)\} \\
&\quad = 0.
\end{aligned}$$

The minimum mean square error criterion allows the simple result of $E(Y|X)$ being best predictor, and since $E(Y|X)$ is analytically more tractable when joint distributions of (X, Y) are provided, the same criterion is used commonly in ordinary least squares econometric estimation methods.

2.6. Modeling Default Correlations

We employ Bernoulli distribution to model correlation of the default events as distinct from correlation of default probabilities which occur in another type of analysis.

Suppose what we are interested about two firms is whether they default or not. Thus, we can represent them by RVs X and Y, each taking the values 0 or 1, with 1 representing default, and 0 otherwise.

Let $P(X = 1) = p > 0$ and $P(Y = 1) = q > 0$. Let $P(X = 1, Y = 1) = p_{11}$, $P(X = 1, Y = 0) = p_{10}$, $P(X = 0, Y = 1) = p_{01}$, and $P(X = 0, Y = 0) = p_{00}$. Note, therefore, $p_{11} + p_{01} + p_{10} + p_{00} = 1$, $p_{10} + p_{11} = p$, and $p_{01} + p_{11} = q$.

Using the generic definition of correlation between X and Y,

$$r = \frac{E(XY) - E(X)E(Y)}{\sigma_X \sigma_Y} = \frac{p_{11} - pq}{\sqrt{p(1-p)q(1-q)}}.$$

Suppose we are considering a basket of returns of the two firms, and the event of first default, that is either one or both of the firms default (during a specified period). Probability of this event is $P(\{X = 1\} \cup \{Y = 1\}) = 1 - p_{00} = \theta$.

Previously

$$\begin{aligned} r &= \frac{p_{11} - pq}{\sqrt{p(1-p)q(1-q)}} \\ &= \frac{(p + q + p_{00} - 1) - pq}{\sqrt{p(1-p)q(1-q)}} \\ &= \frac{p_{00} - (1-p)(1-q)}{\sqrt{p(1-p)q(1-q)}}. \end{aligned}$$

Hence

$$\theta = 1 - p_{00} = 1 - (1-p)(1-q) - r\sqrt{p(1-p)q(1-q)}. \tag{2.4}$$

Thus, holding p and q constant, if the Bernoulli default correlation r increases (decreases), then the event of first-to-default decreases (increases) in probability. In reality, as in the recent 2008 economic crisis, however, both default correlation and individual firms' default probabilities, p, q, tend to increase in tandem, in which case the event becomes even more likely. This observation is similar to a high correlation in p, q, and p_{11}. The latter can be easily explained by Merton-type structural model where asset prices of both firms fall drastically versus their liabilities so as to induce increases in p, q, and also p_{11}, and hence also r.

The above initial implication is based on holding p and q constant while increasing r (or equivalently p_{11}) as seen in Eq. (2.4). In other words, it is a result conditional on holding p and q as constants. In real life, many — if not all — of our decisions and perceptions are conditioned on some things. We go out with an umbrella conditional on the belief that it is likely to rain. We do not go out with an umbrella conditional on the belief that it is not likely to rain, and so on.

Suppose now we fix the primitives or exogenous probabilities as conditioned on the correlation coefficient r itself. The probability table, Table 2.5, is shown below.

Table 2.5: Probability Table

Probability	$X = 0$	$X = 1$
$Y = 0$	p_{00}	$E(X) - E(XY)$
$Y = 1$	$E(Y) - E(XY)$	$E(XY)$

It can be verified that $P(X = 1) = P(X = 1, Y = 0) + P(X = 1, Y = 1) = E(X)$. Similarly, $P(Y = 1) = E(Y)$. Further, $P(X = 1, Y = 1) = E(XY)$.

Now, we model $P(X = 1) = E(X) = rp$, where p is a constant. We should ensure that for all values of r and p, $0 \leq rp \leq 1$. Similarly, let $P(Y = 1) = E(Y) = rp$. Hence, $E(X) = E(Y)$. Since r is

$$r = \frac{E(XY) - [E(X)]^2}{E(X)[1 - E(X)]},$$

then

$$\begin{aligned} E(XY) &= [E(X)]^2 + rE(X)[1 - E(X)] \\ &= (rp)^2 + r^2p(1 - rp) = r^2p(1 + p) - r^3p^2 \in (0, 1). \end{aligned}$$

The probability of the event of first-to-default is

$$\theta = 2E(X) - E(XY) = 2rp - [r^2p(1 + p) - r^3p^2].$$

Then, it can be shown that

$$\frac{d\theta}{dr} = 2p\left(1 - r - rp + \frac{3}{2}r^2p\right) > 0$$

is attainable for some regions of r and p. Thus, a converse result is obtained. Higher default correlation leads to higher probability of first-to-default.

2.7. Problem Set 2

1. Consider (bivariate) jointly normal RVs X and Y with joint PDF.

$$f(x,y) = \frac{1}{2\pi\sigma_X\sigma_Y\sqrt{1-\rho^2}}\exp(-q/2), \qquad -\infty < x, y < \infty$$

   where

$$q = \frac{1}{1-\rho^2}\left[\left(\frac{x-\mu_X}{\sigma_X}\right)^2 - 2\rho\left(\frac{x-\mu_X}{\sigma_X}\right)\left(\frac{y-\mu_Y}{\sigma_Y}\right) + \left(\frac{y-\mu_Y}{\sigma_Y}\right)^2\right].$$

   Show that $E(Y|x) = \mu_Y + \rho\frac{\sigma_Y}{\sigma_X}(x-\mu_X)$.
2. X, Y is distributed as bivariate normal, with $\mu_X = 5$, $\mu_Y = 10$, $\sigma_X^2 = 1$, and $\sigma_Y^2 = 25$. Suppose $P(Y \in (4,16)|x = 5) = 0.95$, find $\rho > 0$.
3. Show that for a normally distributed $X \sim N(\mu, \sigma^2) < v$,

$$\mu - \frac{\sigma^2\phi(v')}{\Phi(v')} < v.$$

4. Show $E(X|\mathcal{F}_b) = X$, and $E(E(X|\mathcal{F}_b)|\mathcal{F}_a) = E(X|\mathcal{F}_a)$.
5. If a firm's revenue may be modelled as $R = S - 100{,}000 + e$ where S is sales, and e is a random variable with mean zero that is independent of S, what is the best forecast of R if sales is expected at $200{,}000$? What is the best forecast of R if sales is registered as $250{,}000$? Which is a better forecast of R?
6. Find $E(g(Y)|Y = y)$ for any non-linear function g.

Chapter 3

LAWS OF PROBABILITY

Mathematics is a beautiful and logical creation of the human mind. It tests and purifies itself, and draws as close to the truth as its very axioms. It is also a great construction for the advancement of the works of science and of scientific investigations. It exists in its own right with laws, of which the central limit theorem (CLT) is one of the key results most widely applied. De Moivre provided the first example of a CLT in 1733, and remarked, "And thus in all cases it will be found, that altho' Chance produces irregularities, still the Odds will be infinitely great, that in the process of Time, those irregularities will bear no proportion to the recurrency of that Order which naturally results from original design."

3.1. Some Important Inequalities

Theorem 3.1 (Markov's Inequality): *If X is a nonnegative RV, then for $a > 0$,*

$$P(X \geq a) \leq \frac{E(X)}{a}. \qquad \blacksquare$$

Proof.

$$E(X) = \int_0^{\infty} xf(x)dx \geq \int_a^{\infty} xf(x)dx \geq a\int_a^{\infty} f(x)dx = a\,P(X \geq a).$$

Then,

$$\frac{E(X)}{a} \geq P(X \geq a). \qquad \blacksquare$$

Another way to prove makes use of the indicator function and its expectation as a probability.

Suppose RV Y is defined on the same probability space $(\Omega, \mathcal{F}, \mathcal{P})$ as X.

$$Y(\omega) = \begin{cases} 1 & \text{if } X(\omega) \geq a \\ 0 & \text{if } X(\omega) < a. \end{cases}$$

Since $X(\omega) \geq 0$, $Y(\omega) \leq \frac{X}{a}$. Thus, $E(Y) \leq \frac{E(X)}{a}$. As $E(Y) = P(X \geq a)$, the inequality is obtained.

Theorem 3.2 (Chebyshev's Inequality): *If X is a RV with mean μ and variance σ^2, then for $a > 0$,*

$$P(|X - \mu| \geq a) \leq \frac{\sigma^2}{a^2}. \quad ■$$

Proof. Using the Markov inequality, if we use $(X - \mu)^2 > 0$ as a nonnegative RV, and note that $(X-\mu)^2 \geq a^2$ if and only if $(\Leftrightarrow)$ $|X-\mu| \geq a$, then we obtain the result. ■

Another version is to use $\frac{(X-\mu)^2}{\sigma^2}$ as the nonnegative RV, then $P(|X - \mu| \geq a\sigma) \leq \frac{1}{a^2}$. From Chebyshev's inequality, we can obtain a one-sided inequality as follows.

$$P(|X - \mu| \geq a) \leq \frac{\sigma^2}{a^2}.$$

So,

$$P(X - \mu \geq a) + P(X - \mu \leq -a) \leq \frac{\sigma^2}{a^2}.$$

Hence, $P(X-\mu \geq a) \leq \frac{\sigma^2}{a^2}$. However, this is not a tight bound, meaning we can find a bound that is smaller, and therefore more effective.

The following is a proof of the Cantelli's inequality (or sometimes called Chebyshev one-tailed inequality by others).

Theorem 3.3 (Cantelli's Inequality): *If X is a RV with mean μ and variance σ^2, then for $a > 0$,*

$$P(X - \mu \geq a) \leq \frac{\sigma^2}{a^2 + \sigma^2}. \quad ■$$

For the proof, we require use of the following variance decomposition lemma. Y and S are jointly distributed.

$$\begin{aligned}\text{var}(Y) &= E(Y^2) - \{E(Y)\}^2 \\ &= E[E(Y^2|S)] - [E[E(Y|S)]]^2 \\ &= E[(\text{var}(Y|S) + [E(Y|S)]^2)] - [E[E(Y|S)]]^2 \\ &= E[\text{var}(Y|S)] + \text{var}[E(Y|S)].\end{aligned}$$

Proof. Let the joint distribution of Y and S be as follows. The form of the distribution is not restricted. $a > 0$.

Prob	p	q	$r = 1 - p - q$
$\tilde{Y}$	$> a$	$= a$	$< a$
$\tilde{S}$	1	0	-1

Note $E(Y|S \geq 0) \geq a$ since Y is either $= a$ or $> a$. Without loss of generality, assume $E(Y) = 0$. So, $E(E(Y|S)) = 0$. Thus, $(p+q)E(Y|S \geq 0) + rE(Y|S = -1) = 0$. Then, $(p+q)a + rE(Y|S = -1) \leq 0$. Hence, $E(Y|S = -1) \leq -\frac{(p+q)a}{r}$.

Now, $[E(Y|S \geq 0)]^2 \geq a^2$ and $[E(Y|S = -1)]^2 \geq \frac{(p+q)^2 a^2}{r^2}$. Therefore,

$$\begin{aligned}\text{var}[E(Y|S)] &= E[E(Y|S)]^2 - [E[E(Y|S)]]^2 \\ &= E[E(Y|S)]^2 \geq (p+q)a^2 + r\frac{(p+q)^2 a^2}{r^2}.\end{aligned}$$

Given the variance decomposition lemma, therefore

$$\begin{aligned}\text{var}(Y) &\geq \text{var}[E(Y|S)] \\ &\geq (p+q)a^2 + (p+q)^2\frac{a^2}{r} \\ &= a^2(1-r) + a^2\frac{(1-r)^2}{r} = a^2(1-r)\left[1 + \frac{(1-r)}{r}\right] \\ &= a^2\frac{(1-r)}{r} = a^2\frac{(p+q)}{1-(p+q)}.\end{aligned}$$

Thus, $\text{var}(Y) \geq a^2(p+q) + \text{var}(Y)(p+q)$.

And $p+q = P(Y \geq a) \leq \frac{\text{var}(Y)}{a^2+\text{var}(Y)}$. Put $Y = X - \mu$, and $\sigma^2 = \text{var}(X)$, then

$$P(X - \mu \geq a) \leq \frac{\sigma^2}{a^2 + \sigma^2}. \qquad \blacksquare$$

For the next inequality result, we define that a function $\Psi(x)$ on real line x is convex if for any $0 \leq \pi \leq 1$, $\pi\Psi(a) + [1-\pi]\Psi(b) \geq \Psi(\pi a + [1-\pi]b)$, for an arbitrary interval $[a, b]$ on x. This can be illustrated in Fig. 3.1 as follows. A convex function also has a nonnegative second derivative. If $\Psi(x)$ is convex, then $-\Psi(x)$ is concave.

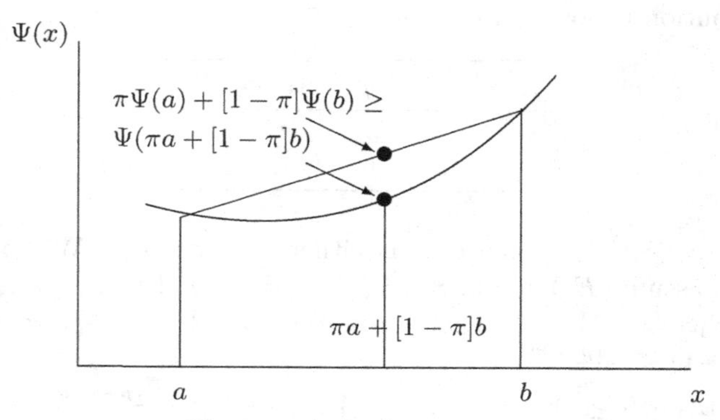

Fig. 3.1: A Convex Function

We can form any arbitrary point z within $[a, b]$ by any arbitrarily large weighted average of points x_i on $[a, b]$, i.e., $z = \sum_i \pi_i x_i$ where $\sum_i \pi_i = 1$. Then, $\sum_i \pi_i \Psi(x_i) \geq \Psi\big(\sum_i \pi_i x_i\big)$. If we use weights π_i that correspond to a probability distribution on x_i, then

$$E[\Psi(X)] \geq \Psi(E[X]).$$

This is called Jensen's inequality, and will be of use in risk aversion theory.

Jensen's inequality shown above applies for any probability distribution, not just the Bernoulli distribution. Let $E[X] = m$. Use Taylor's expansion of $\Psi(X)$ about m. Then,

$$\Psi(X) = \Psi(m) + (X - m)\Psi'(X) + \frac{(X - m)^2}{2!}\Psi''(k),$$

(where k is between X and m)

$$\geq \Psi(m) + (X - m)\Psi'(X) \quad \text{since } \Psi''(k) \geq 0.$$

So, $E[\Psi(X)] \geq \Psi(m) \equiv \Psi(E[X])$.

Next we shall see two of the most celebrated probability laws (results) in mathematical statistics.

3.2. Law of Large Numbers

Theorem 3.4 (Weak Law of Large Numbers, WLLN): *Let $\{X_i\}_{i=1,2,...}$ be a sequence of uncorrelated identical RVs each with mean μ and variance σ^2. Then, for any $\epsilon > 0$,*

$$P\left(\left|\frac{X_1 + X_2 + \cdots + X_n}{n} - \mu\right| \geq \epsilon\right) \downarrow 0, \quad \textit{as } n \uparrow \infty. \qquad \blacksquare$$

Proof. Note that

$$E\left(\frac{X_1 + X_2 + \cdots + X_n}{n}\right) = \mu,$$

and

$$\text{var}\left(\frac{X_1 + X_2 + \cdots + X_n}{n}\right) = \frac{\sigma^2}{n}.$$

From Chebyshev's inequality, therefore,

$$P\left(\left|\frac{X_1 + X_2 + \cdots + X_n}{n} - \mu\right| \geq \epsilon\right) \leq \frac{\sigma^2}{n\epsilon^2}.$$

Thus, as $n \uparrow \infty$, for any given $\epsilon > 0$, no matter how small, the probability that the sample mean deviates from μ can be made ever closer to zero. $\blacksquare$

Hence, the WLLN says that the sample mean in the uncorrelated identical variables case (and obviously also for the special case of i.i.d.) converges in probability to the population mean μ. RV X above is weakly stationary.

Applications

The WLLN gives us some degree of confidence in estimating population parameters such as mean or variance using sample mean or sample variance respectively when the sample size is large. In a laboratory setting, the estimators are reliable and would deviate from the true population parameters by sampling errors due to finite sample size. In finance and economics, using time series data, one hopes to be able to similarly estimate population mean returns, return variances and covariances, and so on. However, in financial markets, economic conditions change all the time, and the assumption that the random variable is stationary over time or even stronger, i.i.d., can be quite brave. One may have to take a really long time series to estimate any moments even if the series is unconditionally stationary.

We revisit the asset price kernel of a traded asset:

$$p_t = E(m_{t+1}p_{t+1}|\Phi_t),$$

where m_{t+1} is the marginal rate of substitution of a representative agent in the economy. Given information Φ_t which contains p_t, the above Euler equation can be rewritten as

$$E\left(m_{t+1}\frac{p_{t+1}}{p_t}\middle|\Phi_t\right) = 1.$$

Taking unconditional expectation on both sides, and noting that m_{t+1} contains population parameters such as risk aversion coefficient ρ, we have

$$E\left(m_{t+1}(\rho)\frac{p_{t+1}}{p_t}\right) = 1.$$

If we take the argument $m_{t+1}(\rho)\frac{p_{t+1}}{p_t}$ as a single RV $Z(\rho)$ across time, assume it is i.i.d. and also existence of the first two moments (so as to apply the WLLN), then there is convergence of sampling average to the population moment in probability. Suppose we pick the correct estimate $\hat{\rho}$, then

$$\frac{1}{T}\sum_{t=1}^{T} m_{t+1}(\hat{\rho})\frac{p_{t+1}}{p_t} \longrightarrow 1$$

when we employ data time series of m_t, p_t, and so on, with sample size T. The convergence to population mean 1 in this case allows a way of estimating ρ which otherwise is elusive given the nonlinear equation.

Convergence Concepts

Let x be a real number on $(0, 1] \in \mathcal{R}$. Let x be represented by a nonterminating (in the sense there are no infinite zeros at the end of the expansion) dyadic expansion, where

$$x = \sum_{n=1}^{\infty}\frac{d_n(x)}{2^n} = \frac{d_1(x)}{2} + \frac{d_2(x)}{2^2} + \frac{d_3(x)}{2^3} + \cdots\cdots$$

and $d_n(x) = 0$ or 1. Thus, $x = 0.d_1(x)d_2(x)\cdots\cdots$ in base 2, where $(d_1(x), d_2(x), d_3(x), \cdots\cdots)$ is an infinite sequence of binary digits 0 or 1.

We always use the infinite expansion when there is a choice, e.g., $1/2 = 0.011111\cdots$ instead of 0.1. Or $3/8 = 0.01011111\cdots$ instead of 0.011.

The probability of drawing a number x from $(0, 1] \in \mathcal{R}$ is thus equivalent to the probability of having an infinite sequence of binary digits where "1" may denote head and "0" may denote tail in an infinite number of coin tosses of a fair coin.

If the sample space is $\Omega = (0, 1]$, a generic sample point is x (or it can be notated as ω) and one specific sample point could be $x_1 = 1 = 0.1111\dot{1}$ representing an infinite number of all heads.

Thus, if we perform an infinite number of coin tosses and count the fraction of head occurrences as $\lim_{n\to\infty} \frac{1}{n}\sum_{i=1}^{n} d_i(x_1)$, this limit is not $1/2$, but in fact 1 for a sequence of all heads. For a sequence that corresponds to a specific sample point x' on $(0, 1]$ which ends with an infinite number of ones, it is not true that

$$\lim_{n\to\infty} \frac{1}{n}\sum_{i=1}^{n} d_i(x') = \frac{1}{2} \quad \text{for every } x' \in \Omega.$$

If $f_n(x) = \frac{1}{n}\sum_{i=1}^{n} d_i(x)$, the function f_n here is not a deterministic one (unlike e.g., $f_n(x) \equiv \frac{x}{n} + x \to x$), but a function of a RV (think of $\frac{1}{n}\sum_{i=1}^{n} d_i(x)$ as a RV that evolves probabilistically as n increases, since each $d_i(x)$ is a Bernoulli RV).

In the above, for each $x' \in M$,

$$\lim_{n\to\infty} \frac{1}{n}\sum_{i=1}^{n} d_i(x') \neq \frac{1}{2}.$$

Such x' are numbers ending with all ones (a number with binary digit 1 in the $(k-1)^{\text{th}}$ binary position and all zeros starting at the kth binary place is equal to a number with all ones starting at the kth binary place and a 0 preceding in the $(k-1)$th position — our number system is chosen to recognize only the latter nonterminating one) such as 1/2, 3/4, 1/4, 7/8, 3/8, 1/8, etc. It can be shown that $P(M) = 0$ using Lesbesgue measure since M is smaller than the set of rational numbers which has zero measure. M is called a set of measure zero or a null set.

Then, for all $x \in A = \Omega \backslash M, d_i(x)$ occurs infinitely often as 0 and also infinitely often as 1, for all i, so:

$$P\left(\left[x \in A : \lim_{n\to\infty} \frac{1}{n}\sum_{i=1}^{n} d_i(x) = \frac{1}{2}\right]\right) = 1.$$

We say that $\frac{1}{n}\sum_{i=1}^{n} d_i(x)$ converges to 1/2 almost everywhere (a.e.), or almost surely (a.s.), or "with probability 1" (w.p.1), or strong convergence or convergence with "almost all $\omega \in \Omega$" to 1/2.

Note the difference of the above strong convergence with convergence in probability discussed earlier via the Chebyshev's inequality and the WLLN, i.e.,

$$\lim_{n\to\infty} P\left(\left|\frac{1}{n}\sum_{i=1}^{n} d_i(x) - \frac{1}{2}\right| > \epsilon\right) = 0,$$

for any $\epsilon > 0$.

In probability, there is another type of convergence in L^p of RVs $X_n \equiv f_n(x) = \frac{1}{n}\sum_{i=1}^{n} d_i(x)$ to $X \equiv f(x) = 1/2$, viz.

$$\lim_{n\to\infty} E(|X_n - X|^p) = 0.$$

X_n is said to converge in pth mean to X. For $p = 2$, X_n is said to converge in mean square to X, or notationally as is commonly seen, $X_n \stackrel{\mathcal{L}^2}{\to} X$.

As an example of convergence in mean square, consider that

$$E\left(\left|\frac{1}{n}\sum_{i=1}^{n} d_i(x) - \frac{1}{2}\right|^2\right) = \text{var}\left(\frac{1}{n}\sum_{i=1}^{n} d_i(x)\right)$$

since $E(\frac{1}{n}\sum_{i=1}^{n} d_i(x)) = \frac{1}{2}$. Now, $\text{var}(\frac{1}{n}\sum_{i=1}^{n} d_i(x)) = \frac{1}{4n}$, then

$$\lim_{n\to\infty} E\left(\left|\frac{1}{n}\sum_{i=1}^{n} d_i(x) - \frac{1}{2}\right|^2\right) \equiv \lim_{n\to\infty} \text{var}\left(\frac{1}{n}\sum_{i=1}^{n} d_i(x)\right) = 0.$$

By the Chebyshev two-tailed inequality,

$$\lim_{n\to\infty} P\left(\left|\frac{1}{n}\sum_{i=1}^{n} d_i(x) - \frac{1}{2}\right| \geq \epsilon\right) \leq \frac{\text{var}\left(\frac{1}{n}\sum_{i=1}^{n} d_i(x)\right)}{\epsilon^2}.$$

Since the numerator on the RHS approaches zero in the limit as $n \uparrow \infty$, therefore

$$\lim_{n\to\infty} P\left(\left|\frac{1}{n}\sum_{i=1}^{n} d_i(x) - \frac{1}{2}\right| > \epsilon\right) = 0,$$

for any $\epsilon > 0$. In the above, we show that convergence in $\mathcal{L}^2 \Rightarrow$ convergence in P. More generally, convergence in $\mathcal{L}^p (1 \leq p < \infty) \Rightarrow$ convergence in P. The converse is not necessarily true.

Convergence a.s. $\Rightarrow$ convergence in P. Hence, convergence a.s. and $\mathcal{L}^p$ convergence are stronger convergence than convergence in P. We show below an example in which for measurable $\omega \in (0,1]$, a sequence of RVs $X_n(\omega)$ converges to 0 in P but not a.e.

$$X_1 = 1_{(0,1]}$$

$$X_2 = 1_{(0,\frac{1}{2}]} \quad X_3 = 1_{(\frac{1}{2},1]}$$

$$X_4 = 1_{(0,\frac{1}{3}]} \quad X_5 = 1_{(\frac{1}{3},\frac{2}{3}]} \quad X_6 = 1_{(\frac{2}{3},1]}$$

$$\cdots\cdots\cdots\cdots\cdots\cdots\cdots\cdots\cdots\cdots$$

In the above function, each line represents a partition of the interval $(0,1]$. For any $\omega \in (0,1]$, there is one X_n value equal to 1 in each line. Hence, as $n \to \infty$, X_n will take the value 1 infinitely often, i.e., event $\{X_n = 1\}$ will recur again and again infinitely often as $n \to \infty$. Thus, X_n does not converge to 0 for any $\omega \in (0,1]$. Neither does it converge to 1. It is like an infinitely alternating series between values 0 and 1. Since

$$P\big(\omega : \lim_{n\to\infty} X_n(\omega) = 0\big) = 0,$$

$X_n \not\to 0$ a.s.

For arbitrarily small $\epsilon > 0$, when $n = 1$,

$$P(|X_1 - 0| > \epsilon) = 1 \text{ since all } X_1(\omega) = 1 \text{ for all } \omega \in (0,1].$$

When $n = 2$,

$$P(|X_2 - 0| > \epsilon) = 1/2 \text{ since } P(\omega : X_2 = 1) = 1/2.$$

When $n = 4$,

$$P(|X_4 - 0| > \epsilon) = 1/3 \text{ since } P(\omega : X_3 = 1) = 1/3.$$

We can see that as $n \to \infty$, $P(\omega : X_n = 1)$ becomes smaller and smaller and tends toward zero. Hence,

$$\lim_{n\to\infty} P(\omega : |X_n(\omega) - 0| > \epsilon) = 0.$$

Thus, in this case, X_n converges to zero in P.

A diagrammatic explanation of the two types of a.s. and probability or P convergence of $X_n(\omega)$ is shown in Table 3.1.

For a.s. or a.e. convergence, $X_n(\omega_i) \to X(\omega_i)$ for a set $\{\omega_i\} \in A$ such that $P(A) = 1$. For convergence in probability, $P(|X_n - X| > \epsilon)$ converges to zero.

Table 3.1: Illustration of Different Convergence Modes

Ω	$X_1(\omega)$	......	$X_n(\omega)$	
ω_1	$X_1(\omega_1)$	...	$X_n(\omega_1)$	$\to X(\omega_1)$
ω_2	$X_1(\omega_2)$	...	$X_n(\omega_2)$	$\to X(\omega_2)$
ω_3	$X_1(\omega_3)$	...	$X_n(\omega_3)$	$\to X(\omega_3)$
$\vdots$	$\vdots$	$\vdots$	$\vdots$	$\vdots$
For each n over all ω	$P(\lvert X_1 - X\rvert > \epsilon)$	......	$P(\lvert X_n - X\rvert > \epsilon)$	$\to 0$

With the concept of a.e. convergence explained, we now proceed with the Strong Law of Large Numbers.

Theorem 3.5 (Strong Law of Large Numbers, SLLN): *Let $\{X_i\}_{i=1,2,\ldots}$ be a sequence of i.i.d. RVs each with finite mean μ. Then, with probability* 1 *or a.s.,*

$$P\left(\omega : \lim_{n\to\infty} \frac{X_1(\omega) + X_2(\omega) + \cdots + X_n(\omega)}{n} = \mu\right) = 1. \qquad \blacksquare$$

There is a weaker type of convergence — convergence in distribution. First, we define the empirical distribution function (edf) and visit the Glivenko–Cantelli theorem.

Suppose $X_1, X_2, \ldots, X_n$ is a random sample from X, with distribution function $F(X)$. For a given value y, we can determine if for each i,

$$\tilde{1}_{[X_i \le y]}(\omega) = \begin{cases} 1 & \text{if } X_i(\omega) \le y \\ 0 & \text{otherwise.} \end{cases}$$

Note that each ω constitutes a set of values of the random sample of size n.

Define an edf of $F(X)$ as

$$\hat{F}_n(y, \omega) = \frac{1}{n} \sum_{i=1}^{n} 1_{[X_i \le y]}(\omega).$$

Theorem 3.6 (Glivenko–Cantelli): *With probability 1 (or for all $\omega \in A \ni$ (such that) $P(A) = 1$),*

$$\lim_{n\to\infty} \sup_{-\infty<y<\infty} |\hat{F}_n(y, \omega) - F(y)| = 0. \qquad \blacksquare$$

Clearly, given y, for all y, each $1_{[X_i \leq y]}(\omega)$ is an i.i.d. Bernoulli RV with probability $F(y)$ of 1 and probability $1 - F(y)$ of 0. Then by the SLLN, its average converges to the population fraction $F(y)$. The "supremum" tells of a stronger, uniform convergence, or convergence at a rate regardless of point y.

The above says that RV X_n which is distributed with edf such that $P(X_n \leq y) = \hat{F}_n(y)$ converges to RV X in distribution or in law, i.e., $X_n \xrightarrow{\mathcal{D}} X$ or $X_n \xrightarrow{\mathcal{L}} X$, if $\hat{F}_n(y, \omega) \xrightarrow{a.s.} F(y)$.

Suppose $f(x)$ is any given continuous bounded function, and if $X_n \xrightarrow{D} X$, then $\lim_{n \to \infty} \int f(x) d\hat{F}_n(x) = \int f(x) dF(x)$. It is equivalent to $\lim_{n \to \infty} E_n[f(x)] = E[f(x)]$, where probability measure $P_n(x)$ converges in distribution to probability measure $P(x)$. This result allows for the convergence of sample moments when the edf converges in distribution.

Convergence in distribution or probability measure is, however, a weak convergence, and generally does not imply the stronger a.s. convergence or convergence in P. For example, we see how given a binomial distribution $X_n \sim B(n, p)$, $X'_n = \frac{X_n - np}{\sqrt{np(1-p)}}$ converges to a standard normal distribution $Z \sim N(0, 1)$ in distribution:

$$\lim_{n \to \infty} P(i \leq X'_n \leq j) = P(i \leq z \leq j) \quad \text{for all } i, j < n.$$

However, if set A is the set of discrete values (for a large n):

$$A = \left\{ \omega : \frac{j - np}{\sqrt{np(1-p)}};\ j = 0, 1, 2, \ldots, n \right\},$$

then $P_X{}'(\omega \in A) = 1$, whereas $P_Z(\omega \in A) = 0$ since the discrete values are just a countable set of points. Thus, RV X does not converge to Z a.s. (w.p. 1). However, convergence *w.p.* 1 and convergence in P imply convergence in distribution.

3.3. Central Limit Theorem (CLT)

Theorem 3.7 *Suppose $\{X_i\}_{i=1,2,\ldots}$ is a sequence of i.i.d. random variables, each having mean μ and variance σ^2. Then, the RV $\frac{\sum_{i=1}^{n} X_i - n\mu}{\sigma\sqrt{n}}$ converges in distribution to the standard normal RV as $n \uparrow \infty$.* ■

(Note that the numerator of the RV has mean 0 and the denominator is the standard deviation of the numerator.)

Proof. We can rewrite the RV $\frac{\sum_{i=1}^{n} X_i - n\mu}{\sigma\sqrt{n}}$ as $Y_n \stackrel{d}{=} \sum_{i=1}^{n} \frac{1}{\sqrt{n}} Z_i$ where $Z_i = \frac{X_i - \mu}{\sigma}$ is i.i.d. with mean 0 and variance 1.

For a given n, the MGF of $\sum_{i=1}^{n} \frac{Z_i}{\sqrt{n}}$ is

$$\begin{aligned} M_Y(\theta) &= E\left(\exp\left(\theta \sum_{i=1}^{n} \frac{Z_i}{\sqrt{n}}\right)\right) \\ &= E\left(\prod_{i=1}^{n}\left[\exp\left(\frac{\theta}{\sqrt{n}} Z_i\right)\right]\right) \\ &= \left[M_Z\left(\frac{\theta}{\sqrt{n}}\right)\right]^n. \end{aligned}$$

Let $t = \theta/\sqrt{n}$ and $Q(t) = \log M_Z(t)$. Taking derivatives w.r.t. the argument, i.e., $\frac{dM(t)}{dt} = M'(t)$, and so on, we have

$$Q'(t) = \frac{M_Z'(t)}{M_Z(t)}, \quad \text{and} \quad Q''(t) = \frac{M_Z(t)M_Z''(t) - [M_Z'(t)]^2}{[M_Z(t)]^2}.$$

Now, putting $t = 0$, $M_Z(0) = 1$. $Q(0) = \log M_Z(0) = \log 1 = 0$.
$M_Z'(0) = E(Z) = 0$, since $E(Z) = 0$, so $Q'(0) = 0$.
$M_Z''(0) = E(Z^2) = 1$, since $E(Z^2) = 1$, so $Q''(0) = 1$.

Now, let $t = m\theta$, where $m = 1/\sqrt{n}$. Using the L' Hôpital's rule,

$$\begin{aligned} \lim_{m\to 0} \frac{Q(t)}{m^2} &= \lim_{m\to 0} \frac{\frac{dt}{dm}\frac{dQ}{dt}}{2m} \\ &= \lim_{m\to 0} \frac{\theta dQ/dt}{2m} \\ &= \lim_{m\to 0} \frac{\theta^2 d^2Q/dt^2}{2} = \frac{\theta^2}{2} \end{aligned}$$

since the argument t in Q goes to zero as $m \to 0$. Re-expressing in terms of n instead of m,

$$\lim_{n\to\infty} nQ\left(\frac{\theta}{\sqrt{n}}\right) = \frac{\theta^2}{2},$$

or,

$$\lim_{n\to\infty} \left[M_Z\left(\frac{\theta}{\sqrt{n}}\right)\right]^n = e^{\frac{\theta^2}{2}}.$$

Therefore, $M_Y(\theta)$ converges to $e^{\frac{\theta^2}{2}}$ which is the MGF of a standard normal RV. ■

The above CLT is about the simplest, and is sometimes called the Laplace–De Moivre theorem. There are many other more complicated versions of the CLT, some dealing with weaker requirements such as independence but allowing heterogeneous variances such as in the famous Lindeberg–Levy–Feller theorem. In such cases, the variances also have to be bounded in some way to enable convergence. The speed toward convergence or rate of convergence, i.e., how large n must be in order to get to within certain probabilistic deviation or "closeness" of the limit, is obviously slower when the conditions for CLT become more relaxed. When $E|X_i|^3 < \infty$, i.e., is bounded, the Berry–Esseen theorem gives an idea of the degree of deviations from convergence for any sample size n.

We have seen how convergence to 0 of $\sum_i^n X_i - n\mu$ is achieved by suitable division (or "normalization") by n, and convergence to RV $N(0,1)$ is achieved by a different normalization by $\sigma\sqrt{n}$. Thus, normalization by anything between n and $\sigma\sqrt{n}$ can produce interesting results. Another related law, the law of iterated logarithm deals with characterizing how the partial sum $\sum_i^n X_i - n\mu$ would behave as n increases.

Applications

The normal distribution can be used as an approximation to both the binomial and the Poisson distributions.

Consider a random sample $\{X_i\}_{i=1,2,\ldots}$ with each RV $X_i = 1$ or 0 from a Bernoulli distribution of an event happening, and $X_i = 1$ has probability p. The mean and variance of X_i are p and $p(1-p)$. Suppose the random sample consists of n such Bernoulli RVs. Then aggregate outcome $Y_n = X_1 + X_2 + \cdots + X_n$ follows the binomial distribution with mean np and variance $np(1-p)$.

By the CLT, therefore $\frac{\sum_{i=1}^n X_i - np}{\sqrt{np(1-p)}}$ or $\frac{Y_n - np}{\sqrt{np(1-p)}}$ converges toward a $N(0,1)$ as $n \to \infty$. For large n, therefore, $Y_n \approx N(\mu, \sigma^2)$ where $\mu = np$, and $\sigma^2 = np(1-p)$.

Consider a consumer example: if the probability of closing a sale in a sales call is 0.2, and 50 sales calls are made, what is the probability of landing 10 sales? Using a normal approximation to the binomial problem,

$$\begin{aligned}
&P(Y = 10; p = 0.2, n = 50) \\
&\quad = \binom{50}{10} (0.2)^{10}(0.8)^{40}
\end{aligned}$$

$$\approx P(9.5 < Y < 10.5)$$
$$= P\left(\frac{9.5-(0.2)(50)}{\sqrt{50(0.2)(0.8)}} < \frac{Y-(0.2)(50)}{\sqrt{50(0.2)(0.8)}} < \frac{10.5-(0.2)(50)}{\sqrt{50(0.2)(0.8)}}\right)$$
$$= P\left(\frac{9.5-10}{\sqrt{8}} < \frac{Y-10}{\sqrt{8}} < \frac{10.5-10}{\sqrt{8}}\right)$$
$$= \Phi(0.5/\sqrt{8}) - \Phi(-0.5/\sqrt{8}) = \Phi(0.177) - \Phi(-0.177) = 0.1398.$$

A Poisson approximation to the binomial problem would yield

$$P(Y = 10; \lambda = np = 10) \approx e^{-10}\frac{10^{10}}{10!} = 0.125.$$

The CLT also affords a way in which the standardized Poisson RV $\frac{X-\lambda}{\sqrt{\lambda}}$ (where X is Poisson(λ)) with mean 0 and variance 1 may approximately be normal for large λ.

To show this, we consider the MGF of $\frac{X-\lambda}{\sqrt{\lambda}}$. We use the fact that the MGF of X, $M_X(\theta)$, is $\exp(\lambda(e^\theta - 1))$. We also employ the Taylor series expansion of an exponential function.

$$E\left(\exp\left[\theta\left(\frac{X-\lambda}{\sqrt{\lambda}}\right)\right]\right)$$
$$= \exp(-\theta\sqrt{\lambda})E\left(\exp\left[\frac{\theta}{\sqrt{\lambda}}X\right]\right)$$
$$= \exp(-\theta\sqrt{\lambda})\exp\left(\lambda(e^{\theta/\sqrt{\lambda}} - 1)\right)$$
$$= \exp(-\theta\sqrt{\lambda})\exp\left(\lambda\left[\theta/\sqrt{\lambda} + \frac{1}{2!}(\theta/\sqrt{\lambda})^2 + \frac{1}{3!}(\theta/\sqrt{\lambda})^3 + \ldots\right]\right)$$
$$= \exp(-\theta\sqrt{\lambda})\exp\left(\theta\sqrt{\lambda} + \frac{1}{2!}\theta^2 + \frac{1}{3!}(\theta^3/\sqrt{\lambda}) + o(\lambda)\right)$$
$$= \exp\left(\frac{1}{2!}\theta^2 + \frac{1}{3!}(\theta^3/\sqrt{\lambda}) + o(\lambda)\right)$$
$$\to \exp\left(\frac{1}{2}\theta^2\right) \quad \text{as } \lambda \uparrow \infty.$$

Thus, CLT enables meaningful statistical testing and inference.

3.4. Convergence of Binomial Model

Having seen the CLT, we now introduce an important workhorse for option pricing — the binomial model, or sometimes called the binomial tree model as it is often depicted in a tree or lattice.

Suppose there is a stock (or asset) with price S_0 at time $t = 0$. Suppose the stock is traded at equal time intervals so that the next time it is traded is at $t = 1$, then $t = 2$, and so on. The period between each successive trades is to be determined later.

At time now, $t = 0$, there is no way of knowing what price the stock will take in future, but the trader may take S_t, for each $t > 0$, as a RV with a probability distribution. The binomial asset pricing model assumes that over one period this distribution is a Bernoulli distribution with two outcomes: an increase (state U) in price by a factor $u > 1$ with probability p, or a decrease (state D) in price by a factor $d < 1$ with probability $q = 1 - p$.

Over two periods, there are 4 possible states: UU (U first then U again), UD (U first then D next), DU, or DD. However, when we use the factor characterization of price evolution, there is recombination in some of the states, e.g., $ud = du$ (i.e., commutative). So, the middle nodes in the evolution of the binomial tree recombine or "meet". Thus, at $t = 1$, there are 2 nodes, at $t = 2$, there are 3 nodes, at $t = 3$, there are 4 nodes, and at $t = n$, there are $n + 1$ nodes. The number of nodes grow linearly (and slow enough to enable fast computations in this lattice method) in what is called a recombining binomial tree/lattice. If the nodes do not "meet", then there will be 2 nodes at $t = 1$, 4 nodes at $t = 2$, 8 nodes at $t = 3$, 2^n nodes at $t = n$, and thus the number of nodes grows exponentially with time t, viz. $e^{n \ln 2}$. The latter is computationally tedious when n gets to be large, e.g., $2^{50} > 10^{15}$ (and poses a more critical problem in terms of requirements on computer memory). The latter is called a nonrecombining binomial tree/lattice.

The possible prices are indicated on the binomial tree/lattice in Fig. 3.2. Consider a binomial distribution $B(n, p)$ with $n = 5$ and probability of U as p. The outcome for x steps up, for $x = 0, 1, 2, 3, 4$, or 5, has attached probability ${}^nC_x p^x (1 - p)^{n-x}$. Here, we map x to a price via price function $P : x \mapsto u^x d^{n-x} S_0$. So, the binomial distribution of price becomes $P(S_n = u^x d^{n-x} S_0) =^n C_x p^x (1 - p)^{n-x}$. For example, at $t = 5$, price is a RV with a binomial distribution $P(S_5 = u^x d^{5-x} S_0) = \binom{5}{x} p^x (1 - p)^{5-x}$.

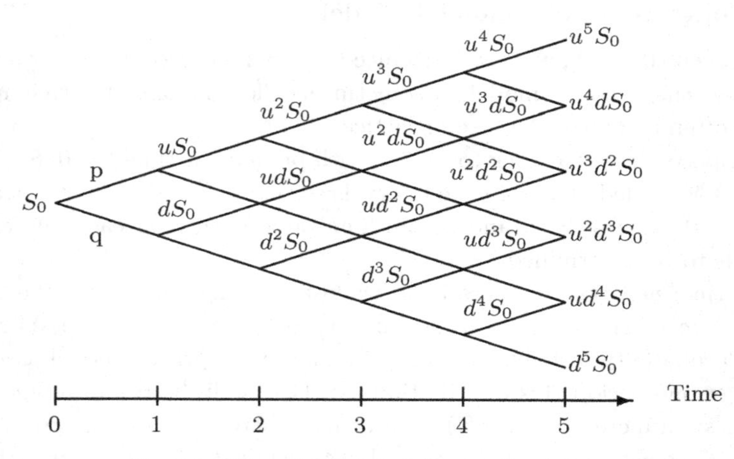

Fig. 3.2: Binomial (Lattice) Tree of Stock Prices

Therefore, the expected value of the stock price at time n, as viewed at $t = 0$, is

$$E_0(S_n) = S_0 \sum_{x=0}^{n} u^x d^{n-x} \binom{n}{x} p^x (1-p)^{n-x}.$$

Suppose the risk-free interest rate factor is R per period. If investors are risk-neutral, then the current stock price trading in the market should be

$$S_0 = \frac{E_0(S_n)}{R^n} = R^{-n} S_0 \left[\sum_{x=0}^{n} u^x d^{n-x} \binom{n}{x} p^x (1-p)^{n-x} \right].$$

We see that indeed the RHS becomes S_0 if we note that under risk neutrality, the one-period stock return is $pu + (1-p)d = R$. Then, $p = \frac{R-d}{u-d} \in (0,1)$. Use this to substitute into the pricing formula above.

At $t = n$, the price of stock is a RV $S_n = u^x d^{n-x} S_0$ depending on x, the number of U states that had occurred. The continuously compounded (or logarithmic) return rate over n periods is then a RV

$$\ln \frac{S_n}{S_0} = x \ln \frac{u}{d} + n \ln d. \tag{3.1}$$

Suppose we have a call option expiring in the actual calendar time of τ year, e.g., 3 months is 1/4 year. This time-to-maturity is modelled using n periods each of which has calendar time length $\triangle = \tau/n$ year. Suppose we keep τ constant, but increase $n \uparrow \infty$, so $\triangle \downarrow 0$.

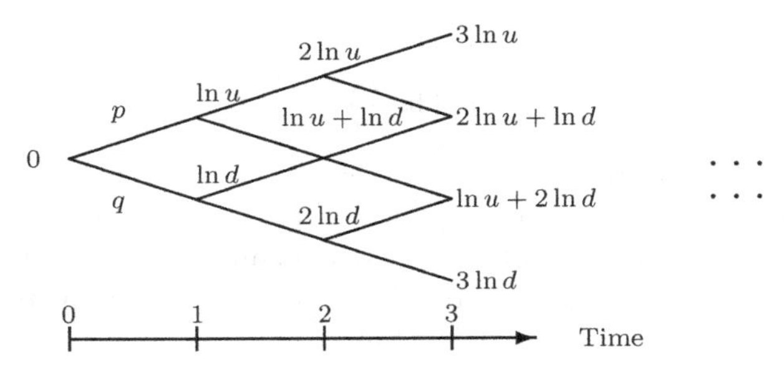

Fig. 3.3: Binomial Tree of Log-Returns

Now in Eq. (3.1), we have an expression of the log-return of stock under the binomial process. In other words, the binomial tree is really also about the evolution of $\ln \frac{S_n}{S_0}$ over time, as shown in Fig. 3.3.

Such a binomial distribution in log-returns or continuously compounded returns converges in the limit to a normal distribution in log-returns. Under risk neutrality, the instantaneous return rate is r. (As a preview of what to come, this comes from the lognormal diffusion process used by Black and Scholes: $\frac{dS}{S} = rdt + \sigma dW$.) The expectation of the log-returns at τ is $(r - \frac{1}{2}\sigma^2)\tau$. The variance of the log-returns at τ is $\sigma^2\tau$.

The expectation of the binomial log-returns at $\tau = n\triangle$ (or abbreviated "$t = n$" on the binomial tree) is, by taking expectation on Eq. (3.1), $E(x) \ln \frac{u}{d} + n \ln d$. The variance of the log-returns at $\tau = n\triangle$, by Eq. (3.1), is $\text{var}(x)\left(\ln \frac{u}{d}\right)^2$.

Since the binomial distribution converges to the normal distribution, the means and variances of the two distributions must equate. Hence in equating means,

$$E(x) \ln \frac{u}{d} + n \ln d = np \ln \frac{u}{d} + n \ln d = \left(r - \frac{1}{2}\sigma^2\right)\tau = \left(r - \frac{1}{2}\sigma^2\right) n\triangle,$$

or

$$p \ln \frac{u}{d} + \ln d = \left(r - \frac{1}{2}\sigma^2\right)\triangle. \tag{3.2}$$

In equating variances,

$$\text{var}(x)\left(\ln \frac{u}{d}\right)^2 = np(1-p)\left(\ln \frac{u}{d}\right)^2 = \sigma^2\tau = \sigma^2 n\triangle,$$

or

$$p(1-p)\left(\ln \frac{u}{d}\right)^2 = \sigma^2 \triangle. \tag{3.3}$$

In Eqs. (3.2) and (3.3), we solve for model parameters p, u, and d in terms of the exogenous market parameters r, σ^2, and $\triangle$. Since there are only two equations and three model parameters to solve, as is done by Cox, Ross, and Rubinstein (1979),[1] we arbitrarily fix one of the model parameters, $d = 1/u$. Solving, we obtain:

$$u = \exp\left(\sqrt{(r - \frac{1}{2}\sigma^2)^2\triangle^2 + \sigma^2\triangle}\right),$$

$$d = \exp\left(-\sqrt{(r - \frac{1}{2}\sigma^2)^2\triangle^2 + \sigma^2\triangle}\right),$$

$$p = \frac{1}{2}\left(1 + \frac{(r - \frac{1}{2}\sigma^2)\triangle}{\sqrt{(r - \frac{1}{2}\sigma^2)^2\triangle^2 + \sigma^2\triangle}}\right).$$

Noting that $\triangle$ should be very small for large n, we treat terms $\triangle^k$, $k \geq 3/2$, as zeros and approximate the above exact solution by the following.

$$u = e^{\sigma\sqrt{\triangle}}, \quad d = e^{-\sigma\sqrt{\triangle}}, \quad p = \frac{1}{2}\left[1 + \left(\frac{r - 1/2\sigma^2}{\sigma}\right)\sqrt{\triangle}\right]. \tag{3.4}$$

We shall also use the following result later. First note that over interval $\triangle$, risk-free interest factor $R = e^{r\triangle}$. Then, using Taylor's expansion of exponentials in the process, and ignoring terms of $\triangle^k$ for $k \geq 3/2$,

$$\begin{aligned}
p' &= \frac{up}{R} \\
&= e^{\sigma\sqrt{\triangle} - r\triangle}p \\
&= \left(1 - \left(r - \frac{1}{2}\sigma^2\right)\triangle + \sigma\sqrt{\triangle}\right)\frac{1}{2}\left[1 + \left(\frac{r - 1/2\sigma^2}{\sigma}\right)\sqrt{\triangle}\right] \\
&= \frac{1}{2}\left[1 + \left(\frac{r - 1/2\sigma^2}{\sigma}\right)\sqrt{\triangle} - \left(r - \frac{1}{2}\sigma^2\right)\triangle + \sigma\sqrt{\triangle} + \left(r - \frac{1}{2}\sigma^2\right)\triangle\right] \\
&= \frac{1}{2}\left[1 + \left(\frac{r + 1/2\sigma^2}{\sigma}\right)\sqrt{\triangle}\right] = p + \frac{1}{2}\sigma\sqrt{\triangle}.
\end{aligned}$$

[1] Cox, JC, SA Roes, and M Rubinstein (1979). Option pricing: A simplified approach. *Journal of Financial Economics*, 7, 229–263.

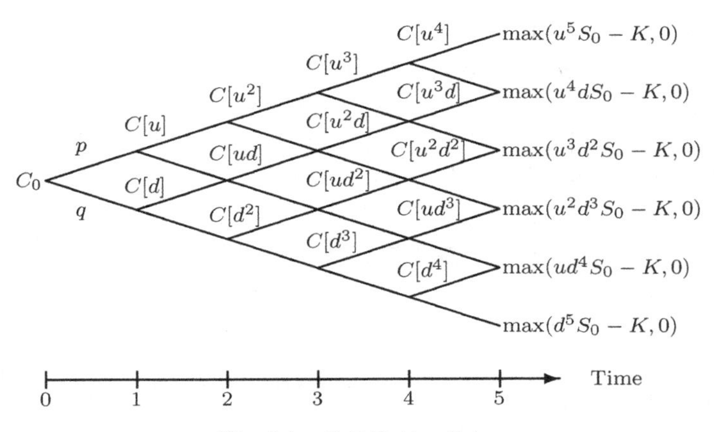

Fig. 3.4: Call Option Prices

We use the binomial lattice in Fig. 3.4 to calibrate a European stock call option price C based on exercise price K. Suppose maturity is at $t = 5$. At $t = 5$, if $S_5 = u^5S_0$, then $C_5 = \max(u^5S_0 - K, 0)$. If $S_5 = u^4dS_0$, then $C_5 = \max(U^4dS_0 - K, 0)$, and so on. Notationally, each call price when exercised at an underlying stock price of $u^xd^{n-x}S_0$ is $C[u^xd^{n-x}] = \max(u^xd^{n-x}S_0 - K, 0)$. Therefore, the current price of the stock call option trading in the market should be

$$C_0 = \frac{E_0(C_n)}{R^n} = R^{-n}\left[\sum_{x=0}^{n}\binom{n}{x}p^x(1-p)^{n-x}\max(u^xd^{n-x}S_0 - K, 0)\right].$$

Suppose $a > 0$ is the smallest integer so that $S_0u^ad^{n-a} - K > 0$, or equivalently the smallest integer so that

$$a > \frac{\ln\left(\frac{K}{S_0d^n}\right)}{\ln\left(\frac{u}{d}\right)} = \frac{\ln\left(\frac{K}{S_0}\right) - n(\ln d)}{\ln\frac{u}{d}}. \tag{3.5}$$

It should be noted that as $n \uparrow \infty$, $a \downarrow \frac{\ln(\frac{K}{S_0}) - n\ln d}{\ln\frac{u}{d}}$.

All nodes associated with more than or equal to "a" number of U states end up in-the-money. The other nodes are out-of-the-money. Thus,

$$\begin{aligned}C_0 &= R^{-n}\left[\sum_{x=a}^{n}\binom{n}{x}p^x(1-p)^{n-x}(u^xd^{n-x}S_0 - K)\right]\\ &= \left[\sum_{x=a}^{n}\left(\frac{n!}{x!(n-x)!}\right)p^x(1-p)^{n-x}\left(\frac{u^xd^{n-x}S_0}{R^n}\right)\right]\end{aligned}$$

$$
\begin{aligned}
&-\frac{K}{R^n}\left[\sum_{x=a}^{n}\left(\frac{n!}{x!(n-x)!}\right)p^x(1-p)^{n-x}\right] \\
&= S_0\left[\sum_{x=a}^{n}\left(\frac{n!}{x!(n-x)!}\right)p'^x(1-p')^{n-x}\right] \\
&\quad -\frac{K}{R^n}\left[\sum_{x=a}^{n}\left(\frac{n!}{x!(n-x)!}\right)p^x(1-p)^{n-x}\right] \\
&= S_0\Phi(a;n,p') - KR^{-n}\Phi(a;n,p) \qquad (3.6)
\end{aligned}
$$

where $p' = \frac{up}{R}$, and $\Phi(a;n,p') = \left[\sum_{x=a}^{n}\left(\frac{n!}{x!(n-x)!}\right)p'^x(1-p')^{n-x}\right]$ is one minus the binomial cdf.

We study convergence of the terms $\Phi(\cdot)$ in the above equation.

$$
1-\Phi(a;n,p) = P(x \le a-1) = P\left[\frac{x-np}{\sqrt{np(1-p)}} \le \frac{a-1-np}{\sqrt{np(1-p)}}\right].
$$

From Eqs. (3.1)–(3.4), we can work out the following.

$$
x = \frac{\ln\frac{S_n}{S_0}}{2\sigma\sqrt{\triangle}} + \frac{n}{2},
$$

$$
np = \frac{n}{2} + \frac{n(r-1/2\sigma^2)\triangle}{2\sigma\sqrt{\triangle}},
$$

and

$$
np(1-p) = \frac{n}{4}\left[1-\left(\frac{r-1/2\sigma^2}{\sigma}\right)^2\triangle\right].
$$

Therefore, by the CLT, as $\Delta \downarrow 0$,

$$
\begin{aligned}
\frac{x-np}{\sqrt{np(1-p)}} &= \frac{\ln\left(\frac{S_n}{S_0}\right) - (r-1/2\sigma^2)n\triangle}{\sigma\sqrt{n\triangle}} \\
&= \frac{\sum_{x=0}^{n-1}\ln\left(\frac{S_{x+1}}{S_x}\right) - n[(r-1/2\sigma^2)\triangle]}{(\sigma\sqrt{\triangle})\sqrt{n}} \xrightarrow{D} Z \sim N(0,1).
\end{aligned}
$$

Meanwhile, using Eqs. (3.1) to (3.5), as $\Delta \downarrow 0$,

$$
\frac{a-1-np}{\sqrt{np(1-p)}} \longrightarrow \frac{\ln\left(\frac{K}{S_0}\right) - (r-1/2\sigma^2)\tau}{\sigma\sqrt{\tau}}.
$$

Hence,

$$\begin{aligned}
\Phi(a; n, p) &\longrightarrow 1 - P\left[Z \leq \frac{\ln\left(\frac{K}{S_0}\right) - (r - 1/2\sigma^2)\tau}{\sigma\sqrt{\tau}}\right] \\
&= P\left[Z \leq -\frac{\ln\left(\frac{K}{S_0}\right) - (r - 1/2\sigma^2)\tau}{\sigma\sqrt{\tau}}\right] \\
&= P\left[Z \leq \frac{\ln\left(\frac{S_0}{K}\right) + (r - 1/2\sigma^2)\tau}{\sigma\sqrt{\tau}}\right]. \qquad (3.7)
\end{aligned}$$

Now, from Eqs. (3.1) to (3.4) and the definition of p', we can work out the following.

$$\begin{aligned}
x &= \frac{\ln\frac{S_n}{S_0}}{2\sigma\sqrt{\triangle}} + \frac{n}{2}, \\
np' &= \frac{n}{2} + \frac{n(r + 1/2\sigma^2)\triangle}{2\sigma\sqrt{\triangle}},
\end{aligned}$$

and

$$np'(1 - p') = \frac{n}{4}\left[1 - \left(\frac{r + 1/2\sigma^2}{\sigma}\right)^2 \triangle\right].$$

Therefore, by the CLT, noting mean of $\ln(S_n/S_0)$ is now $n(r+1/2\sigma^2)\triangle$ under probability p', as $\Delta \downarrow 0$,

$$\begin{aligned}
\frac{x - np'}{\sqrt{np'(1 - p')}} &= \frac{\ln\left(\frac{S_n}{S_0}\right) - (r + 1/2\sigma^2)n\triangle}{\sigma\sqrt{n\triangle}} \\
&= \frac{\sum_{x=0}^{n-1}\ln\left(\frac{S_{x+1}}{S_x}\right) - n[(r + 1/2\sigma^2)\triangle]}{(\sigma\sqrt{\triangle})\sqrt{n}} \xrightarrow{D} Z \sim N(0, 1).
\end{aligned}$$

Using Eq. (3.5), and Eqs. (3.1)–(3.4),

$$\frac{a - 1 - np'}{\sqrt{np'(1 - p')}} \longrightarrow \frac{\ln\left(\frac{K}{S_0}\right) - (r + 1/2\sigma^2)\tau}{\sigma\sqrt{\tau}}.$$

Hence,

$$\begin{aligned}\Phi(a;n,p') &\longrightarrow 1 - P\left[Z \leq \frac{\ln\left(\frac{K}{S_0}\right) - (r + 1/2\sigma^2)\tau}{\sigma\sqrt{\tau}}\right] \\ &= P\left[Z \leq -\frac{\ln\left(\frac{K}{S_0}\right) - (r + 1/2\sigma^2)\tau}{\sigma\sqrt{\tau}}\right] \\ &= P\left[Z \leq \frac{\ln\left(\frac{S_0}{K}\right) + (r + 1/2\sigma^2)\tau}{\sigma\sqrt{\tau}}\right]. \end{aligned} \tag{3.8}$$

From Eqs. (3.7) and (3.8), we substitute into Eq. (3.6) to obtain

$$\begin{aligned}\lim_{n\uparrow\infty} C_0 &= S_0\Phi(a;n,p') - KR^{-n}\Phi(a;n,p) \\ &= S_0\Phi_Z(d_1) - Ke^{-r\tau}\Phi_Z(d_2),\end{aligned} \tag{3.9}$$

where

$$d_1 = \frac{\ln\left(\frac{S_0}{K}\right) + (r + 1/2\sigma^2)\tau}{\sigma\sqrt{\tau}},$$

and

$$d_2 = \frac{\ln\left(\frac{S_0}{K}\right) + (r - 1/2\sigma^2)\tau}{\sigma\sqrt{\tau}},$$

where Φ_Z is the normal cdf.

Equation (3.9) is the celebrated Black–Scholes option pricing formula for European call option. Conditional on S_0, S_n converges to a lognormal distribution. Or binomial $\ln(S_n/S_0)$ for finite n converges in distribution to a normal RV. In the limit, expected value of continuous bounded functions of S_n such as $E[\max(S_n - K, 0)]$ should also converge to expectation of its limit, i.e., $\lim_n E[\max(S_n - K, 0)] = E[\max(\lim_n S_n - K, 0)]$. The discrete version as in the binomial lattice is called the binomial option pricing model or sometimes the Cox–Ross–Rubinstein (CRR) model.

While we find C_0 via the LHS limit above through establishing the convergent values under the binomial distribution, we should be able to obtain the same Black–Scholes formula in (3.9) by the RHS where we obtain the limit distribution as in a lognormal S_n and then find the expected value under the limit lognormal distribution directly. The latter turns out to be easier to solve. Try this in the exercise.

3.5. Remarks on Numerical Methods

In the last sub-section, we showed that the CRR binomial option pricing model provides a European option price once the terminal or boundary conditions at maturity time T is given, e.g., $C_T = \max(S_T - K, 0)$ for a call option, or $P_T = \max(K - S_T, 0)$ for a put option. Other types of options can be priced in the same way on the binomial tree if their terminal boundary condition is explicit and implementable (i.e., forms a stopping time), e.g., a power call option $f_T = \max([S_T - K]^p, 0)$, for $p > 0$, a binary option $f_T = K \times 1_{(S_T > K)}$, and so on. Some like an Asian call option with terminal payoff $f_T = \max(\frac{\sum_{t>0}^{T} S_t}{T-t} - K, 0)$ which is based on an average of underlying prices during a period before maturity, can be priced only with a nonrecombining binomial tree since the payoffs are path dependent.

We see also that as the binomial tree interval $\triangle \downarrow 0$, then the binomial option price converges to that under a continuous diffusion process. In the case of underlying stock price following a geometric Brownian motion, something that will become clearer in later chapters, then the plain vanilla call and put prices converge to the Black–Scholes prices.

Sometimes the binomial option pricing model is treated in its own right as a pricing model for a sufficiently small $\triangle$. Sometimes it is interpreted as an approximation to the Black–Scholes or a continuous diffusion option model. In the latter regard, there is a large variety and number of numerical methods providing approximations to theoretical option pricing models, whether European-style or the more complicated American-style when option exercise by the option holder can take place before option expiry or maturity. One generic method is the use of a finite-difference scheme in the numerical solution of the PDE and initial as well as boundary conditions associated with the option price.

In Chap. 9, we see how the Black–Scholes model can be represented by a fundamental PDE of the parabolic type. By suitable transformation, the PDE can be expressed as a simple heat equation, viz.

$$u_{\tau'} = u_{zz},$$

where the subscripts denote partial derivatives. Once $u(z, \tau')$ is found at τ' or time $t = T - 2\tau'/\sigma^2$, then the call price at t (or else τ') is

$$f_t = \exp(\omega(r, K, S_t, \sigma, T - t))u(z, \tau'),$$

where $\omega(\cdot)$ is a predetermined function.

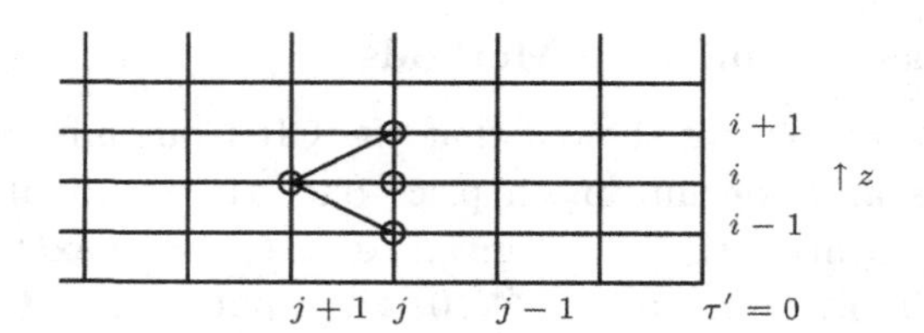

Fig. 3.5: Finite-Difference Method of Option Pricing

Using the finite-difference method to numerically solve for $u_{\tau'} = u_{zz}$, we obtain:

$$\frac{1}{\triangle\tau'}(u_{i,j+1} - u_{i,j}) = \frac{1}{(\triangle z)^2}(u_{i+1,j} - 2u_{i,j} + u_{i-1,j}),$$

where $u_{i,j}$ refers to a value of u on a mesh (Fig. 3.5) with time grids of $\triangle\tau'$ distance apart, and underlying value z-grid of distance $\triangle z$ apart. Rearranging terms:

$$u_{i,j+1} = ru_{i+1,j} + (1-2r)u_{i,j} + ru_{i-1,j},$$

where $r = \frac{\triangle\tau'}{(\triangle z)^2} < \frac{1}{2}$ acts like a pseudo-probability. The mesh in Fig. 3.5 is expressed by τ' increasing to the left from 0.

Thus, it is seen that the finite difference is essentially equivalent to solving the option prices on a trinomial tree once the initial conditions $u(z,0)$ are defined. A binomial tree in turn can be transformed into a trinomial tree if we take a recombining binomial tree and skip every other interval, so that at even intervals, the tree is seen as growing with recombining trinomial branches. Hence, if a sequence converges to a RV based on CLT that is equivalent to another version of the RV based on an analytical solution to a PDE, and that another sequence converges to the latter RV, then the two convergent sequences should be related. In other words, the binomial lattice and the finite-difference grid are related in option pricing. Typically, more elaborate finite-difference methods with faster convergence such as the Crank–Nicolson method are often used, though the convergence principle is the same.

3.6. Problem Set 3

1. Show that if $E(X) = \mu$, $\text{var}(X) = \sigma^2$, and the density distribution of X is symmetrical, then

$$P(|X-\mu| \geq a) \leq \min\left(\frac{\sigma^2}{a^2}, \frac{2\sigma^2}{a^2+\sigma^2}\right).$$

2. Find $P[\omega : d_i(\omega) = u_i, i = 1, 2, \ldots, n]$ for u_i, $i \leq n$, being a given set of numbers that are either 0 or 1.
3. Prove that $|E(XY)| \leq \sqrt{E(X^2)E(Y^2)}$. (Hint: Use the fact that $E(aX + Y)^2 \geq 0$ for any constant a.)
4. If a risk-free bond sells for price $B < 1$ today and matures in 1 period's time with a par value payoff of $1, what is the risk-free rate of interest per period in investing in this bond? Suppose you are risk neutral, which will you prefer — a risk-free bond costing $0.90 to mature in 1 period or a stock with per period expected return of 10%?
5. In the binomial tree model, show that $u > r > d$ if there should be no arbitrage.
6. In the following binomial lattice, given per period interest rate is 1.5%, find the European call price $C[u^2d]$. What is the American call price at node [UUD] if the underlying stock is trading at a price of 2 while the call exercise price is 1? What is the European call price $C[ud]$? Assume p and q are the risk-neutral probabilities.

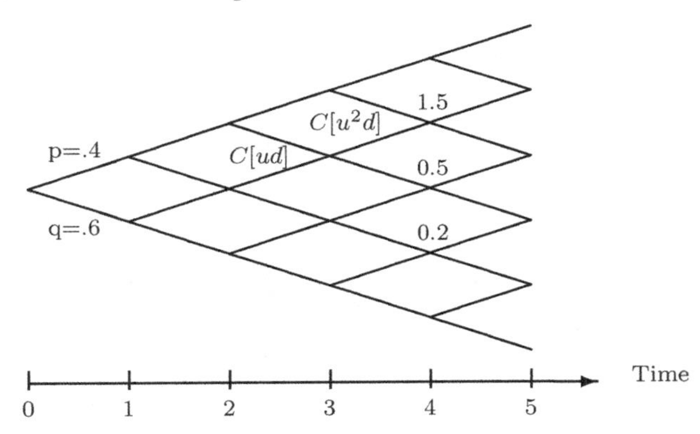

7. $\ln(S_\tau/S_0)$ is distributed as $N((r - 1/2\sigma^2)\tau, \sigma^2\tau)$. $C_\tau = \max(S_\tau - K, 0)$. $C_0 = e^{-r\tau}E[C_\tau]$.
   (a) Find $P(S_T \geq K)$.
   (b) Find $E(S_\tau | S_\tau \geq K)$.
   (c) Show $E(C_\tau) = E(S_\tau | S_\tau \geq K)\ N(d_2) - K\ P(S_\tau \geq K)$. Hence find C_0.
8. Find the Black–Scholes price of a European nondividend-paying stock call option with exercise price of $10. Current stock price is $10. Risk-free rate is 5% p.a. Annualized volatility is 20%. Time to maturity is 1 month.

Chapter 4

THEORY OF RISK AND UTILITY

In the 2001 Hollywood movie, "A Beautiful Mind", Russell Crowe played the role of John Forbes Nash in which he quipped, "See if I derive an equilibrium where prevalence is a non-singular event where nobody loses, can you imagine the effect that would have on conflict scenarios, arm negotiations...." The study of human decision-making under strategic non-cooperative games that Nash had extended in a major way is based upon the pioneering works of game theory by John von Neumann and Oskar Morgenstern. The influence of game theory pervades the social-economic sciences. This chapter provides the background economic theory and analytical framework of strategic decision-making by single agents.[1]

4.1. Expected Utility Theory

A rational framework for decision-making starts with preference — how a consumer would choose to consume among different consumption bundles, each of which is affordable by his/her budget. A consumption vector is $(x_1, x_2, \ldots, x_n)$ where x_i is number of units consumed of good i, and so on. If the consumer strictly prefers bundle $X = (x_1, \ldots, x_n)$ to bundle $Y = (y_1, \ldots, y_n)$, then we write $X \succ Y$. If consumer indeed chooses X over Y, this is called his/her revealed preference or choice. If the consumer has equal preference or is indifferent between bundles X and Y, then we write $X \sim Y$.

Suppose there is a utility function $U(\cdot)$ on the vector of consumption goods bundle such that $X \succ Y$ if and only if (iff) $U(X) > U(Y)$, $X \prec Y$ iff $U(X) < U(Y)$, and $X \sim Y$ iff $U(X) = U(Y)$. The actual number of the utility function, "utils", is just an ordering that indicates which bundle is preferred. Thus $U(\cdot)$ is an ordinal or ordering number, and not a cardinal number (which would have consistent rankings of ratios of the numbers).

[1]Readers can pursue the subject of game theory in excellent works such as the *Handbook of Game Theory with Economic Applications* (1992–1994), Aumann, R and S Hart (eds.), Vol. 1–3. Amsterdam: Elsevier.

Note that the assumption of existence of such a utility function $U : X \mapsto u \in \mathcal{R}$ is not trivial as it projects a consumption vector to a scalar number.

Revealed preferences can in theory be used to build, for example in a two-goods world, a set of indifference curves quantified by the ordinal numbers of utils for each consumer. Then, one can tell if another bundle is preferred to existing ones or not by looking at the indifference curves. But this is as far as it gets; there is nothing else in the cookie jar for understanding choices under uncertainty situations.

Choices Under Uncertainty

In order to build choice theory and decision-making on preferences of risky or uncertain outcomes, the von Neumann–Morgenstern (VM) expected utility representation or framework is very useful and popular in economics and financial research.

A risky outcome is generically represented by a lottery which is characterized as a chance game in which there are two probabilistic outcomes: a probability p of winning X and a probability $(1-p)$ of winning Y. X and Y can be monetary amounts, need not be consumption bundles, and can also be lotteries.

The lottery is expressed as $[p \odot X + (1-p) \odot Y]$. Even though X or Y may be a vector of units of goods, the operation $\odot$ is not a multiplication, but simply denotes the association of probability p with lottery claim X in $p \odot X$ and of probability $(1-p)$ with claim Y in $(1-p) \odot Y$. The idea is represented by the following lottery diagrams.

The simple lottery diagram, Fig. 4.1(a), shows the lottery $[p \odot X + (1-p) \odot Y]$. Entities on the nodes represent consumption bundles or lotteries,

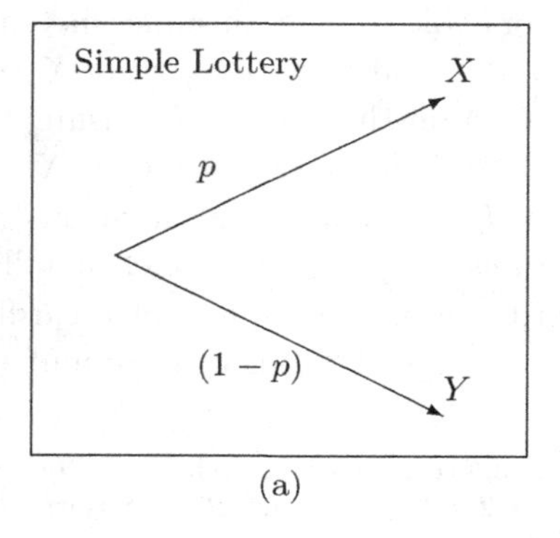

(a)

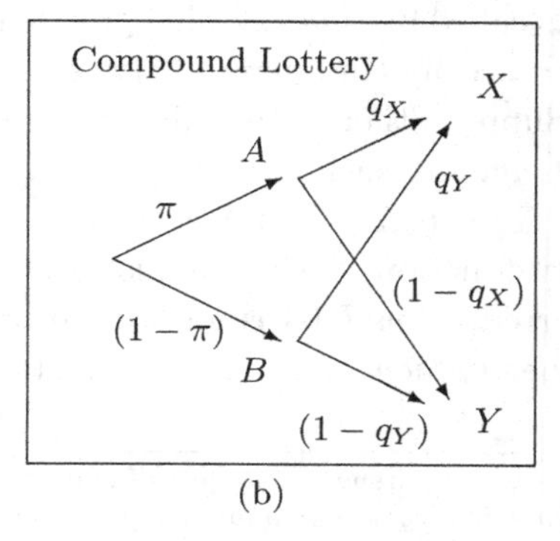

(b)

Fig. 4.1: Simple and Compound Lotteries

while those on the branches represent probabilities. For the compound lottery diagram, Fig. 4.1(b), A and B are themselves lotteries, with A being $[q_X \odot X + (1-q_X) \odot Y]$, and B being $[q_Y \odot X + (1-q_Y) \odot Y]$. This compound lottery is the same as $[(\pi q_X + (1-\pi) q_Y) \odot X + (\pi(1-q_X) + (1-\pi)(1-q_Y)) \odot Y]$ when we consider that the probability of winning X is now $p = \pi q_X + (1-\pi) q_Y$.

We require a few axioms to construct the useful class of VM utility functions. We suppose there is a primitive ordinal utility function $U(\cdot)$ that is continuous and increasing in the sense that $U(X) \geq (\leq)\, U(Y)$ iff $X \succeq (\preceq)\, Y$. Without loss of generality, we define $U : X \to [0,1] \in \mathcal{R}$. Unit utility is associated with the best or most preferred lottery B, i.e., $U(B) = 1$, while zero utility is associated with the worst or least preferred lottery W, i.e., $U(W) = 0$. Use of these two extreme reference points simplify the construction, but are not really necessary. Probability p lies in $[0,1]$. "$\succ$" means strict preference while "$\succeq$" means preference and includes indifference "$\sim$".

There are three assumptions as follows.

(A1) Any two lotteries X, Y can be put into one or both of the preference relations: $X \succeq Y$, $X \preceq Y$.
(A2) If $X \succeq Y$, then for any other lottery Z, $[p \odot X + (1-p) \odot Z] \succeq [p \odot Y + (1-p) \odot Z]$, where $p \in [0,1]$.
(A3) Suppose $X \succ Y \succ Z$ are any 3 lotteries, then $p, q, r \in [0,1]$ can be found such that $[p \odot X + (1-p) \odot Z] \succ Y \sim [q \odot X + (1-q) \odot Z] \succ [r \odot X + (1-r) \odot Z]$.

(A1) is called the completeness axiom. It includes the case $X \sim Y$ when both $X \succeq Y$ and $X \preceq Y$.

(A2) is sometimes called the substitution axiom. This is intuitive, but is not some natural fixture, so it has to be axiomatized. If I prefer a China holiday to a European holiday, it may also be that I prefer an even lottery of a European versus Mediterranean holiday to an even lottery of a China versus Mediterranean holiday, perhaps because of the lesser anxiety in the locational differences of the lottery outcomes. The axiom does compel some rationality onto the probability structure so as to make it a bit like a physical fraction p of outcome 1 and fraction $(1-p)$ of outcome 2 in the lottery. In addition, this axiom yields (a) the reflexivity principle, i.e., put $p = 0$ in (A2), and for any Z, $Z \succeq Z$; and (b) the transitivity principle, i.e., $X \succeq Y \Longrightarrow [p \odot X + (1-p) \odot Z] \succeq [p \odot Y + (1-p) \odot Z]$; $Y \succeq Z \Longrightarrow [p \odot Y + (1-p) \odot Z] \succeq [p \odot Z + (1-p) \odot Z] = Z$, hence $[p \odot X + (1-p) \odot Z] \succeq Z$, and $X \succeq Z$ by putting $p = 1$.

(A2) also implies that if $X \sim Y$ (or $\{X \succ Y$ and $X \prec Y\}$), then for any other lottery Z, $[p \odot X + (1-p) \odot Z] \sim [p \odot Y + (1-p) \odot Z]$ (or $\{[p \odot X + (1-p) \odot Z] \succeq [p \odot Y + (1-p) \odot Z]$ and $[p \odot X + (1-p) \odot Z] \preceq [p \odot Y + (1-p) \odot Z]\}$).

(A3) is a continuity axiom and is sometimes called the Archimedean axiom. It buys a lot of things. First, it allows one to put a lottery of X and Z in equal preference with possibly a non-lottery Y. Second, using B and W, (A3) allows any lottery to be put into equal preference with a lottery on B and W, i.e., any lottery $X \sim [\pi \odot B + (1-\pi) \odot W]$ for a $\pi \in [0,1]$.

(A3) also gives rise to the following lemma.

Lemma 4.1 $[p \odot B + (1-p) \odot W] \succ [q \odot B + (1-q) \odot W]$ iff $p > q$.

■

Proof. Suppose $X \succ Y$. By (A3), $\exists$ (there exists) $p_X, p_Y \ni$ (such that)

$$X \sim [p_X \odot B + (1-p_X) \odot W] \succ [p_Y \odot B + (1-p_Y) \odot W] \sim Y.$$

Since $X \succ Y \succ W$, by (A3) again, $\exists\ \pi \ni$

$$[\pi \odot X + (1-\pi) \odot W] \sim Y.$$

Left-hand side (LHS) is

$$[\pi \odot X + (1-\pi) \odot W] \sim [\pi \odot [p_X \odot B + (1-p_X) \odot W] + (1-\pi) \odot W]$$
$$\sim [\pi p_X \odot B + (1-\pi p_X) \odot W].$$

Therefore, $[\pi p_X \odot B + (1-\pi p_X) \odot W] \sim Y \sim [p_Y \odot B + (1-p_Y) \odot W]$. Thus, $\pi p_X = p_Y$. As $0 < \pi < 1$, we have $p_X > p_Y$, which is the proof of the "only if" part when we put $X = B \succ W = Y$.

Conversely, if $p_X > p_Y$, and $X \sim [p_X \odot B + (1-p_X) \odot W]$, while $Y \sim [p_Y \odot B + (1-p_Y) \odot W]$, we can find $0 < \pi < 1$, such that $\pi p_X = p_Y$. Thus, $Y \sim [\pi \odot X + (1-\pi) \odot W]$. By (A3), $X \succ Y$. ■

Theorem 4.1 (VM Expected Utility Representation): *There is a utility function on the lottery space X, $U : X \to [0,1] \in \mathcal{R}$, such that*

$$U(p \odot X + (1-p) \odot Y) = pU(X) + (1-p)U(Y) \tag{4.1}$$

where p is the probability of outcome X, and $1-p$ is the probability of outcome Y.

Proof. By (A3), we can characterize lotteries X and Y as $X \sim [p_X \odot B + (1-p_X) \odot W]$ and $Y \sim [p_Y \odot B + (1-p_Y) \odot W]$ for $p_X, p_Y \in [0,1]$. By

Lemma 4.1, $X \succ Y$ iff $p_X > p_Y$. Let $U : X \to [0,1] \in \mathcal{R}$ be an ordinal function. Using Lemma 4.1, we can thus fix $U(X) = p_X$ and $U(Y) = p_Y$ without loss of generality. This preserves the ordinal ranking of $X \succ Y$ given $p_X > p_Y$, and vice-versa, $X \succ Y \Rightarrow U(X) > U(Y) \Rightarrow p_X > p_Y$. Indeed, for any general $Z \sim [p_Z \odot B + (1 - p_Z) \odot W]$, $p_Z \in [0,1]$, we can fix $U(Z) = p_Z$.
Next

$$\begin{aligned}
&p \odot X + (1-p) \odot Y \\
&\quad \sim p \odot [p_X \odot B + (1-p_X) \odot W] + (1-p) \odot [p_Y \odot B + (1-p_Y) \odot W] \\
&\quad \sim [(pp_X + (1-p)p_Y) \odot B + (1 - pp_X - (1-p)p_Y) \odot W] \\
&\quad \sim [(pU(X) + (1-p)U(Y)) \odot B + (1 - pU(X) - (1-p)U(Y)) \odot W].
\end{aligned}$$

Since the LHS lottery is again expressed as a compound lottery of B and W, we can assign its utility as the probability of B in the compound lottery, i.e.,

$$U(p \odot X + (1-p) \odot Y) = pU(X) + (1-p)U(Y). \qquad \blacksquare$$

The utility function $U(\cdot)$ that satisfies axioms (A1), (A2), (A3) and thus the relationship $U(p \odot X + (1-p) \odot Y) = pU(X) + (1-p)U(Y)$ is called the VM utility function. It adds more properties to the primitive ordinal utility function $U(\cdot)$. It has a strong advantage over the primitive ordinal utility in that this VM utility function is able to provide a cardinal number in terms of expectation and rank preferences by the expectation outcome. It is also characterized as an expected utility function since any utility can be expressed as an expected utility.

Is $U(X)$ function limited to characterization by p_X in $X \sim [p_X \odot B + (1 - p_X) \odot W]$? We see what happens when we broaden it to be $U(X) = ap_X + b$ where a and b are constants. This includes the case when $U(X) = p_X$. Now

$$\begin{aligned}
&p \odot X + (1-p) \odot Y \\
&\quad \sim p \odot [p_X \odot B + (1-p_X) \odot W] + (1-p) \odot [p_Y \odot B + (1-p_Y) \odot W] \\
&\quad \sim [(pp_X + (1-p)p_Y) \odot B + (1 - pp_X - (1-p)p_Y) \odot W] \\
&\quad \sim \left[\left(p\left\{\frac{U(X)-b}{a}\right\} + (1-p)\left\{\frac{U(Y)-b}{a}\right\}\right) \odot B\right. \\
&\qquad \left. + \left(1 - p\left\{\frac{U(X)-b}{a}\right\} - (1-p)\left\{\frac{U(Y)-b}{a}\right\}\right) \odot W\right].
\end{aligned}$$

And so

$$\begin{aligned} &U(p \odot X + (1-p) \odot Y) \\ &= a\left(p\left\{\frac{U(X)-b}{a}\right\} + (1-p)\left\{\frac{U(Y)-b}{a}\right\}\right) + b \\ &= pU(X) + (1-p)U(Y). \end{aligned}$$

Thus, the expected utility representation is preserved and $U(\cdot)$ is unique up to an affine transformation. Any expected utility function or VM function $cU(\cdot) + d$, where c and d are constants, is equivalent to expected utility function $U(\cdot)$, producing the same preference outcomes.

There have been refutations of the VM representation via offered empirical paradoxes showing inconsistencies of VM utility implications, e.g., the Ellsberg and Allais paradoxes, being two of the most famous. However, the camp of rationalists remains very strong, and VM framework remains a major tool in economics and financial-theoretic modeling.

Hirshleifer and Riley commented,[2] "The dissident literature claims that the discrepancies revealed by these results refute the economist's standard assumption of rationality, or at least the expected utility hypothesis as a specific implication of that assumption. We reject this interpretation. A much more parsimonious explanation, in our opinion, is that this evidence merely illustrates certain limitations of the human mind as a computer. It is possible to fool the brain by the way a question is posed, just as optical illusions may be arranged to fool the eye."

4.2. Utility Functions

Under market mechanism where goods $i = 1, 2, \ldots, n$ are traded, suppose the market prices $\{p_i\}_{i=1,2,\ldots,n}$ are competitive and strictly positive (taken as given; in other words, individual consumer choices cannot affect the prices), a representative individual's demand on the goods $\{x_i\}_{i=1,2,\ldots,n}$ is restricted by his/her income Y accordingly. Denote X as the vector $(x_1, x_2, \ldots, x_n)$. The budget constraint is

$$\sum_i p_i x_i \leq Y.$$

[2] J Hirshleifer and JG Riley (1992). *The Analytics of Uncertainty and Information*, p. 34. Cambridge: Cambridge University Press.

As a consumer, he/she chooses x_i's to

$$\max_X U(X) \quad \text{subject to} \quad \sum_i p_i x_i \leq Y \quad \text{and} \quad X > 0.$$

Note that vector $X \geq 0 \Rightarrow$ any element is either > 0 or $= 0$. Vector $X > 0 \Rightarrow$ at least one element $x_i > 0$ while the others are ≥ 0. $X \gg 0 \Rightarrow$ all x_i's > 0. U as a function of direct consumption goods amounts is called a direct utility function and the optimal demands x_i as a solution is called the Marshallian demand function.

How do we solve this constrained optimization problem? Let us consider necessary conditions for a maximum. In Figs. 4.2(a) and (b), it is seen that keeping all other variables constant while varying x_i, if a maximum point occurs in the closed interior set where $x_i \geq 0$, as in Fig. 4.2(a), then in this case, $U_i' = 0$ at the maximum point. However, in Fig. 4.2(b), the maximum stationary point occurs outside the constraint set $x_i \geq 0$, and hence the constrained maximum occurs at $x_i = 0$. At this point, the slope U_i' is clearly negative.

Hence, under constraint $x_i \geq 0$, $U_i' \leq 0$ at the maximum point. This is one necessary condition. In fact, either $\{U_i' = 0, x_i \geq 0\}$ or $\{U_i' < 0, x_i = 0\}$, so we can use another necessary condition for maximum, i.e., $U_i' \times x_i = 0$.

Form the "Lagrangian" function with Lagrange multiplier λ; $L \equiv U(X) + \lambda(Y - \sum_i p_i x_i)$. Then, maximize the objective function

$$\max_{X,\lambda} U(X) + \lambda \left(Y - \sum_i p_i x_i \right) \qquad \text{s.t.} \quad X > 0.$$

Note that in L, if the maximum occurs within the constraint $\left(Y - \sum_i p_i x_i\right) > 0$, or in its interior, then the solution should be as if solving

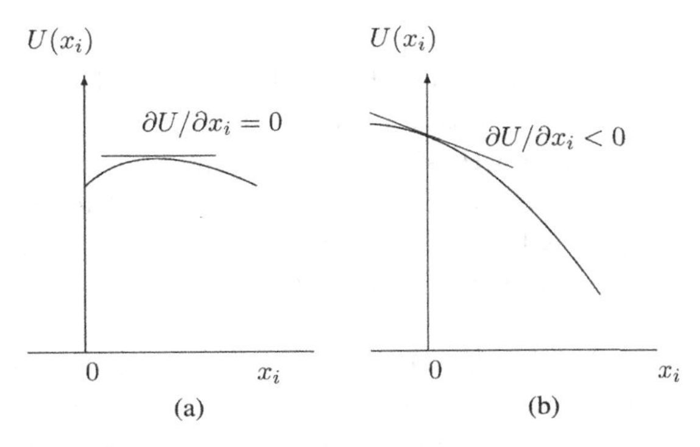

Fig. 4.2: Unconstrained and Constrained Maximums

$\max U(X)$ without the constraint, hence λ necessarily equals zero. If the maximum is right at the boundary where $\left(Y - \sum_i p_i x_i\right) = 0$, then $\lambda > 0$ since any increase of Y in the constraint set would increase L by the shadow price λ, which must necessarily be strictly positive in this case.

In fact, either $\{(Y - \sum_i p_i x_i) = 0, \lambda > 0\}$ or $\{(Y - \sum_i p_i x_i) > 0, \lambda = 0\}$, so we can use another necessary condition for maximum, i.e., $\left(Y - \sum_i p_i x_i\right) \lambda = 0$.

The first-order necessary conditions (FOC) are:

$$
\begin{array}{llll}
\text{(C1)} & \dfrac{\partial L}{\partial x_i}: & U_i(X) - \lambda p_i \le 0, & \forall i \\
\text{(C2)} & & (U_i(X) - \lambda p_i)\, x_i = 0, & \forall i \\
\text{(C3)} & \dfrac{\partial L}{\partial \lambda}: & Y - \sum_i p_i x_i \ge 0, & \\
\text{(C4)} & & (Y - \sum_i p_i x_i)\, \lambda = 0 & \\
\text{(C5)} & & x_i \ge 0, & \forall i, \text{ and} \\
\text{(C6)} & & \lambda \ge 0. &
\end{array}
$$

Second-order conditions for maximum are met as $U(c)$ is assumed to be strictly concave.

Conditions (C1) and (C2) follow the same arguments laid out for diagrams in Figs. 4.2(a) and (b) when $x_i \ge 0$. (C4) is sometimes called the complementary slackness condition.

Suppose we have an interior solution in (C1)–(C6). Then, $\lambda > 0$, and $x_i > 0 \quad \forall i$. (C2) $\Rightarrow U_i = \lambda p_i$. (C4) $\Rightarrow Y = \sum_i \frac{U_i x_i}{\lambda}$, so $\lambda = \frac{\sum_i U_i x_i}{Y}$. Then, we have

$$U_i = \frac{p_i}{Y}\left(\sum_i U_i x_i\right) \qquad \forall i,$$

and a solution $x_1^*, x_2^*, \ldots, x_n^*$ can be found. Each x_i^* is a function of p_i's and Y, or $x_i^*(Y;p)$. When expressed in terms of given prices and income Y, the demand function is called a Hicksian demand function.

Utility function $U(X(Y;p))$ or $U(Y;p)$ based on Hicksian demand or income or wealth Y is called an indirect utility function. It is this indirect form of utility function in terms of available income or wealth that is most often used in finance theory.

Under preference relations that can be represented as VM expected utility formulated above, an individual determines the probabilities of lottery payoffs, assigns an index U to each possible payoff, and then makes a decision to maximize the expected value of the index. The index or sometimes "utils" is a function of income or wealth. We shall refer to the use of utility under the representation as VM utility function.

Suppose there is an investment A that leads to final wealth W_A which is a RV with probability distribution $P(W_A)$. Another investment B leads to final wealth W_B with probability distribution $P(W_B)$. A is preferred to B iff $E(U(W_A)) > E(U(W_B))$ or $\sum U(W_A)P(W_A) > \sum U(W_B)P(W_B)$. Henceforth, we shall use money in the argument of $U(\cdot)$. Even if we use a certain good (say, gold) and its amount x in the argument, we can treat it as "money" or as a numéraire, so all other goods can be denominated in terms of the amount of gold.

4.3. Risk Aversion

First, we note that by the basic axiom of nonsatiation in human economic behavior (generically, excepting some self-sacrificial souls or instances), $U(x) > U(y)$ iff $x > y$.

This is because for fixed x greater than fixed y, $x \succ y$, so by VM expected utility (which after this point we shall always assume, unless it is otherwise indicated), $E[U(x)] > E[U(y)]$, hence $U(x) > U(y)$ as x, y are constants in this case. If for any $x > y$, $U(x) > U(y)$, then $U(\cdot)$ is a strictly increasing function of its argument — money. Assuming continuous function $U(\cdot)$ with existence of at least the first and second derivatives, for purposes of easy analysis and exposition of basic theoretical results, then clearly, $U'(\cdot) > 0$. Similarly, if $U(x) > U(y)$ and hence $E[U(x)] > E[U(y)]$, then $x > y$.

We shall define that a person (or agent or investor) is risk neutral if he (or she) is indifferent between doing nothing or value 0 and an actuarially (probabilistically in the expectations sense) fair amount $E(X) = 0$ where X is a RV or a lottery:

$$\tilde{X} = \begin{cases} \pi r & \text{with probability} \quad (1-\pi) \\ -(1-\pi)r & \text{with probability} \quad \pi. \end{cases}$$

The agent is defined to be risk averse if he or she prefers doing nothing to accepting the gamble, i.e., prefers certainty to an actuarially fair game. He/she is defined to be risk loving if he or she prefers the gamble to certainty.

Theorem 4.2 *An agent is risk averse iff $U(W)$, where W is his or her wealth, is a strictly concave function.* ■

Proof. By Jensen's inequality, if $U(\cdot)$ is strictly concave (which means $-U(\cdot)$ is strictly convex), $E[U(W+X)] < U(E[W+X]) = U(W)$ since $E(X) = 0$ for an actuarially fair lottery. Thus, the agent always prefers certainty to the actuarially fair gamble X, and is thus risk averse.

For the "only if" part, suppose the agent is risk averse, then

$$U(W) > \pi U(W - [1-\pi]r) + (1-\pi)U(W + \pi r)$$

for all r and $\pi \in (0,1)$. Since

$$U(W) = U(\pi(W - [1-\pi]r) + (1-\pi)(W + \pi r))$$

strict concavity of $U(\cdot)$ is shown. ■

We shall henceforth assume all agents are risk averse unless otherwise specified.

Suppose an agent faces a risky lottery of RV X, with $E(X) = 0$, so his/her final wealth is $W + X$, where W is a constant. An insurance company charges him/her an insurance amount I to remove any uncertainty in his/her final wealth. Then

$$E[U(W + \tilde{X})] = U(W - I).$$

Using Taylor's expansion, considering I is small relative to variance in X, then

$$E\left[U(W) + \tilde{X}U'(W) + \frac{1}{2}\tilde{X}^2 U''(W) + o(U)\right] = U(W) - IU'(W) + o(U).$$

Thus, $\frac{1}{2}E[\tilde{X}^2]U''(W) \approx -IU'(W)$, or $I \approx [-\frac{U''(W)}{U'(W)}]\ \frac{1}{2}\text{var}(X)$.

Intuitively, for any agent, $[-\frac{U''(W)}{U'(W)}] \stackrel{d}{=} A(W)$ is a positive number such that insurance premium I increases with this number $A(W)$ for a given risk X. If the agent is willing to pay a higher risk premium, he/she is more risk averse. Thus, $A(W)$ is a measure of risk aversion: the higher the $A(W)$, the higher the risk aversion. $A(W) > 0$ since concave $U(\cdot)$ implies $U''(\cdot) < 0$.

$A(W) = [-\frac{U''(W)}{U'(W)}]$ is called the absolute risk aversion function; $T(W) = 1/A(W)$ is called the risk-tolerance function; $R(W) = W \times A(W)$ is called the relative risk aversion function.

In some VM utility, when these functions become constants, we have the associated "coefficients."

For example, $U(W) = -e^{-aW}$, where a is a constant, is a negative exponential utility function.

$U' = -aU$, and $U'' = a^2 U$, so $A(W) = -U''/U' = a$, and $a > 0$ is called the constant absolute risk aversion coefficient.

Another example is $U(W) = \frac{W^{1-\gamma}}{1-\gamma}$, where γ is a constant, and U is a power utility function.

$U' = W^{-\gamma}$, and $U'' = (-\gamma)W^{-\gamma-1}$. So, $R(W) = -WU''/U' = \gamma$, and $\gamma > 0$ is called the constant relative risk aversion coefficient.

$U(W) = \frac{t}{1-t}(\frac{cW}{t} + d)^{1-t}$, $d > 0$, is a class of hyperbolic absolute risk aversion (HARA) utilities.

$U' = c(\frac{cW}{t} + d)^{-t}$, and $U'' = -c^2(\frac{cW}{t} + d)^{-t-1}$.

So, $A(W) = -U''/U' = c(\frac{cW}{t} + d)^{-1}$.

Thus, $T(W) = 1/A(W) = \frac{W}{t} + \frac{d}{c}$. HARA utility functions are linear risk tolerance functions (in wealth).

The logarithmic utility function $\log(W)$ is a special case of the power utility, as seen below, when $\gamma \to 1$. Apply L'hôpital's rule:

$$\lim_{\gamma \to 1} \frac{W^{1-\gamma}}{1-\gamma} = \log(W).$$

Note $\frac{d}{dW}\log W = 1/W > 0$, so log utility is increasing in W. Next, $\frac{d^2}{dW^2}\log W = -1/W^2 < 0$, so log utility exhibits strict concavity and thus risk aversion.

It is interesting to note that since the 1980s there have been increasing attempts to model non-standard utility, i.e., non-VM utility that does not necessarily obey Theorem 4.1. A non-standard utility may have the advantage of higher flexibility to explain some of the aberrations or anomalies occurring that could not be satisfactorily explained by VM-type preferences. An example of non-standard utility is the recursive Epstein–Zin utility function[3] that is used frequently in life-cycle modeling where intertemporal substitutional issues are significant. In the Epstein–Zin utility, an extra elasticity of intertemporal substitution parameter allows its disentanglement from the coefficient of risk aversion so that both intertemporal substitution or resolution of uncertainty in near versus far risks and also risk aversion in degree of overall uncertainty can be separately considered.

[3]Epstein, LG and SE Zin (1989). Substitution, risk aversion, and the temporal behavior of consumption growth and asset returns I: A theoretical framework. *Econometrica*, 57(4), 937–969.

Application of VM Expected Utility

Nicholas Bernoulli in 1713 posed the "St. Petersburg Paradox" as follows. It was commonly accepted at that time that the price of a lottery would be its expected value. The lottery pays 2^{n-1} if the first head occurs in the nth toss of a coin.

The probability of a head at the nth toss follows a geometric distribution (special case of negative binomial) $p(1-p)^{n-1}$ where $p = 1/2$ is the probability of a head. The expected payoff of the lottery is then

$$\sum_{i=1}^{\infty} \frac{1}{2}\left(\frac{1}{2}\right)^{i-1} 2^{i-1} = \sum_{i=1}^{\infty} \frac{1}{2} = \infty!$$

The paradox is that no one would pay ∞ to buy this lottery!

Now suppose people are risk averse and not risk neutral (who would then play actuarially fair games). Suppose they have log utility $\log(W)$. Then, a fair price π is $\ni$:

$$\log(\pi) = \sum_{i=1}^{\infty} \frac{1}{2}\left(\frac{1}{2}\right)^{i-1} \log\left(2^{i-1}\right) = \sum_{i=1}^{\infty}\left(\frac{1}{2}\right)^{i} (i-1)\log 2 = \log 2.$$

Hence, $\pi = 2$ which is a much smaller sum to pay given the risk aversion, and can thus explain the paradox!

4.4. Value of Information

Suppose a market offers state shares that give rise to a payoff for every \$1 invested in state i share as follows. There are N finite states of the world, thus N types of shares:

$$\$\text{payoff} = \begin{cases} X_i & \text{with probability} \quad p_i \\ 0 & \text{with probability} \quad (1-p_i). \end{cases}$$

An investor or individual has a budget of \$1 with which to allocate amount $a_i \geq 0$ to the ith share such that $\sum_{i=1}^{N} a_i = 1$. His/her payoff if state j occurs next period is $\$a_j X_j$. He/she maximizes his/her expected utility subject to budget constraints, i.e.,

$$\max_{a_i} \sum_{i=1}^{N} p_i\, U(a_i X_i) \qquad \text{s.t.} \sum_i a_i = 1 \quad \text{and} \quad a_i > 0\ \forall i$$

If Lagrangian function $L \equiv \sum_{i=1}^{N} p_i \; U(a_i X_i) + \lambda(1 - \sum_i a_i)$ is maximized, assuming the budget is satisfied, and assuming for simplicity of exposition that all $a_i > 0$, then the necessary condition is

$$p_i X_i \, U'(a_i X_i) = \lambda.$$

For log utility $U(\cdot)$, the necessary conditions are

$$p_i X_i \, \frac{1}{a_i X_i} = \frac{p_i}{a_i} = \lambda.$$

Solving, $\lambda = 1$, so $a_i = p_i$. Then, the maximum value of the objective function is

$$\sum_{i=1}^{N} p_i \log(p_i X_i) = \sum_{i=1}^{N} p_i \log p_i + \sum_{i=1}^{N} p_i \log X_i. \tag{4.2}$$

Now, the uncertainty in the problem has to do with which state i will eventually occur. Suppose the investor is given an information set on which state occurred before he or she makes an investment decision about a_i. We saw how to interpret this in Chap. 2. In particular, we saw $E(\tilde{X}|\tilde{X}) = \tilde{X}$. Then, the investor's *ex ante* objective function, given this full information each time before he or she decides, is

$$\max_{a_i} E\left(U(a_i X_i)|\tilde{X}\right) = \sum_{i=1}^{N} p_i \; \log(1 \times X_i) = \sum_{i=1}^{N} p_i \; \log(X_i). \tag{4.3}$$

Maximum log utility is obtained in (4.2) without information while that in (4.3) is obtained with full information from distribution $\tilde{X}$. The information value (in util sense, not $ sense here) is therefore

$$\left[\sum_{i=1}^{N} p_i \log X_i\right] - \left[\sum_{i=1}^{N} p_i \log p_i + \sum_{i=1}^{N} p_i \log X_i\right] = -\sum_{i=1}^{N} p_i \log p_i.$$

In statistical information theory, $-\sum_{i=1}^{N} p_i \log p_i$ is also called the entropy of RV $\tilde{X}$ and is a measure of the uncertainty embodied in the randomness of $\tilde{X}$. For example, $X = 1$ with probability 0.5 and $X = 0$ with probability 0.5 is a lot more uncertain than $X = 1$ with probability 0.1 and $X = 0$ with probability 0.9, in that one is more able to predict the latter with the higher probability of which value will occur.

4.5. Economics of Information

We shall briefly develop here, some ideas of economics of information and show how they have entered as ideas in finance. Microeconomics and finance theory are especially closely connected in many ways.

Adverse Selection

Gresham's law has its intention of saying that cheap money will drive out the good ones, in the following sense. Suppose good coins are minted with pure silver weighing 10 grams, and cheaper metals such as brass are also mixed to produce counterfeit coins also weighing 10 grams and are passed off as silver coins. A smart investor who finds a way of detecting the genuine silver coins would retrieve them, melt down the silver and sell them for more money. Thus, bad coins will drive out the good ones eventually under this scenario.

In this same sense, Akerlof's[4] second-hand car market will see lemons drive out peaches if buyers have no way of telling which are lemons (junks) and which are peaches (good running cars). Sellers will not put their peaches on the market because buyers will tend to offer a lower price as they are more likely to get a lemon that is not worth that much.

Adverse selection is about a situation where the bad will be adversely selected with a higher chance because the good will be less forthcoming. This phenomenon occurs frequently in real economic situations.

Suppose there are many insurance companies offering insurance to clients against car accidents, paying \$1 per unit of insurance purchased. Price per unit of insurance is \$p. There are two types of car drivers, the good ones who are safe drivers with probability of accident $\pi_G < \pi_B$, the probability of accident of the other type of bad drivers. Assume they all have identical utility function U. If an accident happens, the loss by the driver is \$K. For a good driver, with original wealth W, he chooses to buy Q units of the insurance in the following optimization

$$\max_Q \ \pi_G U(W - K + Q[1-p]) + (1-\pi_G)U(W - Qp).$$

[4]Akerlof, GA (1970). The market for 'lemons': Quality uncertainty and the market mechanism. *The Quarterly Journal of Economics*, 84(3), 488–500.

Type G's FOC is

$$\frac{U'(W-K+Q_G[1-p])}{U'(W-Q_Gp)}=\frac{(1-\pi_G)p}{\pi_G(1-p)}. \tag{4.4}$$

Similarly, type B's FOC is

$$\frac{U'(W-K+Q_B[1-p])}{U'(W-Q_Bp)}=\frac{(1-\pi_B)p}{\pi_B(1-p)}. \tag{4.5}$$

Since $\frac{(1-\pi_G)}{\pi_G}>\frac{(1-\pi_B)}{\pi_B}$ in (4.4) and (4.5), the LHS of (4.4) > the LHS of (4.5). As $U''<0$, and thus U' is decreasing in Q, therefore $Q_G<Q_B$. In other words, the G type will demand less insurance than the B type for the same price per unit of insurance.

In fact, (4.4) and (4.5) are quite instructive. If $p=\ (>)\ \pi_G$, then $Q_G=\ (<)\ K$. If $p=\ (<)\ \pi_B$, then $Q_B=\ (>)\ K$. If $\pi_G<p<\pi_B$, then G underinsures ($<K$) and B overinsures ($>K$).

Suppose the proportion of G is θ. Then, insurance firms being in a perfectly competitive market will fetch zero (normal) expected profit. The zero expected profit condition per capita is

$$(\theta Q_G+(1-\theta)Q_B)p=\theta\pi_GQ_G+(1-\theta)\pi_BQ_B. \tag{4.6}$$

LHS is per person insurance revenue to firm. RHS is expected per person payout by firm. Competitive condition (4.6) simplifies to

$$Q_G\theta(p-\pi_G)=Q_B(1-\theta)(\pi_B-p),\quad \text{for } \pi_G<p<\pi_B. \tag{4.7}$$

Equation (4.7) shows that when the insurance firm cannot identify and thus discriminate between type G and type B, it can offer only one price p to both G and B, and in this case, the gains it makes from type G (LHS) (if firms can identify G, its price should be lower at π_G) are used to subsidize for the losses it incurs from type B (RHS) (if firms can identify B, its price should be higher at π_B).

However, (4.7) may not be consistent with the implication of $Q_G<Q_B$ from (4.4) and (4.5). If $\frac{(1-\theta)(\pi_B-p)}{\theta(p-\pi_G)}>1$, or if $\theta<\frac{\pi_B-p}{\pi_B-\pi_G}$, then $Q_G\not<Q_B$. In other words, if there are too few G types (θ small) to help subsidize B, or if the single price p is too low, then there are too many B's to subsidize. In such a situation, there is no equilibrium (hence no equilibrium price solution) where demand by B's and G's equal supply of insurance by firms. Since the equilibrium with a single price would have been a "pooling" equilibrium, there is thus no pooling equilibrium. In

general, there are plenty of economic forces to cause breakaway from any temporary pooling equilibrium. For example, the G types could organize themselves into a cooperative (assuming they can identify themselves) to self-insure at a lower price. As less ($\theta \downarrow 0$) G types buy insurance, the equilibrium gives way. Eventually, it could be left with just the B type and then the price of insurance would have risen to $p = \pi_B$. Thus, B "drives out" G.

Another way in which markets work to produce equilibrium is when the good type can find a way to signal to the firms they are the G type, even if *a priori* the firms cannot identify them.

Signalling

Suppose the G type can enrol in a car driving test and get a pass certificate. Assume a G-type's disutility (think of it as inconvenience cost e) enters its utility function as

$$V_G(x, e) = U(x) - e$$

where U is the function in the last section, V is the new VM utility function, and $e > 0$. Note that such a test or not does not affect the outcome x. If G does not decide to go for a driving test, $e = 0$. Similarly, B type can enrol in the test and get a pass certificate, but he/she has a greater disutility and inconvenience cost in doing so (perhaps expending a lot more effort). B's VM utility function is now

$$V_B(x, e) = U(x) - ne$$

where $e > 0$ and $n > 1$. If B does not decide to go for a driving test, $e = 0$.

Suppose in this case, the insurance firms figure out a way to reach price equilibrium by offering two types of insurance contracts: (a) pay π_G if client shows proof of a driving test certificate; and (b) pay higher π_B if client does not show proof of a test certificate. Suppose the resulting situation is that G-types all take up the test and also contract (a) and pay π_G, so $Q_G = K$. Further, all B-types choose not to take up the test and opt for contract (b) and pay π_B, so $Q_B = K$. In this situation, the insurance firms also face zero expected profit, since for each type of contracts, revenue $\pi_{\{.\}} Q_{\{.\}}$ exactly equals expected payout $\pi_{\{.\}} Q_{\{.\}}$.

Although the optimality (demand) equations (4.4) and (4.5), and the competitive (supply) condition Eq. (4.7), are satisfied, the situation may still not be an equilibrium if B decides to switch to the cheaper contracts in (a), or if G decides to switch to the more expensive contracts in (b). To keep G and B to their respective contracts, we have to ensure another condition exists. This is called incentive compatibility (IC) condition to ensure there is an incentive to stick to a certain choice.

For B to stick to (b), the IC condition is that his/her expected utility $E[V]$ under (b) must be greater than that under (a). Hence

$$\begin{aligned} &\pi_B[U(W-K+K[1-\pi_B])]+(1-\pi_B)[U(W-K\pi_B)] \\ &\quad > \pi_B[U(W-K+K[1-\pi_G])]+(1-\pi_B)[U(W-K\pi_G)]-ne. \end{aligned} \tag{4.8}$$

The LHS is the resulting expected utility under choice (b) while the RHS is resulting expected utility under choice (a). In the latter, B pays a cheaper price π_G, but suffers a lower utility due to $-ne$ in the certificate-signal.

For G to stick to (a), the IC condition is

$$\begin{aligned} &\pi_G[U(W-K+K[1-\pi_G])]+(1-\pi_G)[U(W-K\pi_G)]-e > \\ &\pi_G[U(W-K+K[1-\pi_B])]+(1-\pi_G)[U(W-K\pi_B)]. \end{aligned} \tag{4.9}$$

IC condition (4.8) becomes

$$\begin{aligned} &\pi_B[U(W-\pi_B K)]+(1-\pi_B)[U(W-K\pi_B)] \\ &> \pi_B[U(W-\pi_G K)]+(1-\pi_B)[U(W-K\pi_G)]-ne. \end{aligned}$$

Or,

$$e > \frac{1}{n}[U(W-\pi_G K)-U(W-\pi_B K)] > 0.$$

IC condition (4.9) becomes

$$\begin{aligned} &\pi_G[U(W-\pi_G K)]+(1-\pi_G)[U(W-K\pi_G)]-e \\ &> \pi_G[U(W-\pi_B K)]+(1-\pi_G)[U(W-K\pi_B)]. \end{aligned}$$

Or

$$e < [U(W-\pi_G K)-U(W-\pi_B K)].$$

IC (4.9) is interesting as it shows that G can (unintuitively) opt for the more expensive (b) if the signal cost e is too high. IC (4.8) shows that B will stick to (b), provided ne is sufficiently high.

Thus the IC conditions require

$$\begin{aligned}&[U(W-\pi_G K)-U(W-\pi_B K)]\\ &>e>\frac{1}{n}[U(W-\pi_G K)-U(W-\pi_B K)]>0.\end{aligned}$$

Clearly, we can find n such that the IC conditions will be satisfied and there will be a separating equilibrium under the signal.

Signalling theory in economics became popular modeling tools after Michael Spence published, among others, his Nobel-winning paper[5] on the signalling equilibrium illustrating a classic case of how more able workers can use education as a signal which is less costly to them than to less able workers.

In finance, signalling is sometimes suggested as an explicit mechanism used by corporate listed firms to signal good news when they issue more dividends (weaker firms find it too costly to do so as they will go bankrupt), or finance by issuing more debt than using equity (weaker firms again will find too much debt too costly to service). Except for tax benefits of debt financing, Modigliani and Miller[6] showed in an early path-breaking paper that debt-equity financing is irrelevant in firm value, although under an implicit assumption of riskless debt or debt without bankruptcy cost.

Moral Hazard and Principal Agency Theory

Apart from not being able to observe types in adverse selection problems, decision-makers face many situations in the real world in which they cannot observe the actions by others after some mutual contracting is done. For example, an employer (called the principal) may agree to pay a wage of $5,000 to a worker (called an agent) to perform a task worth $5,000. However, because the principal cannot observe the performance of the worker, the worker may shirk (idle) and perform only a substandard job worth $2,000.

This behavior of shirking or deviating in action from what is implicitly or explicitly agreed, just because it is not observable by the principal, is called moral hazard. A related problem in public goods (e.g., provision

[5]Spence, M (1973). Job market signalling. *The Quarterly Journal of Economics*, 87(3), 355–374.

[6]See Modigliani, F and M Miller (1958). The cost of capital, corporation finance, and the theory of investment. *American Economic Review*, June, 261–297.

of leisure parks, public gardens, etc.) is called "negative externality", in which one litters and is unobserved, or if a firm pumps toxic waste down the river and is unobserved. Negative externality can be reduced by policing and imposing a hefty fine for bad behavior. If not, public goods may become greatly underprovided (e.g., poorly maintained parks). Thus unobservability and nonenforceability of good behavior cause great deadweight loss (loss to all relative to a situation where all behave and enjoy a nice park).

We now return to the worker-employer scenario where many scholars such as Bengt Holmstrom have done excellent research. Suppose we have a principal with utility function $V(x_i - w(x_i))$, where x_i is the \$ outcome of an agent's job performance with effort level $e \in (0,1)$ which is not observable by the principal. $w(\cdot)$ is a wage function offered by the principal.

Higher effort e by the agent will result in a better payoff distribution to the principal. Assume $x_i(e')$ stochastically dominates $x_i(e)$ iff $e' > e$. To be clear, effort level e affects the probability distribution of x_i, but not the discrete values of X.

We explain stochastic dominance below. Distribution $G(X)$ is said to stochastically dominate (first-order stochastic dominance) $F(X)$ if for every x, $G(x) < F(x)$, or $P_G(X \leq x) < P_F(X \leq x)$. Another way to look at it is that for f, g satisfying $P_G(X \leq g) = P_F(X \leq f)$, then $g > f$. Or yet another way is, $P_G(X > x) > P_F(X > x)$, or for any x, the probability of G having a larger probability of exceeding x is higher than F. The CDF curves are shown in Fig. 4.3.

The way we shall model the outcome distributions X — which are dependent both on agent effort and also exogenous noise — follows Hart

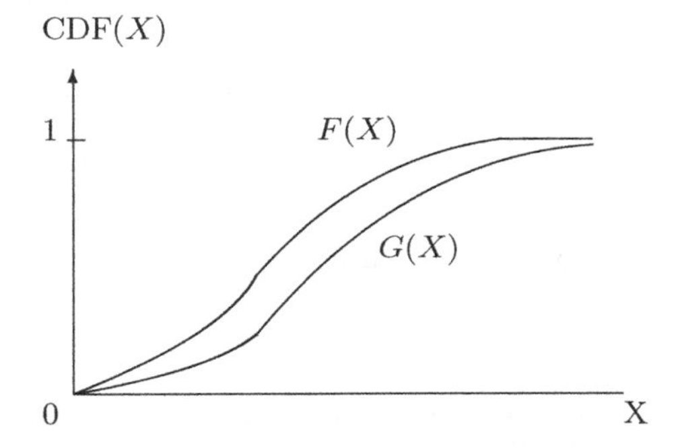

Fig. 4.3: Illustration of First-Order Stochastic Dominance

Table 4.1: Agent's Dollar Output Distribution

State j	Payoff X	Prob π_j
1	x_1	$eL_1 + (1-e)H_1$
2	x_2	$eL_2 + (1-e)H_2$
$\vdots$	$\vdots$	$\vdots$
M	x_M	$eL_M + (1-e)H_M$

and Holmstrom (1987).[7] Outcomes are described by not just values x_j of RV X, but also its attendant probabilities. Specifically, where e is the agent's unobserved effort once the wage contract is committed, $x_1 < x_2 < \cdots < x_M$, and $0 < L_i < H_i, \forall i \leq M-1$. The payoff distributions are as shown in Table 4.1.

For $e \in (0,1)$, in order to ensure $\sum_{j=1}^{M} \pi_j = 1$, for the Mth state, $\pi_M = e[1 - \sum_{j=1}^{M-1} L_j] + (1-e)[1 - \sum_{j=1}^{M-1} H_j]$. Or we can define $L_M = [1 - \sum_{j=1}^{M-1} L_j] > [1 - \sum_{j=1}^{M-1} H_j] = H_M$. Then, clearly, $P(X; e_1)$ stochastically dominates $P(X; e_2)$ iff $e_1 > e_2$.

Given wage function $w(x_i)$, the agent maximizes his/her own VM utility defined as $U[X,e] = u[w(X)] - \phi(e)$ by choosing optimal effort, where u is continuous and strictly concave in $w(X)$ and ϕ is continuous and strictly convex in e. The latter is a disutility, being the cost of extra effort:

$$\begin{aligned}\max_e \; E[U] &= \sum_{j=1}^{M} [eL_j + (1-e)H_j] \; u[w(x_j)] - \phi(e) \\ &= \sum_{j=1}^{M} H_j u[w(x_j)] + e \sum_{j=1}^{M} [L_j - H_j] u[w(x_j)] - \phi(e).\end{aligned}$$

FOC where $\frac{\partial E[U]}{\partial e} = 0$ gives

$$\sum_{j=1}^{M} [L_j - H_j] u[w(x_j)] = \phi'(e). \tag{4.10}$$

[7] Hart, O and B Holmstrom (1987). The theory of contracts. In *Advances in Economic Theory*, T Bewley (ed.), Fifth World Congress. Cambridge: Cambridge University Press.

The principal in designing his/her optimal wage contract function must solve:

$$\max_{w(x_1),w(x_2),\ldots,w(x_M)} \sum_{j=1}^{M} [eL_j + (1-e)H_j]\, V[x_j - w(x_j)]$$

subject to

$$\sum_{j=1}^{M} [L_j - H_j] u[w(x_j)] = \phi'(e),$$

and

$$E[U] = \sum_{j=1}^{M} [eL_j + (1-e)H_j]\, u[w(x_j)] - \phi(e) \geq \underline{U}.$$

The first constraint is the IC constraint to ensure the wage function $w(\cdot)$ chosen is compatible with the agent's choice in Eq. (4.10), and the second constraint is to ensure the wage function meets some minimal condition to induce the agent to enter the wage contract. It must meet the agent's minimum expected utility of $\underline{U}$.

In the solution of the principal-agent optimal contract or wage function $w(x_j)$ for given constants $x_1 < x_2 < \ldots < x_m$, the constraint in Eq. (4.10) can be solved once continuous functions $U(w)$ and $\phi(e)$ are explicitly specified. The agent's optimal effort choice e^* is then a function of $w(x_j)$ for $j = 1, 2, \ldots, M$. The second IC constraint can be satisfied when $\underline{U}$ is sufficiently low. Then the objective function is the maximum of $\sum_{j=1}^{M} [e^* L_j + (1-e^*)H_j] V[x_j - w(x_j)]$ and can be solved with respect to the principal's optimal choices of $w_1 < w_2 < \ldots < w_M$, once continuous function $V(X, w)$ is specified.

Information can become critical in determining choices and actions of players in a transaction setting, especially when one set of players has asymmetrically superior information relative to another set of players. The understanding of preference and choice theory as well as value of information above will help one to appreciate the subtleties involving such asymmetric information situations. For example, in English auctions (with buyers openly bidding up prices under a reserve price system), very often the uninformed or those with inferior information will pay too high and suffer what is called a winner's curse. In initial public offer (IPOs) tenders, investors usually bid and obtain offerings at a discount to the opening price of the stock due to a risk premium being paid to them, as on average, they will not have superior information but are necessary to soak up all the offerings. Asymmetric information also extends to game settings in small groups of players.

4.6. Problem Set 4

1. Suppose an individual has a quadratic utility function $U(W) = aW - bW^2$, where a and b are constants. Is there any constraint on a, b to make this a feasible utility function?
2. Suppose subject X is presented with a choice between prospects A and B.
   A: receive \$100 K with certainty
   B: receive \$1 M with 10% probability, \$100 K with 89% probability, \$0 with 1% probability.
   X chooses A. Next, X is presented with a choice between prospects C and D.
   C: receive \$1 M with 10% probability, \$0 with 90% probability.
   D: receive \$100 K with 11% probability, \$0 with 89% probability.
   X chooses C. Show if there is inconsistency in terms of VM expected utility.
3. In the separating equilibrium case for the insurance contracts, where price per unit of insurance is π_G if a test certificate is shown, and π_B otherwise, prove that optimal $Q_B = Q_G = K$.
4. Suppose there are N stocks and an investor holds x_i shares of each stock i. Let investment vector $X = (x_1, x_2, \ldots, x_N)^T$. There are $j = 1, 2, \ldots, M$ states of the world. At the end of the investment period, per share values are p_{ij}, or a price vector of $P_j = (p_{1j}, p_{2j}, \ldots, p_{Nj})^T$ in state j. His/her portfolio wealth becomes $\sum_{i=1}^{N} x_i p_{ij}$ or $X^T P_j$ if the jth state occurs.

   The jth state occurs with (*ex ante*, or "before-the-fact") probability of π_j where $\sum_{j=1}^{M} \pi_j = 1$. His/her expected utility is $V(X) = \sum_{j=1}^{M} \pi_j U(X^T P_j) = \sum_{j=1}^{M} \pi_j U(\sum_{i=1}^{N} x_i p_{ij})$. If U is increasing and strictly concave, in X (given P_j), show that V is also strictly concave in X.
5. Show that if RV X first-order stochastically dominates RV Y, then it is equivalent to an equal-in-distribution relationship $Y \stackrel{d}{=} X + e$, where e is a non-positive RV.

Chapter 5

STATE PRICE AND RISK-NEUTRAL PROBABILITY

By the 1950s, the existence of the Walrasian equilibrium — the idea that in free competitive markets, a set of prices clears demands and supplies so that all agents have optimally allocated their initial endowments and budgets — was proven in great generality by Arrow and Debreu (1954).[1] Radner (1972) added information sets and rational expectations to this framework to show existence of sequential market equilibrium.[2] Neo-Walrasian economics, as part of neo-classical economics (the study of the determination of prices, outputs, and incomes through individual or consumer choice theory and market clearing) in the post-WW II years moved into issues of optimal risk sharing,[3] social choice theory, and sometimes normative economics of welfare theory. These were the forerunners of the intensive rational agent decision-making approach and game-theoretic equilibrium modeling that followed with path-breaking findings. At the same time, as a consequence of the rational equilibrium paradigm, the idea of state securities or Arrow–Debreu certificates became a springboard for development of financial economics into a rich set of asset pricing theories and models well into the 1980s.

5.1. Uncertainty and State Prices

In modeling a single period world from time $t = 0$ to $t = 1$, i.e., over interval $[0, 1]$, an investor at time $t = 0$ faces uncertainty in the future at $t = 1$, but nevertheless has to make investment decisions at $t = 0$. In the modeling, a current consumption C_0 at $t = 0$ may also be added. There are only two *de facto* time points in a one-period model, $t = 0$ and $t = 1$. The uncertainties are fully resolved and revealed at the end of period, $t = 1$.

[1] Arrow, K and G Debreu (1954). Existence of equilibrium for a competitive economy. *Econometrica*, 22, 265–290.

[2] Radner, R (1972). Existence of equilibrium of plans, prices and price expectations in a sequence of markets. *Econometrica*, 40(2), 289–303.

[3] This is one of the many major contributions of Kenneth Arrow; see his classic *Essays in the Theory of Risk-Bearing* (1974) Chicago: Markham Publishing Co.

At $t = 1$, the investor's investment returns are realized and he or she consumes the final wealth and ends investment (and presumably dies). However, rather than being overly prescriptive about what happens to the individual at $t = 1$, the model leaves off. Sometimes, modelers implicitly assume that the story can then repeat itself with the investor again making new decisions in another period. However, in economic modeling, there is connectivity between uncertainties at $t = 1$ and at $t = 2$ in this case. Therefore, rational investors would make decisions at $t = 0$ by considering not only their information about uncertainties at $t = 1$ but also at $t = 2$. This is where single period models differ from dynamic multi-period models which we consider later. The situation would be quite different, say, in playing poker where each hand or game is independent of the next, and players can work out their strategies as if in a single period model and then repeat it in the next period, provided obviously that capital is not a constraint. Nevertheless, one-period models capture many of the financial economic ideas and concepts that are shared with multi-period models, and these are discussed in this and the next chapter.

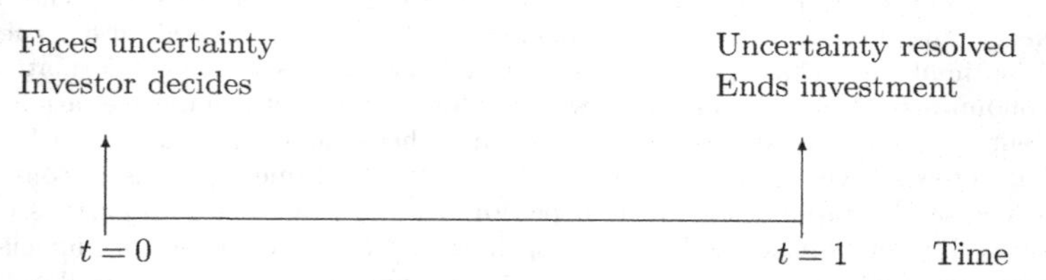

Suppose uncertainties that matter to the investor, i.e., enter his/her utility or preference function, are S states of the world denoted by ω_i for each state $i = 1, 2, \ldots, S$. This is a state representation of future uncertainties. Note that S is at most countably infinite (less dense than a continuous real line segment).

We begin with a simple example for illustration. An investor starts with a wealth of \$1 at $t = 0$. Suppose $S = 2$, so there are 2 states of the world ending in up (U) state with \$payoff or return u, and down (D) state with \$payoff or return d on a \$1 stock purchase at $t = 0$. Note that this is a parsimonious setup: we could start with price S_0 and end with uS_0, and dS_0. Buying $1/S_0$ unit, we end up similarly with the state-contingent (state-dependent) payoffs equal to u and d. (However, in the latter, we save one variable and notation, S_0. This is different when S_0 is explicitly needed as in option pricing given exercise price K, where it then becomes material how large S_0 is, relative to K.) We assume $u > d$.

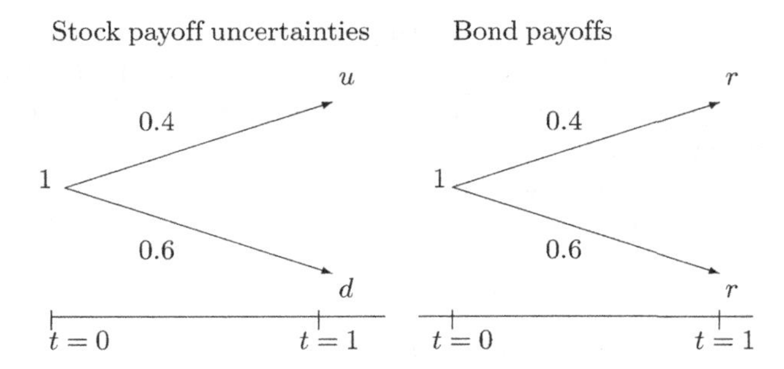

Fig. 5.1: Security Payoffs in a One-Period Model

We suppose investors know the structure of the uncertainties to be resolved in the end at $t = 1$, into u with actual or empirical probability 0.4, and d with actual or empirical probability 0.6. However, at $t = 0$, investors do not know which state will eventually occur. Besides the risky stock, there is one risk-free bond with risk-free return or payoff r, where $u > r > d$, or there will be riskless arbitrage opportunities. (Note that a return or payoff is equal to one plus the rate of return.) See Fig. 5.1.

Define a state price (or sometimes called elementary claim, or state-contingent claim, or originally called Arrow–Debreu certificate price) as the price of a simple security which pays \$1 iff that state occurs, and 0 otherwise. Assume there is a market state price for U-state. Buying this simple security would provide a payoff of \$1 when U occurs, but a zero payoff when D occurs. Similarly, assume there is a market state price for D-state. Buying this simple security would provide a payoff of \$1 when D occurs, but a zero payoff when U occurs. In this simple economy with two states and two market instruments viz. the risky stock and the risk-free bond, we can show that the state U and state D securities can exist and can be replicated.

Let us try to find the state security (or contingent claim) prices here. First, note an interesting property depicted below which says that any linear combination of traded payoff-price vectors leads to another traded payoff-price vector, i.e.,

$$A\begin{pmatrix} u \\ d \\ -1 \end{pmatrix} + B\begin{pmatrix} r \\ r \\ -1 \end{pmatrix} = \begin{pmatrix} Au + Br \\ Ad + Br \\ -(A+B) \end{pmatrix}. \tag{5.1}$$

The upper two elements of the 3×1 vector are the payoffs on the U and the D states, respectively. The third element is the cost (negative in sign) of the security giving rise to the state-contingent payoffs. The security with the state-contingent payoffs and the cost is represented by the traded payoff-price vector. A and B are the number of units of stocks and bonds that are purchased respectively.

If we choose $A > 0$ (long A units or buy \$$A$ of stock) and $B < 0$ (short B units or sell \$$B$ of bond) such that

$$Au + Br = 1 \quad \text{and}$$
$$Ad + Br = 0,$$

two equations in two unknowns A, B, then the solution is $A = \frac{1}{(u-d)}$ and $B = -\frac{d}{r(u-d)}$. Hence, the price of this state-U contingent claim is

$$A + B = \frac{1}{(u-d)} - \frac{d}{r(u-d)} = \frac{r-d}{r(u-d)}. \tag{5.2}$$

If we choose A and B such that

$$Au + Br = 0$$
$$Ad + Br = 1,$$

then, the solution is $A = \frac{1}{(d-u)}$ and $B = \frac{u}{r(u-d)}$. Hence, the price of this state-D contingent claim is

$$A + B = -\frac{1}{(u-d)} + \frac{u}{r(u-d)} = \frac{u-r}{r(u-d)}. \tag{5.3}$$

The state prices that are derived from the traded securities are arbitrage-free prices and should be strictly positive. In fact, they are unique given the structure of the price processes, i.e., they are expressed in terms of u, d, and r.

We can provide a bit more characterization of the state prices. If in the two-state model, there are only two securities, none of which is redundant (a redundant security is just a scalar multiple of an existing payoff-price vector), then, the market is said to be complete. In a complete market, every state is attainable in the sense that all the state prices can exist and can be replicated as shown above. In a complete market, the state prices are also unique as seen in (5.2) and (5.3). We also note "strangely" thus far that the actual or empirical probabilities of the states do not enter into the picture. This is because the state security prices in (5.2) and (5.3) are dependent on the parameters u, r, and d that are defining the stock and bond prices. In actual fact, under general equilibrium in a market with investors having risk-return preferences or utility functions (as seen in Chap. 4), the stock and bond prices would have reflected or embedded parameters defining the utility function or preference, as well as the empirical probabilities defining the uncertain future payoffs. This notion shall be seen more fully in this and the next chapter. However, these preference parameters and empirical probabilities do not show up in replicating non-arbitrageable derivatives such as the state securities here (derivable from the stock and the bond) because the stock price and the bond price as represented by u, r, and d are sufficient statistics for the price of the state securities (no other parameters such as preferences or empirical probabilities provide any more information on the state prices than those of the stock and bond prices). We can show the reverse, that if Arrow–Debreu state-U and state-D prices are the primitive building blocks in the economy, and are given as prices c^u and c^d, respectively, then the stock and bond prices of Fig. 5.1 are $uc^u + dc^d$ and $r(c^u + c^d)$, respectively. Again, note that here the stock and bond are derivative prices on the state securities, and the prices do not show any empirical probabilities or preference parameters.

Application

Suppose there is a European binary call option traded in the market such that if the underlying stock price at maturity time T, S_T, is larger than the strike price K, then the option can be exercised by the owner or holder to receive \$$B$. If the underlying $S_T \leq K$, then the binary call option expires out-of-the-money, and becomes worthless. Furthermore, suppose there are two traded Arrow–Debreu certificates. One costs c^u and pays \$1 in the state $S_T > K$, or else nothing. The other costs c^d and pays \$1 in the state $S_T \leq K$ or else nothing. Then, in a no-arbitrage situation, clearly the price of the binary call option is \$ Bc^u. A European binary put in this case would cost \$ Bc^d. A portfolio of one binary call and one binary put yields identical future time T payoff of certain \$$B$ as a riskless T-maturity bond paying \$$B$ at maturity. If the effective risk-free return is r over $[0, T]$, then current $t = 0$ bond price is B/r. By the no-arbitrage condition, this is also the price of the call-put portfolio, $B(c^u + c^d)$. Hence, $c^u + c^d = \frac{1}{r}$.

No-Arbitrage Conditions

Now we define more rigorously what is arbitrage. Arbitrage is basically being able to make positive payoff or return without bearing risk and net of time premium. The latter is seen from the bond — while it is risk-free in the sense of no default or bankruptcy, there is positive return r when the investor carries the bond (deposits money on it) over time from $t = 0$ to $t = 1$; r is the time premium or time value of money in the economy.

A security is said to be dominant over another when both can be purchased at the same price at $t = 0$ (could be different number of shares but the same dollar amount), but the dominant security will yield a higher return at t in every state of the world. In the above example, if $u > r$ in U-state, and $d > r$ in D-state, i.e., for any state, the stock always provides a higher return, then the stock is said to be dominant over the bond. If $r > u > d$, then the bond is said to dominate the stock. A security is said to be weakly dominant over another when they can be purchased at the same price at $t = 0$ (could be different number of shares but the same dollar amount), but the weakly dominant security will yield a higher return at $t = 1$ in at least one state of the world, and at least the same return in all other states of the world. Obviously, a dominant security is always weakly dominant.

In the case of dominance, one can make endless money by buying the dominant security and selling the dominated security, if they are traded at the same price. This is thus a riskless arbitrage opportunity. In Fig. 5.1, when $u > d > r$, sell X units of bond and use the $\$X$ to buy $\$X$ of stock, resulting in zero outlay at $t = 0$. At $t = 1$, if it is U-state, the investor makes $\$X(u - r) > 0$. If it is D-state, the investor makes $\$X(d - r) > 0$. Hence, with zero outlay, the investor ends up making positive money. In the case of weak dominance, e.g., $u > d \geq r$, there is still arbitrage opportunity (making something valuable such as a lottery out of zero cost) as with zero outlay, state-U yields $\$X(u - r) > 0$ while state-D yields $\$X(d - r) \geq 0$.

We define an arbitrage opportunity or arbitrage to be present when
(1a) a portfolio can be purchased at zero cost (self-financing) and its state-dependent future payoffs are all positive with at least one state having strictly positive payoff. A zero outlay portfolio as described above is called an arbitrage portfolio. In the case where $u > d > r$ for example, an arbitrage portfolio consists of long $\$X$ units of stock and short $\$X$ units of bond.

An arbitrage opportunity is also present when
(1b) the investor can purchase a portfolio with strictly negative value, i.e., short-selling more than buying, and yet achieve positive (can be zero) returns in every state of the world. Purchasing with strictly negative value is the same as being paid to hold the portfolio.

From Eq. (5.1), any portfolio in the two-state two-security economy can be written, for arbitrary A and B, as a payoff-price vector:

$$\begin{pmatrix} Au + Br \\ Ad + Br \\ -(A+B) \end{pmatrix}$$

Suppose we have $c_1 > 0, c_2 > 0, c_3 > 0$. Consider the scalar

$$(c_1\ c_2\ c_3) \begin{pmatrix} Au + Br \\ Ad + Br \\ -(A+B) \end{pmatrix}$$
$$= c_1(Au + Br) + c_2(Ad + Br) - c_3(A + B) = H.$$

By (1a), when an arbitrage occurs, $A+B = 0$, and $\max(Au+Br, Ad+Br) > 0$ while $\min(Au + Br, Ad + Br) \geq 0$, thus $H > 0$. Or $-A - B = 0$, and $\max(-Au - Br, -Ad - Br) > 0$ while $\min(-Au - Br, -Ad - Br) \geq 0$, thus $H < 0$.

By (1b), when an arbitrage occurs, $-(A + B) > 0$, and $\min(Au + Br, Ad + Br) \geq 0$, so $H > 0$ again. Or, $(A + B) > 0$, and $\min(-Au - Br, -Ad - Br) \geq 0$, so $H < 0$ again.

Note that if we take $(-A, -B)$ positions instead of $(+A, +B)$ positions, then when arbitrage occurs, $H < 0$. Hence, in equilibrium, when no arbitrage occurs, then $H = 0$.

Conversely, consider if $H = 0$. Under (1a), when $A + B = 0$, either $Au+Br = Ad+Br = 0$ or else $Au+Br$ and $Ad+Br$ are of opposite signs. However, since A, B are arbitrary in the situation when $H = 0$, the latter is not always the case. Hence, $A+B = Au+Br = Ad+Br = 0$ which means no arbitrage of type (1a). Under (1b), when $A + B < (>)0$, then either $Au+Br, Ad+Br \leq 0(\geq 0)$ with at least one $< 0(>0)$, or else, $Au+Br$ and $Ad+Br$ are of opposite signs such that $H = 0$. Again, the latter is ruled out because of the arbitrariness of A and B. Thus, $Au + Br, Ad + Br \leq 0(\geq 0)$ with at least one $< 0(>0)$ which means no arbitrage of type (1b).

Therefore, no arbitrage occurs iff $H = 0$. Now, $H = 0$ can be written as

$$H = A(c_1 u + c_2 d - c_3) + B(c_1 r + c_2 r - c_3) = 0. \tag{5.4}$$

Since A and B are arbitrary (while u, d, and r are given) in Eq. (5.4), and without loss of generality we can fix $c_3 = 1$, then

$$c_1 u + c_2 d = 1 \tag{5.5}$$

$$c_1 r + c_2 r = 1 \tag{5.6}$$

from which we can solve to obtain $c_1 = \frac{r-d}{r(u-d)}$ and $c_2 = \frac{u-r}{r(u-d)}$. Checking with (5.2) and (5.3), c_1 is the state price of U-state while c_2 is state price of D-state.

We see that there is no arbitrage iff there exist strictly positive state prices according to Eqs. (5.5) and (5.6).

Let us extend to a three-state world as follows. Suppose $S = 3$ and the three states of the world end in up (U) state with return u, middle (M) state with return m, and down (D) state with return d.

We suppose investors know the structure of the uncertainties to be resolved in the end at $t = 1$ into u with probability 0.2, m with probability 0.5, and d with probability 0.3 (Fig. 5.2). However, at $t = 0$, investors do not know which state will eventually occur.

Besides the stock there is another bond security traded in the economy with payoff r which is risk-free, and shown in Fig. 5.3.

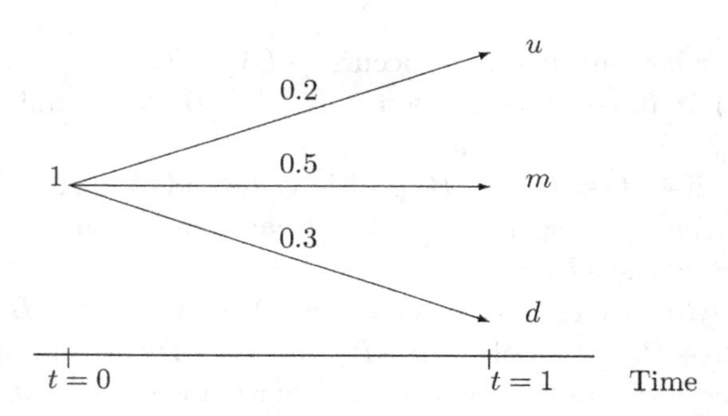

Fig. 5.2: Stock Payoffs in a Three-State World

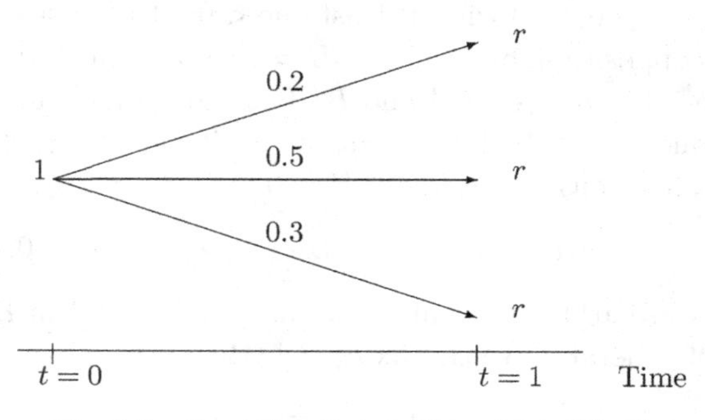

Fig. 5.3: Bond Payoffs in a Three-State World

As in Eq. (5.1), any portfolio in the three-state two-security economy can be written, for arbitrary A and B, as a payoff-price vector:

$$\begin{pmatrix} Au + Br \\ Am + Br \\ Ad + Br \\ -(A + B) \end{pmatrix}.$$

Suppose we have $c_1 > 0, c_2 > 0, c_3 > 0, c_4 > 0$ and consider the scalar

$$(c_1\ c_2\ c_3\ c_4) \begin{pmatrix} Au + Br \\ Am + Br \\ Ad + Br \\ -(A + B) \end{pmatrix}$$

$$= c_1(Au + Br) + c_2(Am + Br) + c_3(Ad + Br) - c_4(A + B) = H.$$

In the same way as argued earlier, there is no arbitrage iff $H = 0$. When $H = 0$

$$A(c_1u + c_2m + c_3d - c_4) + B(c_1r + c_2r + c_3r - c_4) = 0. \tag{5.7}$$

Since A and B are arbitrary in Eq. (5.7), and without loss of generality we can fix $c_4 = 1$, then

$$c_1u + c_2m + c_3d = 1 \tag{5.8}$$

$$c_1r + c_2r + c_3r = 1. \tag{5.9}$$

In Eqs. (5.8) and (5.9) there are 3 unknowns c_1, c_2, c_3, so it is not possible to obtain unique solutions to the unknowns. However, it is always possible to find a solution, e.g., $c_1 = (2r - m - d)/[r(2u - m - d)]$, $c_2 = c_3 = (u - r)/[r(2u - m - d)]$. Moreover, c_1, c_2, c_3 can be interpreted as state prices. Having non-unique state prices is the case in incomplete market when the number of securities is less than the number of states, i.e., the securities do not adequately span the states of the world. In other words, not every state is attainable by a unique state security.

Thus, again we see that there is no arbitrage iff there exist strictly positive state prices according to Eqs. (5.8) and (5.9).

A couple of comments are in order here. We just saw a case in which the number of states of the world is more than the number of securities traded in the market, so not all uncertain states of the future are spanned. Reasonably assuming no arbitrage opportunities in the market, the results indicate that the limited number of traded securities are priced as if there exists state prices (or similarly risk-neutral probabilities, excepting some normalizing constant). However, since the states are not spanned, these phantom state securities are not actually traded. Because of this, their no-arbitrage prices are also not unique. It means there are an infinite number of solutions to the state prices, or in the example of the three-state world, an infinite number of solutions to c_1, c_2, and c_3.

The problem in the real world is that it is more often the case that the market is incomplete and not all states are spanned. Yet, obtaining (some estimates of the) state prices is typically required in order to provide analytical no-arbitrage pricing to new securities that are being issued or sold by investment bankers. Or, they are required in order to perform price valuation to check if a security is temporarily over- or under-priced and hence to be sold or bought. One practical way the industry operates is to use a time series of the security prices e.g., the stock price in the three-state world over time. Assuming statistical deviations in the real world, the price of the stock is not exactly 1 but $1 + \epsilon_t$ where ϵ_t is a small random error representing market frictions or transaction costs or observational errors. Over time, the price series $p_t \approx 1$ is obtained. Assuming the state prices c_1, c_2, and c_3 exist and remain constant over $[0, T]$, $T > 3$, then some estimation criterion such as minimizing least squares, i.e., $\min_{c_1,c_2,c_3} \sum_{t=1}^{T}[p_t - (c_1 u + c_2 m + c_3 d)]^2$ is employed to estimate c_1, c_2, and c_3. Here, we recognize that $(c_1 u + c_2 m + c_3 d)$ in the objective function is the theoretical stock price as indicated in Eq. (5.8). Obtaining the state prices or the risk-neutral probabilities this way is said to be "implying out" the prices or parameters. In an incomplete market, the state prices will reflect preference parameters and empirical probabilities. This will be seen more clearly in the next chapter.

Another way to find the state prices is to search for more securities that may complete the market and then to utilize their prices as additional equations to solve uniquely for the state prices.

5.2. S-State World

More generally, for S number of states ω_i, and for $N \leq S$ number of securities, each with price at $t = 0$ of p_j, the payoff-price matrix of the securities (concatenated payoff-price vectors of securities) is given by

$$Q_{S+1\times N} = \text{State } i \overset{\text{Security } j}{\begin{pmatrix} q_{11} & q_{12} & \cdots & q_{1j} & \cdots & q_{1N} \\ q_{21} & q_{22} & \cdots & q_{2j} & \cdots & q_{2N} \\ \cdots & \cdots & \cdots & \cdots & \cdots & \cdots \\ q_{i1} & q_{i2} & \cdots & q_{ij} & \cdots & q_{iN} \\ \cdots & \cdots & \cdots & \cdots & \cdots & \cdots \\ q_{S1} & q_{S2} & \cdots & q_{Sj} & \cdots & q_{SN} \\ -p_1 & -p_2 & \cdots & -p_j & \cdots & -p_N \end{pmatrix}}.$$

Element q_{ij} is the payoff at $t = 1$ of \$ q_{ij} of one unit of security j in state i. The last row of Q is $-1\times$ the security price vector

$$P_{N\times 1} = \begin{pmatrix} p_1 \\ p_2 \\ \vdots \\ p_N \end{pmatrix}.$$

A portfolio is a vector with elements as number of shares in each security

$$A_{1\times N} = (a_1, a_2, \ldots, a_N).$$

Hence, the market price or value of portfolio A is

$$p_A = A\,P.$$

The payoff-price vector of A is

$$Q_{S+1\times N} A^T_{N\times 1} = \begin{pmatrix} \sum_{j=1}^N q_{1j}a_j \\ \sum_{j=1}^N q_{2j}a_j \\ \vdots \\ \sum_{j=1}^N q_{Sj}a_j \\ -p_A \end{pmatrix} \tag{5.10}$$

where the ith $< S + 1$ element in the vector is the state-i payoff of portfolio A.

Having set up the structure, we pause to provide an important theorem which will be used in our formal proof of the relationship of no-arbitrage with state prices.

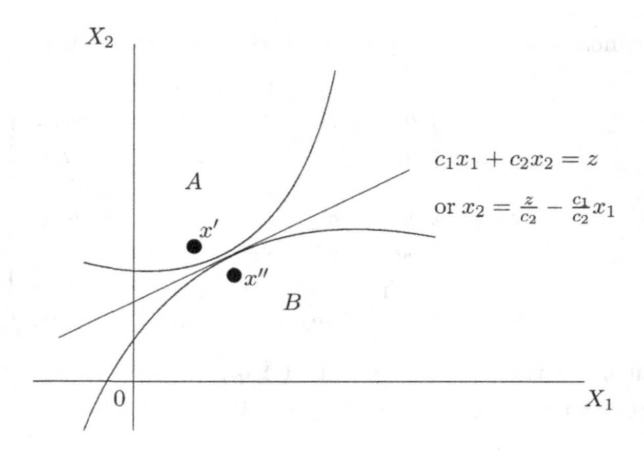

Fig. 5.4: Separating Hyperplanes

Theorem 5.1 (Separating Hyperplane Theorem): *If A and B are convex disjoint subsets of $\mathcal{R}^{S+1}$, then there is some (at least one) non-zero linear functional $F : \mathcal{R}^{S+1} \to \mathcal{R}$ such that $F(x) < F(y)$ for every $x \in A$ and $y \in B$.* ■

In other words, suppose vectors $x^T = (x_1, x_2, \ldots, x_{S+1})$ and $y^T = (y_1, y_2, \ldots, y_{S+1})$, then $F(x) = c^T x = c_1x_1 + c_2x_2 + \cdots + c_{S+1}x_{S+1}$, $F(y) = c^T y = c_1y_1 + c_2y_2 + \cdots + c_{S+1}y_{S+1}$, and $F(x) < F(y)$ for some $c^T = (c_1, c_2, \ldots, c_{S+1})$.

If $S + 1 = 2$, we can show the above separating hyperplane theorem intuitively via geometry as in Fig. 5.4.

Any two points x' in A and x'' in B will have $F(x') = c_1x_1' + c_2x_2' = z' > F(x'') = c_1x_1'' + c_2x_2'' = z''$. Any point $x*$ along the straight line separating the two convex regions A and B will have $F(x*) = z$, where $z' > z > z''$.

The given general payoff-price vector is a $S+1$ dimensional space $\mathcal{R}^S \times \mathcal{R}$. Existence of arbitrage opportunities means that for a portfolio, either:

(2a) last element $=0$ and every ith $< S+1$ element that is a payoff number is ≥ 0 (≤ 0) with at least one element > 0 (< 0), or

(2b) last element > 0 (< 0), $p_A < 0$ ($p_A > 0$) and every ith element $\geq 0 (\leq 0)$, for a portfolio A.

The subspace in $\mathcal{R}^S \times \mathcal{R}$ described by the arbitrage opportunity is a convex cone in the intersection of all positive orthants excepting the origin

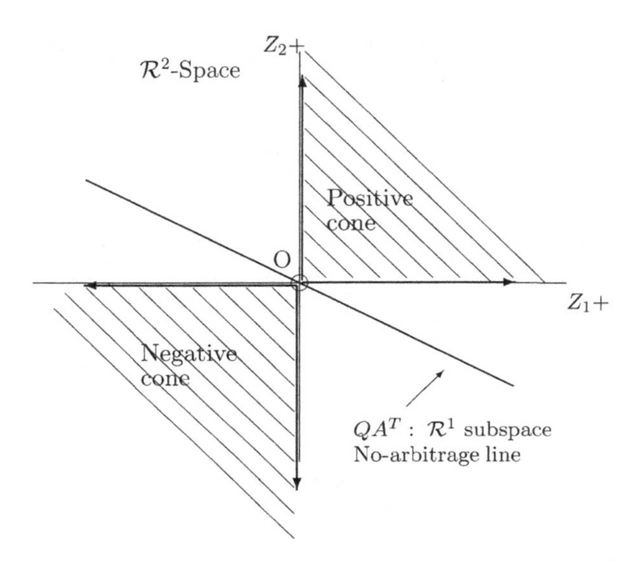

Fig. 5.5: Arbitrage Opportunity in Shaded Cones Less "O" in $\mathcal{R}^2$-Space

"O" when all elements $= 0$, or else a convex cone in the intersection of all negative orthants excepting "O".

From Eq. (5.10), given Q which is a $S+1 \times N$ matrix of fixed numbers, multiplying by a smaller dimension $A^T_{N\times 1}$ where $N \leq S$, but allowing elements of A to vary as in a parameter, produces a linear subspace (hence also convex) QA^T that is of dimension $S+1 \times 1$.

It is easier to see the above idea of a convex cone and a subspace of lower N dimension as follows in a one-state one-security world (Fig. 5.5).

Since choosing $A = 0$-vector is possible, the subspace spanned by QA^T representing no arbitrage passes through origin "O" between the two cones where A is an arbitrary portfolio. For there to be no arbitrage in portfolios formed given Q, the linear subspace QA^T must intersect with the cones only at point "O" (note the arbitrage opportunity are the cones less "O"). Hence, the arbitrage cones and the no-arbitrage subspace QA^T are disjoint subspaces. Therefore, by the separating hyperplane theorem, there exists a linear functional, with coefficients $c \gg 0$ such that $c^T QA^T = 0$, since "O" is on this subspace and $c^T \times 0_{S+1\times 1} = 0$. Moreover, $c^T Z > (<) 0$ for any $Z \in \mathcal{R}^{S+1}$ in the positive (negative) cones less "O".

Now, for arbitrary A,

$$c^T QA^T = 0 \Rightarrow c^T_{1\times S+1} Q_{S+1\times N} = 0_{1\times N}.$$

Without loss of generality, let $c = (c_1, c_2, \ldots, c_S, 1)$, then there exists such a $S+1 \times 1$ vector $c \gg 0$, so

$$(c_1 c_2 \cdots c_S 1) \begin{pmatrix} q_{11} & q_{12} & \cdots & q_{1j} & \cdots & q_{1N} \\ q_{21} & q_{22} & \cdots & q_{2j} & \cdots & q_{2N} \\ \cdots & \cdots & \cdots & \cdots & \cdots & \cdots \\ q_{i1} & q_{i2} & \cdots & q_{ij} & \cdots & q_{iN} \\ \cdots & \cdots & \cdots & \cdots & \cdots & \cdots \\ q_{S1} & q_{S2} & \cdots & q_{Sj} & \cdots & q_{SN} \\ -p_1 & -p_2 & \cdots & -p_j & \cdots & -p_N \end{pmatrix} = 0. \tag{5.11}$$

Equation (5.11) $\Rightarrow$ that for every security $j = 1, 2, \ldots, N$,

$$p_j = \sum_{i=1}^{S} c_i q_{ij}. \tag{5.12}$$

Hence, $\psi_{1\times S} = (c_1, c_2, \ldots, c_S)$ are the state prices. We just showed that if there is no arbitrage, state prices exist.

Let $Q'_{S\times N}$ be Q less the bottom row. Conversely, if state prices exist (all strictly positive), then for any portfolio A, we have its price $p_A = AP = (\psi Q')A^T$. Thus, when $Q'A^T > 0$ (for a vector, "> 0" means each element "≥ 0" but at least one element "> 0"), $p_A > 0$. This does not satisfy (2a), so there is no arbitrage opportunity of type (2a).

Similarly, when $Q'A^T \geq 0$ (For a vector, "≥ 0" means each element "≥ 0". Recall that "$\gg 0$" means every element of the vector "> 0".), $p_A \geq 0$ and not $p_A < 0$. This does not satisfy (2b), so there is also no arbitrage opportunity of type (2b).

Hence, we establish the following:

Theorem 5.2 *There is no arbitrage iff there exists an equilibrium state-price vector. (State prices are by definition strictly positive.)* ■

5.3. Risk-Neutral Probabilities

Equation (5.12) includes as special cases (5.5), (5.6), (5.8), and (5.9), and shows that every traded security price in the market can be expressed in terms of strictly positive scalars c_i for $i = 1, 2, \ldots, S$ given the payoff structure q_{ij}s. Note that in Eq. (5.12), N may be $< S$, hence the market needs not be complete, nor is there a need for a risk-free bond to be traded.

For each security j in Eq. (5.12), divide both sides of the equation by $\sum_{k=1}^{S} c_k$

$$\frac{p_j}{\sum_{k=1}^{S} c_k} = \frac{c_1}{\sum_{k=1}^{S} c_k} q_{1j} + \frac{c_2}{\sum_{k=1}^{S} c_k} q_{2j} + \cdots + \frac{c_S}{\sum_{k=1}^{S} c_k} q_{Sj}. \tag{5.13}$$

In (5.13), let $\pi_i = \frac{c_i}{\sum_{k=1}^{S} c_k}$, then a normalized security price $p'_j = \frac{p_j}{\sum_{k=1}^{S} c_k}$, becomes

$$p'_j = \pi_1 q_{1j} + \pi_2 q_{2j} + \cdots + \pi_S q_{Sj} \tag{5.14}$$

where $\sum_{i=1}^{S} \pi_i = 1$. Thus, π_i's act like pseudo-probabilities > 0 which sum to one over all the states. Moreover, normalized price in (5.14) is as if it is equal to the expected value of future payoffs under the pseudo-probabilities.

Now, suppose among the securities is a risk-free discount bond with bond value B at $t = 0$ and value 1 at $t = 1$, so that risk-free return is $r = 1/B$. For this bond, from (5.12)

$$\sum_{k=1}^{S} c_k = B = \frac{1}{r}.$$

Use this in Eq. (5.14)

$$p_j = \frac{1}{r} \sum_{i=1}^{S} \pi_i q_{ij}. \tag{5.15}$$

We can then interpret Eq. (5.15) as setting price equal to expected future payoffs discounted by the risk-free factor. In finance parlance, this is tantamount to a risk-neutral investor taking expectation and then discounting by the risk-free rate to obtain current price. For this reason, the no-arbitrage equilibrium price in (5.15) is said to be computed based on risk-neutral probabilities π_i's. Again, we see that actual or empirical probabilities do not show explicitly in these price computations.

Comparing Eqs. (5.12) and (5.15), it is seen that state prices are closely related to the risk-neutral probabilities, $rc_i = \pi_i$. Intuitively, this makes sense. Given fixed r, the higher the state price — more demand, the higher the investor is betting for the state to occur, and so its (despite being risk-neutral) probability of state occurrence is higher. Thus, Theorem 5.2 also says that there is no arbitrage iff there exists non-zero risk-neutral probabilities.

5.4. Creating State Securities

We have explained the ideas of state securities and their connection with more complex securities such as stocks and bonds, and how in a no-arbitrage equilibrium condition, there exists state prices and risk-neutral probabilities. In an informal way, we also show how less securities than states (incomplete market) lead to non-unique but still positive state prices and non-unique risk-neutral probabilities. A complete market, when all states are spanned or attainable via existence of state security in each state, has unique positive state prices as well as unique risk-neutral probability of each state. We also see the close connection between state prices and risk-neutral probabilities.

In this section, we first show how completing the market can improve welfare defined as improving VM expected utility for risk-averse investors. Then, we show how states can be spanned using options.

Suppose in a one-period world with three states as discussed in the earlier section, an investor has VM utility $U(W_1)$, $U' > 0$, $U'' < 0$, where his or her end-of-period wealth W_1 is an outcome of investing initial wealth at $t = 0$, W_0 in state securities that are traded. Suppose that initially there are only two traded state securities, those of states U and M, and not D. Assume the state prices are c_1 and c_2 respectively. Then, the investor's expected utility is maximized by

$$\max_{x_1} EU(x_1 1_u + [W_0 - x_1 c_1] 1_m / c_2),$$

where x_i is the number of state securities purchased for state i, and $W_1 = x_1 1_u + x_2 1_m$, 1_i being an indicator variable which takes value 1 only if the state i occurs at $t = 1$. We let $1_1 \equiv 1_u, 1_2 \equiv 1_m$, and $1_3 \equiv 1_d$. The current budget constraint $W_0 = x_1 c_1 + x_2 c_2$ has been imposed in the objective function. Let the empirical probability of state i be θ_i. A solution x_1^*, hence also x_2^*, exists in the following FOC

$$\theta_1 U'(x_1^*) - \frac{c_1}{c_2} \theta_2 U' \left[\frac{W_0 - x_1^* c_1}{c_2} \right] = 0. \tag{5.16}$$

Now, consider the situation where the third state security is also traded at price c_3. Then, the investor's optimization is

$$\max_{x_1, x_3} EU(x_1 1_u + x_3 1_d + [W_0 - x_1 c_1 - x_3 c_3] 1_m / c_2),$$

where now $W_1 = x_1 1_u + x_2 1_m + x_3 1_d$, and $W_0 = x_1 c_1 + x_2 c_2 + x_3 c_3$.

The FOCs are

$$\theta_1 U'(x_1^*) - \frac{c_1}{c_2}\theta_2 U'\left[\frac{W_0 - x_1^* c_1 - x_3^* c_3}{c_2}\right] = 0, \tag{5.17}$$

and

$$\theta_3 U'(x_3^*) - \frac{c_3}{c_2}\theta_2 U'\left[\frac{W_0 - x_1^* c_1 - x_3^* c_3}{c_2}\right] = 0. \tag{5.18}$$

All their second-order conditions (SOCs) checked out to be < 0 and hence maximization is assured.

The objective function in both cases of two-securities or three-securities world is $\theta_1 U(x_1) + \theta_2 U(x_2) + \theta_3 U(x_3)$ s.t. $x_1 c_1 + x_2 c_2 + x_3 c_3 = W_0$. In the three-securities world, the FOC (5.17) can be solved by constraining $x_3 = 0$ in which case we obtain (5.16). However, this is suboptimal as Eq. (5.18) should be solved in conjunction with (5.17) to produce x_3^* that is not necessarily 0. Thus, it is seen that the suboptimal solution with $x_3 = 0$ in the three-securities world yields the same VM expected utility as the maximum in the two-securities world. Therefore, the maximized VM expected utility in the three-securities world is indeed larger than that in the two-securities world. This proves that being able to trade the additional state d security increases the investor's welfare. We say that with complete markets such that all states are spanned (attainable), the investor's security allocation, (x_1^*, x_2^*, x_3^*) is efficient for his or her given preference and the probability distribution underlying the security prices. Any other allocation is inefficient. Therefore, in an incomplete market setting, the allocation is not efficient.

We can enlarge the market or increase the spanning or complete the market by trading more of linearly independent securities (whose state-contingent payoff vector is linearly independent or cannot be replicated by other existing traded security payoff vectors). These securities can be stocks. They can also be options.

Consider a S-state world with S number of states of uncertainty over the next one period. Suppose there is one underlying stock in the market with payoffs in the S different number of states that are not pairwise equal. In a one-period world, the stock's end-of-period payoff is its final price. Let the payoff monetary unit be 1 cent. We shall refer to this as 1 unit. Without loss of generality, the stock's payoffs in the different states may be expressed as a vector:

$$(u_1, u_2, u_3, \ldots\ldots, u_S)^T$$

where $u_1 < u_2 < u_3 < \cdots < u_S$.

Let the exchange-traded price of the stock be in constant increments of 1 monetary unit. Thus, the traded prices of the stock move up from u_1 to u_1+1, u_1+2 and so on, or move down from u_1 to u_1-1, u_1-2, and so on.

Suppose call options on the stock with different strike or exercise prices are now traded, each with maturity equal to the end of the period. Let the first call have exercise price u_1. Its payoff vector is then

$$(0, u_2 - u_1, u_3 - u_1, \ldots\ldots, u_S - u_1)^T.$$

Note that in the first state, the call is out-of-the money, but it is in-the-money for all the other states.

Let the second call have exercise price u_2. Its payoff vector is then

$$(0, 0, u_3 - u_2, u_4 - u_2, \ldots, u_S - u_2)^T.$$

Let the third call have exercise price u_3. Its payoff vector is then

$$(0, 0, 0, u_4 - u_3, u_5 - u_3, \ldots, u_S - u_3)^T.$$

Proceeding with the completion, the $(S-1)^{\text{th}}$ call has exercise price u_{S-1} with payoff vector:

$$(0, 0, 0, 0, \ldots, 0, u_S - u_{S-1})^T.$$

Assembling the payoff vectors of the one stock and $S-1$ number of calls, the market's $Q_{S\times S}$ payoff matrix is

$$\text{State } i \overset{\text{Security } j}{\begin{pmatrix} u_1 & 0 & 0 & \vdots & 0 & \vdots & 0 \\ u_2 & u_2 - u_1 & 0 & \vdots & 0 & \vdots & 0 \\ u_3 & u_3 - u_1 & u_3 - u_2 & \vdots & 0 & \vdots & 0 \\ u_4 & u_4 - u_1 & u_4 - u_2 & \vdots & 0 & \vdots & 0 \\ \vdots & \vdots & \vdots & \vdots & \vdots & \vdots & \vdots \\ u_i & u_i - u_1 & u_i - u_2 & \vdots & (u_i - u_{j-1})^+ & \vdots & 0 \\ \vdots & \vdots & \vdots & \vdots & \vdots & \vdots & \vdots \\ u_S & u_S - u_1 & u_S - u_2 & \vdots & u_S - u_{j-1} & \vdots & u_S - u_{S-1} \end{pmatrix}}.$$

The square payoff matrix is lower triangular and is clearly of full rank. Hence, it provides a complete efficiency market for the S-state world. Similarly, we can use puts with exercise prices u_2, u_3, u_4, and so on to u_S (and by putting the stock vector payoff on the RHS of the Q-matrix instead) to form a market upper triangular payoff matrix which is again of full rank S.

If European calls can be traded on strike prices $1, 2, 3, \ldots, k-1, k, k+1, \ldots$, then we can construct a state-k security with payout of unit money iff state k occurs and zero otherwise, by a portfolio consisting of 1 long call at strike $k-1$, 2 short calls at strike k, and 1 long call at strike $k+1$.

5.5. Problem Set 5

1. Suppose the world exists in a single period when investments are made at $t = 0$ and payoffs are collected at $t = 1$ from the investments at $t = 0$. In the United States U-Arrow–Debreu certificates are sold at $\$a_0^u$ with payoff \$1 iff U occurs, or else zero. D-Arrow–Debreu certificates are sold at $\$a_0^d$ with payoff \$1 iff D occurs, or else zero. In another nearby country called Utopia using "utopi" as its currency, simple U-securities are sold at π_0^u utopi with payoff X^u utopi iff U occurs, or else zero. If the currency exchange rate is fixed at \$1 to y utopi, and there is free trade, then under the no-arbitrage condition, how are π_0^u and X^u related to a_0^u and y? Hence, what is the price of a simple D-security in Utopia and its D-state contingent payoff in utopi?
2. If in the single-period world, portfolio $(x_1, x_2, \ldots, x_M)$ with x_j shares in jth security dominates portfolio $(y_1, y_2, \ldots, y_M)$, show how you would create an arbitrage portfolio.
3. State prices are $e_1 = 0.5, e_2 = 0.3, e_3 = 0.07, e_4 = 0.03$, in a four-state world, what is the price of a security that has payoffs in the various states as follows: $q_1 = 0, q_2 = 1, q_3 = -1, q_4 = 5$? What is the risk-free interest rate per period?
4. The following is a payoff matrix Q in which one unit of security j of S number of securities $j = 1, 2, \ldots, S$ yields \$ q_{ij} in the ith state out of S number of states in the world over the next period.

$$\begin{pmatrix} q_{11} & q_{12} & \cdots & q_{1j} & \cdots & q_{1S} \\ q_{21} & q_{22} & \cdots & q_{2j} & \cdots & q_{2S} \\ \cdots & \cdots & \cdots & \cdots & \cdots & \cdots \\ q_{i1} & q_{i2} & \cdots & q_{ij} & \cdots & q_{iS} \\ \cdots & \cdots & \cdots & \cdots & \cdots & \cdots \\ q_{S1} & q_{S2} & \cdots & q_{Sj} & \cdots & q_{SS} \end{pmatrix}$$

Find the S portfolios (each portfolio is a $S \times 1$ vector $(y_1, y_2, \ldots, y_S)^T$ where y_j is the number of units of the jth security held in the portfolio) such that each is replicating a k-state security with payoff \$1 only in the kth state and zero otherwise.

Suppose the number of securities is instead $M > S$ and the payoff matrix is Q_M of dimension $S \times M$. If there exists a $M \times S$ matrix X such that $Q_M X = I_S$ where I_S is an $S \times S$ identity matrix, are there any redundant securities? (Assume the rank of matrix Q and Q_M is S.)

5. In a three-state world, there is a stock with payoffs $(1, 2, 3)^T$ in each of the states. A call written on the stock at strike price 1 has payoff vector $(0, 1, 2)^T$. Show how the state securities can be derived if a put with strike price 2 is now traded as well. What are the state prices of states 1, 2, 3 if the stock, call, and put prices are S, C, and P respectively?
6. In a three-state world with states U, M, and D, there are three securities being traded. One is a state-D security with price \$ c_d. It pays \$1 iff state D occurs, and zero otherwise. Another security is a dual-state security that pays \$1 if either state U or state M occurs, and zero otherwise. There is also a risk-free bond that pays interest rate R at the end of the period. What is the no-arbitrage price of the dual security?
7. In a three-state world, would an investor be better off in a situation when there are only two state-securities or in a situation when there are only two securites, viz. a stock and a bond as in Fig. 5.2?

Chapter 6

SINGLE PERIOD ASSET PRICING MODELS

William Sharpe's capital asset pricing model (CAPM) was published in 1964, and built on the foundation of portfolio theory as first put forward by Harry Markowitz, and also on the two-fund separation idea between risky assets and risk-free assets by James Tobin. The CAPM was a breakthrough in general equilibrium analysis that provided an intuitive explanation as to why some assets are priced higher than others because they are less risky, in turn fetching lower expected return. The nature of the risk-return tradeoff is made very explicit and clear by the idea of systematic versus diversifiable risks and the measure of systematic risk through beta. What was fuzzy became crystal clear in the understanding that if a security should provide insurance or high returns when the rest of the market does badly in some states of the world, and vice-versa, then the negative correlation implied a lower expected rate of return due to its lesser systematic risk. Such insurance ideas or insurance against adverse states of nature had been studied even earlier as seen in the last chapter. However, equilibrium asset pricing in incomplete market had a path-breaking start with the CAPM.

6.1. State Preference Model

We use state prices to characterize single-period consumption and investment problems.

Suppose actual or empirical probability of state i happening is θ_i. Final or end-period wealth for consumption is a RV that takes value $\sum_{j=1}^{N} q_{ij}a_j$ or ith element of $Q'A^T$ in state i with probability θ_i. Q' is the $S \times N$ payoff matrix on the universe of N securities and S states of the world, and $A_{1\times N}$ is the portfolio of number of shares in each security. The number of securities N could be smaller than the number of uncertain states S, so that the market can be incomplete.

A VM utility function $u\big(\sum_{j=1}^{N} q_{ij}a_j\big)$ is assumed to be strictly increasing and concave in its argument which could be wealth or consumption depending on the particular model. Since almost all VM utility optimization will involve the marginal utility u', we assume throughout that the Inada condition holds, i.e., $\lim_{x\uparrow\infty} u'(x) = 0$. This includes the more restrictive case where u is bounded from above. Original or starting wealth at $t = 0$ is $W_0 = \sum_{j=1}^{N} p_j a_j$ or AP, given P, where $P_{N\times 1}$ is the current share price vector.

The investor's optimal consumption and investment problem is

$$\max_{\{a_j,\lambda\}} E\left[u\left(\sum_{j=1}^{N} q_{ij}a_j\right)\right] + \lambda\left(W_0 - \sum_{j=1}^{N} p_j a_j\right) \tag{6.1}$$

where λ is a Lagrange multiplier on the budget constraint. FOC yields

$$E[q_{ij}\, u'] - \lambda p_j = 0 \qquad \forall j. \tag{6.2}$$

From Eq. (6.2)

$$p_j = \sum_{i=1}^{S} \frac{\theta_i\, u'}{\lambda} q_{ij}, \quad \lambda > 0, \quad \forall j, \tag{6.3}$$

hence, $\left(\frac{\theta_i u'}{\lambda}\right)$ is seen as a state-i price. Thus, given that u' exists, a solution to the optimal consumption and investment problem exists iff there exists a state-price vector. By Theorem 5.2 in Chap. 5, a solution to the optimal consumption and investment problem exists iff there is no arbitrage.

In Eq. (6.2), suppose a risk-free bond exists in the economy with current price 1 and risk-free return or payoff r. Then, one of the N number of first-order conditions is

$$r\, E[u'] = \lambda.$$

Substituting the above back into Eq. (6.3), we obtain

$$p_j = \sum_{i=1}^{S} \frac{\theta_i u'}{rE[u']} q_{ij}, \quad \forall j, \tag{6.4}$$

so the state-i price is more specific as $\left(\frac{\theta_i u'}{rE[u']}\right)$. The risk-neutral probability of state i is $r\times$ the state-i price, or

$$\pi_i = \frac{\theta_i u'}{E[u']}. \tag{6.5}$$

The state-i risk-neutral probability in (6.5) is seen to be increasing in empirical probability of the state, θ_i, and the marginal utility or preference u'.

We have seen two types of probabilities above for every state i, the empirical probability $P_P(\omega_i) = \theta_i$, and the risk-neutral probability

$P_Q(\omega_i) = \pi_i$. We use subscripts to the probability functions to indicate "P" for empirical probability measure, and "Q" for risk-neutral probability measure.

Now, we define for state ω_i in the probability space,

$$d_i(\omega_i) \triangleq \frac{\pi_i(\omega_i)}{\theta_i(\omega_i)} \tag{6.6}$$

to be the state-i price density (or state-price kernel or state-price deflator). Since for every ω_i, $\pi_i(\omega_i) > 0$ and $\theta_i(\omega_i) > 0$, then $d_i(\omega_i) > 0$. This gives it a "probability density" flavor. It is basically a ratio of two different probability measures on the same σ-algebra.

We saw $W_0 = \sum_{j=1}^N p_j a_j$. Using Eq. (6.4) with empirical probability θ_i appearing in the state-i price, we can then express W_0 as

$$\begin{aligned} W_0 &= \sum_{j=1}^N \left[\sum_{i=1}^S \frac{\theta_i u'}{E[u']} \frac{q_{ij}}{r} \right] a_j \\ &= \frac{1}{r} \sum_{i=1}^S \left(\frac{\theta_i u'}{E[u']} \right) \sum_{j=1}^N q_{ij} a_j \\ &= \frac{1}{r} \sum_{i=1}^S \left(\frac{\theta_i u'}{E[u']} \right) W_{1i} \\ &= \frac{E_Q(W_1)}{r}, \end{aligned}$$

where W_{1i} is final wealth in state i, and the expectation in the last line is taken w.r.t. the risk-neutral Q-probability measure.

When there is no arbitrage and thus risk-neutral probability measures exist, we can now rewrite the investor's objective function as

$$\max_{\{a_j, \lambda\}} E_P \left[u \left(\sum_{j=1}^N q_{ij} a_j \right) \right] + \lambda \left(W_0 - \frac{E_Q(W_1)}{r} \right),$$

where we know the constraint under the Lagrange multiplier will be met exactly. Since W_0 and solved λ are constants, this simplifies to

$$\max_{\{a_j\}} E_P \left[u \left(\sum_{j=1}^N q_{ij} a_j \right) \right] - \lambda \left(\frac{E_Q(W_1)}{r} \right),$$

conditional on λ and W_0.

Introducing the state price density $d_i(\omega)$, we have

$$\max_{\{a_j,\lambda\}} E_P\left[u\left(\sum_{j=1}^{N} q_{ij}a_j\right)\right] - \frac{\lambda}{r}\left(\sum_{i=1}^{S} \pi_i W_{1i}\right)$$

$$\text{or} \quad \max_{\{W_{1i}\}} \sum_{i=1}^{S} \theta_i[u(W_{1i})] - \frac{\lambda}{r}\left(\sum_{i=1}^{S} \theta_i d_i W_{1i}\right)$$

$$\text{or} \quad \max_{\{W_{1i}\}} E_P\left[u(\tilde{W}_1) - \frac{\lambda}{r}\tilde{d}\tilde{W}_1\right]. \tag{6.7}$$

Note that in the process, we use

$$E_Q(\tilde{W}_1) = \sum_{i=1}^{S} \pi_i W_{1i} = \sum_{i=1}^{S} \theta_i\left(\frac{\pi_i}{\theta_i}\right) W_{1i} = E_P(\tilde{d}\tilde{W}_1).$$

The FOC is thus

$$u' = \frac{\lambda}{r} d_i \quad \forall i. \tag{6.8}$$

Or $\theta_i u' = \pi_i E[u']$, which is the same as in Eq. (6.5). Hence also, $I(u') = I(d_i E[u'])$, where $I(\cdot)$ is the inverse function corresponding to $u'(W_{1i})$, hence $W_{1i} = I(\tilde{d}\, E[u'])$, and $E_Q\left[\frac{I(\tilde{d}\, E[u'])}{r}\right] = W_0$.

6.2. Two-Solution Approaches

Up to this point, it is interesting to note that the optimal one-period consumption and investment problem represented in (6.1) and also in (6.7) can be solved in two ways:

$$\max_{\{a_j\}} E\left[u\left(\sum_{j=1}^{N} q_{ij}a_j\right)\right] + \lambda\left(W_0 - \sum_{j=1}^{N} p_j a_j\right)$$

or

$$\max_{\{W_1\}} E\left[u(\tilde{W}_1) - \frac{\lambda}{r}\tilde{d}\tilde{W}_1\right],$$

together with constraint $W_0 = \Sigma_{j=1}^{N} p_j a_j = E_Q(W_1)/r$.

The first solution is the classical approach where we differentiate directly w.r.t. the portfolio weights a_js to obtain the first-order conditions for solution. The second is more subtle, and uses the fact that W_1 is itself a control variable, related to controls $a_j s$ in the classical approach, that can be chosen. The budget constraint based on endowment W_0 can be nicely subsumed under the empirical probability measure in terms of a state-price density and discounted wealth, i.e., $\tilde{d}W_1/r$, and thus we can solve it in two stages after first obtaining conditions on the optimal W_{1i}, $\forall i$.

We shall provide a numerical example (Tables 6.1 and 6.2) below to show how both approaches will yield the same solution as expected. The first approach via (6.1) is the classical approach where we solve for portfolio weights a_js explicitly, and the second approach via (6.7) is the state-price density (or martingale) approach where we solve to obtain (6.8) relating marginal utility to state-price density, obtaining the state-price density uniquely if it is a complete market, and then solving for the portfolio weights. The latter is a kind of a two-step approach, but can be more intuitive and simpler in some ways.

Table 6.1: Security Prices and Probabilities

Security prices				
	t=0	ω_1	ω_2	ω_3
Empirical probability		1/4	1/2	1/4
Risk-neutral probability		1/3	1/3	1/3
State-price density		4/3	2/3	4/3
Stock 1	1	1	1	5/2
Stock 2	1	1	3/2	2
Risk-free bond	1	3/2	3/2	3/2

Table 6.2: Two Solution Approaches

Classical approach	State-price density approach
Equation (6.2)	Equation (6.8)
$p_j = \sum_{i=1}^{S} \frac{\theta_i u'\left(\sum_{j=1}^{N} q_{ij} a_j\right)}{\lambda} q_{ij}, \forall j$	$u'\left(\sum_{j=1}^{N} q_{ij} a_j\right) = \frac{\lambda}{r} d_i, \forall i$
$W_0 = \sum_{j=1}^{N} p_j a_j$	$W_0 = \sum_{j=1}^{N} p_j a_j$

Note that without loss of generality, the prices at $t = 0$ can all be fixed at \$1. For example, if it were \$5, we just divide all payoffs by the same factor 5. Suppose investors have (CRRA) negative exponential VM utility function $u(W) = -e^{-W}$, so $u'(W) = e^{-W} = -u(W)$. The solution approaches are tabulated in Table 6.2 for easy reference.

Using Eq. (6.2), we obtain for each of the 3 securities

$$\lambda = \frac{1}{4}e^{-W_1(\omega_1)}(1) + \frac{1}{2}e^{-W_1(\omega_2)}(1) + \frac{1}{4}e^{-W_1(\omega_3)}\left(\frac{5}{2}\right)$$

$$\lambda = \frac{1}{4}e^{-W_1(\omega_1)}(1) + \frac{1}{2}e^{-W_1(\omega_2)}\left(\frac{3}{2}\right) + \frac{1}{4}e^{-W_1(\omega_3)}(2)$$

$$\lambda = \frac{1}{4}e^{-W_1(\omega_1)}\left(\frac{3}{2}\right) + \frac{1}{2}e^{-W_1(\omega_2)}\left(\frac{3}{2}\right) + \frac{1}{4}e^{-W_1(\omega_3)}\left(\frac{3}{2}\right).$$

Solving, we obtain

$$W_1(\omega_1) = W_1(\omega_3) \tag{6.9}$$

$$W_1(\omega_2) = W_1(\omega_3) + \ln 2, \quad \text{and} \tag{6.10}$$

$$W_1(\omega_3) = -\ln\lambda - \ln\frac{8}{9}. \tag{6.11}$$

Let a_1, a_2, a_3 be the number of shares invested in stock 1, 2, and the risk-free bond, respectively. Then, starting with an initial budget of $W_0 = \$1$, we have

$$W_1(\omega_1) = a_1 + a_2 + \frac{3}{2}a_3 \tag{6.12}$$

$$W_1(\omega_2) = a_1 + \frac{3}{2}a_2 + \frac{3}{2}a_3 \tag{6.13}$$

$$W_1(\omega_3) = \frac{5}{2}a_1 + 2a_2 + \frac{3}{2}a_3 \tag{6.14}$$

$$W_0 = 1 = a_1 + a_2 + a_3. \tag{6.15}$$

From Eqs. (6.9)–(6.15), we obtain: $a_1 = -0.924$, $a_2 = 1.386$, and $a_3 = 0.538$. The maximal $W_1(\omega_1) = 1.269$, $W_1(\omega_2) = 1.962$, $W_1(\omega_3) = 1.269$, and thus maximized expected utility is $E(-e^{-\tilde{W}_1}) = -0.211$.

If we use the state-price density approach, maximal conditions in (6.8) yields

$$\exp(-W_1(\omega_1)) = \frac{8}{9}\lambda \tag{6.16}$$

$$\exp(-W_1(\omega_2)) = \frac{4}{9}\lambda \quad \text{and} \tag{6.17}$$

$$\exp(-W_1(\omega_3)) = \frac{8}{9}\lambda. \tag{6.18}$$

It can immediately be verified that Eqs. (6.16)–(6.18) are exactly the same equations in (6.9)–(6.11) under the classical approach. In the second step, we then use similar portfolio conditions in (6.12)–(6.15) to solve the problem completely.

We now look at a theorem which is quite deep in its full generality, but which in a specialized simple version aptly represents what we have done in this section regarding state pricing via the state-price density.

Let $q_j = (q_{1j}, q_{2j}, \ldots, q_{Sj})^T$ be the $S \times 1$ vector of payoffs in various states for security j.

Theorem 6.1 (Simplified Riesz Representation Theorem): *If $p(q_j)$ (Eq. (5.12) in Chap. 5.) is a security price linear functional mapping a payoff vector in $\mathcal{R}^S$ to a price, and if p is bounded, then there exists a monotone increasing function $G(z) = \sum_{i=1}^{z} c_i$, $1 \leq z \leq S$, such that $p(q_j) = \sum_{i=1}^{S} q_{ij} \triangle G = \sum_{i=1}^{S} q_{ij} c_i$ (one can write in continuous distributional form $p(q \in \mathcal{L}^S) = \int q dG$ for a monotone increasing function G) for every q_j.* ■

Clearly, c_i are the state prices $\gg 0$ as in Eq. (5.12) in our case. The Riesz representation theorem guarantees the *existence* of such state prices if prices are bounded. (A lot of deep mathematics theorems worry about existence and uniqueness issues.) The discrete case here is simple as the existence can also be obtained just by the separating hyperplane theorem we saw earlier. When the case becomes continuous distributions and continuous spaces, then the Riesz theorem becomes more powerful in its usage.

Applications

The application of the Riesz representation theorem here becomes more interesting if we define the argument of the price function as vector $q_j \cdot d$ (pointwise product) where $d = (\frac{\pi_1}{\theta_1}, \frac{\pi_2}{\theta_2}, \ldots, \frac{\pi_S}{\theta_S})^T$. Then by the theorem, there exists a monotone increasing function $G(z) = \sum_{i=1}^{z} \frac{\theta_i}{r}$, $1 \leq z \leq S$, such that $p(q_j \cdot d) = \sum_{i=1}^{S} q_{ij} d_i \triangle G = \frac{1}{r} \sum_{i=1}^{S} q_{ij} d_i \theta_i = \frac{1}{r} \sum_{i=1}^{S} q_{ij} \pi_i$, which is precisely Eq. (5.15) in the last chapter, under the risk-neutral pricing. Hence

$$p_j = \frac{1}{r} E_P(\tilde{q}_j \tilde{d}) = \frac{1}{r} E_Q(\tilde{q}_j). \tag{6.19}$$

Another angle of application of the theorem is even more interesting if we define the argument of the price function as vector $q_j \cdot \theta$, where $\theta = (\theta_1, \theta_2, \ldots, \theta_S)^T$. Then by the theorem, there exists a monotone increasing function $G(z) = \sum_{i=1}^{z} \frac{d_i}{r}$, $1 \leq z \leq S$, such that $p(q_j \cdot \theta) = \sum_{i=1}^{S} q_{ij}\theta_i \triangle G = \frac{1}{r}\sum_{i=1}^{S} q_{ij}\theta_i d_i = \frac{1}{r}\sum_{i=1}^{S} q_{ij}\pi_i$ which is the same Eq. (5.15) under the risk-neutral pricing. However, we buy another thing here: the Riesz theorem says there exists a monotone increasing G. Since $G(z) = \sum_{i=1}^{z} \frac{d_i}{r}$, $1 \leq z \leq S$, clearly d_i must ≥ 0 for G to be monotone increasing. But since we do not allow for states with infinitely large empirical probabilities in our economy, we use strict monotonicity, and $d \gg 0$, (and thus also $\pi \gg 0$) which proves the existence of strictly positive state-price densities. From the equation $p_j = E_P(\tilde{q}_j \frac{\tilde{d}}{r})$, it is clear that p_j is strictly increasing in q_j iff $d \gg 0$.

Note that since $d_i = \frac{\pi_i}{\theta_i} \gg 0$ for each state ω_i, we can treat $\tilde{d}$ as a RV with a probability $P(d_i) = P(\omega_i) > 0$. Note $E_P(\tilde{d}) = \sum_{i=1}^{S} \theta_i d_i = \sum_{i=1}^{S} \pi_i = 1$.

From Eq. (6.19),

$$1 = E_P\left[\left(\frac{\tilde{q}_j}{p_j}\right)\left(\frac{\tilde{d}}{r}\right)\right].$$

Or

$$E_P[\tilde{R}_j \tilde{d}'] = 1, \tag{6.20}$$

where R_j is return to security j, and $d' \gg 0$ is state-price density normalized by r. $E_P[d'] = E_P[d/r] = 1/r$. Hence

$$\mathrm{cov}_P(R_j, d') = E_P(R_j d') - E_P(R_j)E_P(d').$$

Or

$$\mathrm{cov}_P(R_j, d') = 1 - \frac{E_P(R_j)}{r}.$$

Or

$$E_P(R_j) - r = -r\mathrm{cov}_P(R_j, d'). \tag{6.21}$$

If R_j is positively correlated with d', i.e., when return is up, its probability θ is down, and vice-versa, then this means a lower expected return, which is indeed the case. The result applies in both a complete market and an incomplete market setting.

Without loss of generality, we can express $R_j = a + bd' + e$, where e is an independent random (disturbance) error, and $R_s = a^* + b^*d'$, where

R_s is a portfolio whose return is perfectly correlated with the state-price density. Then,

$$\mathrm{cov}_P(R_j, d') = b\,\mathrm{var}_P(d'), \text{ and}$$
$$\mathrm{cov}_P(R_s, d') = b^*\,\mathrm{var}_P(d').$$

So

$$\frac{\mathrm{cov}_P(R_j, d')}{\mathrm{cov}_P(R_s, d')} = \frac{b}{b^*}.$$

Now

$$\begin{aligned}\mathrm{cov}_P(R_j, R_s) &= bb^*\mathrm{var}_P(d') = bb^*\left(\frac{\mathrm{var}_P(R_s)}{b^{*2}}\right) = \frac{b}{b^*}\mathrm{var}_P(R_s)\\ &= \frac{\mathrm{cov}_P(R_j, d')}{\mathrm{cov}_P(R_s, d')}\mathrm{var}_P(R_s).\end{aligned}$$

But

$$\frac{E_P(R_j) - r}{E_P(R_s) - r} = \frac{\mathrm{cov}_P(R_j, d')}{\mathrm{cov}_P(R_s, d')} = \frac{\mathrm{cov}_P(R_j, R_s)}{\mathrm{var}_P(R_s)} \equiv \beta_j.$$

Hence

$$E_P(R_j) - r = \frac{\mathrm{cov}_P(R_j, R_s)}{\mathrm{var}_P(R_s)}(E_P(R_s) - r).$$

The last equation may be called a state-price beta asset pricing model for securities. It shows that state-price beta β_j is a scaled correlation, $\mathrm{corr}_P(R_j, R_s) \times \frac{\sigma_j}{\sigma_s}$, of the security j return with return on a portfolio with maximal correlation with the state-price density. The expected excess return of security j is proportional to β_j by the expected excess return on the maximal correlation portfolio. In this case, $(E_P(R_s) - r)$ could be negative. The key implication is that if security j has a high state-price beta, or a high correlation of return with state-price density, then high returns in low probability states and low returns in high probability states is an insurance on a market portfolio whereby most high returns of stocks occur in high probability states and low returns of stocks occur in low probability states, provided the expected premium of R_s is positive. Negative premium would imply the converse.

In the following section, we discuss equilibrium asset pricing under incomplete market settings.

6.3. Single-Period CAPM

The Sharpe–Lintner (sometimes Sharpe–Lintner–Mossin) capital asset pricing model[1] is a single-period model. From about 1964 until the early 1980s, it was a predominant model for understanding the pricing of stocks as well as projects through extensive use of its concept of systematic risk in beta. It is still relevant, although there have been much improvements in the understanding of how stocks are priced, including behavioral aberrations and extensions to multi-factor models. The CAPM is a single-factor model relying on the market index to explain systematic variations in stock returns.

There are N stocks and 1 risk-free bond with risk-free rate r_f. Utility function is strictly increasing and concave. We shall assume either:

(1a) continuously compounded stock returns r_i's are jointly normally distributed[2] (which means any portfolio or linear combination of returns is normally distributed) and/or
(1b) the investor has quadratic utility functions.

Investor k maximizes expected utility based on current wealth W_0 and investment decisions or portfolio weights (percentage investment) on the stocks, x_i:

$$\max_{\{x_i\}_{i=1,2,\ldots,N}} E[U(W_1)]$$

where $W_1 = W_0(1 + r_P)$, r_P being the portfolio return rate, and

$$\begin{aligned} r_P &= \sum_{i=1}^{N} x_i r_i + (1 - \sum_{i=1}^{N} x_i) r_f \\ &= r_f + \sum_{i=1}^{N} x_i (r_i - r_f) \\ &= r_f + x^T (r - r_f 1). \end{aligned}$$

Note that the weights of the N stocks and the risk-free bond sum to 1. In the last step, we switch to matrix notations, where $x^T = (x_1, x_2, \ldots, x_N)$, $r^T = (r_1, r_2, \ldots, r_N)$, and $1^T = (1, 1, \ldots, 1)_{1\times N}$.

---

[1]Sharpe, W (1964). Capital asset prices: A theory of market equilibrium under conditions of risk. *Journal of Finance*, 19, 425–442.

[2]A more general class that includes the normal distribution is the class of elliptical distributions.

For the portfolio return r_P which is normally distributed, note that

$$E(r_P) = \mu_P = r_f + x^T(\mu - r_f 1),$$
$$\text{var}(r_P) = \sigma_P^2 = x^T \Sigma x,$$

and

$$\text{cov}(r, r_P) = \begin{pmatrix} \text{cov}(r_1, r_P) \\ \text{cov}(r_2, r_P) \\ \vdots \\ \text{cov}(r_N, r_P) \end{pmatrix} = \Sigma x,$$

where Σ is the covariance matrix of the N stock returns.

Either assumption (1a) or (1b) has the key effect of ensuring that the investor's preference ultimately depends only on the mean and variance of the return distribution.

For (1a), the third and higher wealth level moments in a Taylor expansion of $E[U(\cdot)]$ (assuming the expansion is convergent) based on normally distributed wealth W_1 at end of period, can be expressed as functions of the first two moments of the normal distribution. Hence, for any arbitrary preferences, the VM expected utility $E[U(W_1)]$ of end-of-period wealth depends only on the mean and variance. However, there has been strong empirical evidence that stock return distributions tend to be skewed and have tails fatter than those of a normal distribution.

For (1b), investor's preference depends only on the mean and variance of returns because third and higher orders of derivatives of $E[U(\cdot)]$ in a Taylor expansion of $U(\cdot)$ (assuming the expansion is convergent) are zeros. Thus, only the first and second moments of return enter into the VM expected utility $E[U(W_1)]$ of end-of-period wealth. However, quadratic utility has the disadvantage that at some wealth level that is sufficiently high, expected utility may decrease, thus violating a standing axiom of nonsatiability. Moreover, there is increasing absolute risk aversion with respect to wealth, which is not very intuitive.

Despite the caveats discussed above where we see that mean–variance optimization (as in Markowitz's portfolio optimization[3]) may be inconsistent at times with expected utility maximization, it is still a very powerful

[3]Markowitz, H (1952). Portfolio Selection. *Journal of Finance*, 7, 77–91. See also CF Huang and RH Litzenberger (1988). *Foundations for Financial Economics*. North-Holland Publishing, for a rigorous discussion of portfolio optimization mathematics.

framework to derive meaningful and positive financial economic theories and understanding. We express $E(U)$ as a function of only μ_P and σ_P^2 for the mean–variance analysis. Thus, we maximize VM expected utility, $E[U(W_1)] \equiv V(\mu_P, \sigma_P^2)$.

The FOC becomes (in vector notation)

$$\frac{\partial V}{\partial x}_{N\times 1} = 0 = \frac{\partial V}{\partial \mu_P}(\mu - r_f 1) + \frac{\partial V}{\partial \sigma_P^2}(2\Sigma x). \tag{6.22}$$

Rearranging,

$$\begin{aligned} xW_0 &= -\frac{W_0}{2}\left(\frac{\frac{\partial V}{\partial \mu_P}}{\frac{\partial V}{\partial \sigma_P^2}}\right)\Sigma^{-1}(\mu - r_f 1) \\ &= t\Sigma^{-1}(\mu - r_f 1), \end{aligned} \tag{6.23}$$

where $t = -\frac{W_0}{2}\left(\frac{\frac{\partial V}{\partial \mu_P}}{\frac{\partial V}{\partial \sigma_P^2}}\right) > 0$ since $\frac{\partial V}{\partial \mu_P} > 0$ due to nonsatiation, and $\frac{\partial V}{\partial \sigma_P^2} < 0$ due to risk aversion.

We see that investors optimally invest in only two funds. The first fund is a portfolio of stocks as given by x in Eq. (6.23), while the second fund is $\$W_0(1 - x^T 1)$ in risk-free bonds. This is sometimes called a two-fund separation theorem, and implies that under mean–variance optimization, investors can achieve optimality by simply investing in two properly construed funds rather than having to worry about deciding weights for every stock. This has in the past been used as an argument for passive market index fund investment, such as in Vanguard.

The story is actually not quite complete yet. The optimization above is for a kth investor in the market. Suppose there are Z number of non-homogeneous or heterogeneous investors in the market. When all their stock demands are aggregated, the total vector dollar demand on stocks is given by

$$\sum_{k=1}^{Z} x_k W_0^k,$$

where superscript k denotes association of the quantity with the kth investor. We allow different investors to have different original wealth, hence W_0^k. They may also have different utility functions $V(\mu_P, \sigma_P^2)$, hence different t^k values although they all have the same information of market return parameters, μ and Σ. Suppose the total market wealth is $M = \sum_{k=1}^{Z} W_0^k$, we can then define $N \times 1$ vector of weights $x_M = \frac{\sum_{k=1}^{Z} x_k W_0^k}{M}$. Thus, x_M acts like the optimal portfolio of stocks of a representative investor (or the aggregated average of all non-homogeneous investors), when aggregate demand is equated to aggregate supply of dollars for each security, i.e., the ith element of $\sum_{k=1}^{Z} x_k W_0^k$ equated to the ith element of $x_M M$. This equation of demand and supply is called an equilibrium condition and is necessary for any good solution to a problem involving the whole set of investors in the market.

Then

$$x_M M = \sum_{k=1}^{Z} x_k W_0^k = \sum_{k=1}^{Z} t^k \Sigma^{-1}(\mu - r_f 1).$$

Now $\frac{x_m}{x_m^T 1}(x_m^T 1 M) = W_m M'$, where $W_m = x_m/(x_m^T 1)$ and $M' = (x_m^T 1)M$.

This implies

$$W_m = \left(\frac{\sum_{k=1}^{Z} t^k}{M'}\right) \Sigma^{-1}(\mu - r_f 1)$$
$$\Rightarrow \Sigma W_m = \mathrm{cov}(r, r_M) = t^M(\mu - r_f 1), \qquad (6.24)$$

where $r_m = W^T r$ is the market portfolio return, and $t^M = (\frac{\sum_{k=1}^{Z} t^k}{M'}) > 0$. Note that $W_m^T 1 = 1$.

Multiplying by W_M^T, Eq. (6.24) becomes

$$W_M^T \Sigma W_M = \sigma_M^2 = t^M(\mu_M - r_f)$$
$$\text{or} \quad t^M = \frac{\sigma_M^2}{(\mu_M - r_f)}, \qquad (6.25)$$

where $W_M^T \mu = \mu_M = E(r_M)$ is the market portfolio return.

Substituting Eq. (6.25) into Eq. (6.24), we obtain

$$\mu - r_f 1 = \frac{\Sigma W_M}{\sigma_M^2}(\mu_M - r_f). \qquad (6.26)$$

Equation (6.26) is the securities market line (SML) of the CAPM. Its ith element is

$$E(r_i) - r_f = \beta_i(E(r_M) - r_f),$$

where $\beta_i = \frac{\mathrm{cov}(r_i, r_M)}{\sigma_M^2}$.

Equation (6.26) says that the expected excess return of any security is equal to a risk premium (required compensation by a risk-averse investor for holding risky stock) which is proportional to its beta, β_i, a measure of its systematic risk. Since systematic risk cannot be diversified away, the

CAPM shows importantly that only diversifiable risk does not cost, but non-diversifiable risk fetches a positive risk premium.

6.4. Arbitrage Pricing Theory

There is another important single-period asset pricing model based on no-arbitrage arguments.[4] This may (although not necessarily so) be used to construct an understanding of multi-factor models in empirical asset pricing which is prevalent in finance.

Suppose there are N assets each with return r_i, and

$$r_i = a_i + \sum_{k=1}^{M} b_{ik} F_k + e_i, \tag{6.27}$$

where a_i, b_{ik} for $k = 1, 2, \ldots, M$ are constants specific to the ith asset, F_ks are M factor RVs with zero means (the zero mean condition is not necessary but is without loss of generality, and will simplify the notations) and zero correlations with each other and with all residual errors e_i's. Residual errors, e_i's, are uncorrelated with the M factors and with each other. $E(e_i) = 0\,\forall i$, and $E(e_i^2) = \sigma_i^2 < \infty$, $\forall i$.

Apart from linearity, the structure above is quite general. However, we must assume there is an almost infinite number of different assets ($N \to \infty$).

Suppose we can form a large portfolio of n_1 assets with weight $x_i \approx \frac{1}{n_1}$ in each of the assets in this group such that $\sum_{i=1}^{n_1} x_i = 1$, and $\sum_{i=1}^{n_1} x_i e_i \approx 0$ by WLLN. Then, this portfolio return is

$$R_1 \approx A_1 + \sum_{k=1}^{M} B_{1k} F_k,$$

where $A_1 = \sum_{i=1}^{n_1} x_i a_i$, and $B_{1k} = \sum_{i=1}^{n_1} x_i b_{ik} \quad \forall k$.

Suppose we can form a second large portfolio of n_2 assets with weight $x_i \approx \frac{1}{n_2}$ in each of the assets in this second group such that $\sum_{i=1}^{n_2} x_i = 1$, and $\sum_{i=1}^{n_2} x_i e_i \approx 0$ by WLLN. Then, this portfolio return is

$$R_2 \approx A_2 + \sum_{k=1}^{M} B_{2k} F_k,$$

where $A_2 = \sum_{i=1}^{n_2} x_i a_i$, and $B_{2k} = \sum_{i=1}^{n_2} x_i b_{ik} \quad \forall k$.

---

[4] Ross, AS (1976). The arbitrage theory of capital asset pricing. *Journal of Economic Theory*, 13, 341–360.

We form T number of such portfolios. Clearly, $\sum_{j=1}^{T} n_j = N$. We have

$$R_j \approx A_j + \sum_{k=1}^{M} B_{jk} F_k, \quad \forall j = 1, 2, \ldots, T.$$

Let

$$\begin{aligned} w &= (w_1, w_2, \ldots, w_T)^T, R = (R_1, R_2, \ldots, R_T)^T, l_{T\times 1} = (1, 1, \ldots, 1)^T, \\ A &= (A_1, A_2, \ldots, A_T)^T, \\ B_1 &= (B_{11}, B_{21}, B_{31}, \ldots, B_{T1})^T, \\ B_2 &= (B_{12}, B_{22}, B_{32}, \ldots, B_{T2})^T, \text{ and so on to} \\ B_M &= (B_{1M}, B_{2M}, B_{3M}, \ldots, B_{TM})^T. \end{aligned}$$

Now, from these large T number of portfolios, we can find weights w_i's to invest in each of these (i.e., form a portfolio of portfolios, which becomes a portfolio of securities across all T groups) such that as $N \uparrow \infty$ (also $T \uparrow \infty$):

(C1) $w^T l = 0$ i.e., a zero cost portfolio
(C2) $w^T B_1 = 0$ i.e., cancels out first-factor effect
(C3) $w^T B_2 = 0$ i.e., cancels out second-factor effect
....
(Cm1) $w^T B_M = 0$ i.e., cancels out the Mth factor effect.

Then, $w^T R = w^T A + 0$. This may be construed as convergence in distribution. Take expectation, so $w^T\ E(R) = w^T A$ since the RHS is a constant, and its expectation is the same constant. Hence, we treat R as expected return from this point.

But since there is zero outlay, then no-arbitrage (that is how the name of this theory comes about) implies this portfolio return $w^T A$ where A is constant vector, must be zero. Therefore, we obtain the $(m+2)^{\text{th}}$ condition, viz.

(Cm2) $w^T R = 0$.

Next, consider the following lemma.

Lemma 6.1 *If vector $X_{N\times 1}$ is orthogonal to vectors $Z_{N\times 1}, Y^1_{N\times 1}$, $Y^2_{N\times 1}, \ldots,$ and $Y^K_{N\times 1}$, such that $X^T Z = 0$, and $X^T Y^j = 0$ for $j = 1, 2, \ldots, K$, then for $\lambda_j \neq 0$,*

$$Z = \sum_{j=1}^{K} \lambda_j Y^j. \quad \blacksquare$$

Proof. $X^T Z = 0$, and $-\lambda_j X^T Y^j = 0$ for $j = 1, 2, \ldots, K$. So, adding them up,

$$X^T Z - \sum_{j=1}^{K} \lambda_j X^T Y^j = 0, \quad \text{or}$$

$$X^T \left(Z - \sum_{j=1}^{K} \lambda_j Y^j \right) = 0.$$

Since X is arbitrary, then $Z - \sum_{j=1}^{K} \lambda_j Y^j = 0$. ■

From conditions (C1)–(Cm2), clearly vector $w_{T\times 1}$ is orthogonal to vectors $R_{T\times 1}$, $l_{T\times 1}$, B_1, B_2, and so on to B_M. By the lemma, for $\lambda_j \neq 0$, for $j = 0, 1, 2, \ldots, M$, then

$$R = \lambda_0 l + \sum_{j=1}^{M} \lambda_j B_j. \tag{6.28}$$

In the arbitrage pricing model expressed in Eq. (6.28), if we put all $\lambda_j = 0$, $j > 0$, then there is no risk and $\lambda_0 = r_f$, the risk-free rate. If $M = 1$, i.e., only one factor and suppose this factor is the market factor, then $\lambda_1 = E(R_M - r_f)$. Thus, (6.28) becomes

$$R_{T\times 1} = r_f \, l_{T\times 1} + E(R_M - r_f) \, B_{T\times 1}$$

which is the Sharpe CAPM model we saw earlier.

Thus, the APT as in (6.28) is quite a general result. The values λ_j, $j > 0$ are the prices of risks (or risk premia or factor prices) associated with the different factors $j = 1, 2, \ldots, M$. B_j is the jth sensitivity to the jth factor risk by the securities or assets.

However, APT alone does not quite provide a positive theory about the number of factors or what they are. The sensitivities or exposures to factors are also not obviously characterized as in the CAPM.

6.5. Multi-Factor Asset Pricing

Many empirical finance papers in the last two decades attempted to find the real factors behind expected returns. There is a famous paper by Fama and French (1992)[5] proposing 3 factors: the market return as factor, the

[5]Fama, E and KR French (1992). The cross-section of expected stock returns. *The Journal of Finance*, 47(2), 427–465. See also Fama, EF and KR French (1993). Common

returns of small stocks less returns of big stocks as firm capitalization or size factor, and the returns of stocks with high book-to-market (B/M) values less returns of stocks with low book-to-market values or value factor. In operational terms, based on Eq. (6.28), $M = 3$, $\lambda_1 = E(R_M - r_f)$, $\lambda_2 = E(SMB)$, and $\lambda_3 = E(HML)$, where SMB is return from a portfolio of the smallest stocks minus biggest stocks (long small short big), and HML is return from a portfolio of the highest book-to-market ratio stocks minus lowest book-to-market ratio stocks (long value short growth).

It is empirically found that using time series sample averages, typically $E(R_M - r_f) > 0$. Now, $E(SMB) > 0$ because of the higher risk associated with smaller stocks, and $E(HML) > 0$ because value stocks have their market prices temporarily depressed, but would on average bounce back. Hence, stocks with higher correlation (B_1) with market return tend to have higher expected returns; stocks with higher correlation (B_2) with SMB, e.g., smaller stocks, tend to have higher returns, and stocks with higher correlation (B_3) with B/M (or sometimes similar to low P/E) tend to have higher returns.

Another significant empirically "discovered" factor is the momentum factor that can be constructed as returns to a dynamic portfolio of long positions in short-term winner stocks, and short positions in short-term loser stocks. An investor may attempt to take an exposure to this systematic risk or factor by investing in a portfolio that is long winners and short losers in the last 3 months. However, the period to track momentum may change. Sometimes, the stocks exhibit reversals in less than 3 months.

6.6. Problem Set 6

1. Suppose in a single period model, an investor is able to select consumption and an investment portfolio to achieve a finite maximum in his/her expected utility based on a strictly increasing concave utility function, is he/she able to find arbitrage opportunity?
2. Show how in the CAPM, if portfolio returns are normally distributed, then the expected utility is a function of only the portfolio return mean and variance.
3. Find the second-order condition in Eq. (6.22) and show that a maximum is indeed obtained.

---

risk factors in the returns on stocks and bonds. *Journal of Financial Economics* 33, 3–56.

4. In CAPM, show that $E(r_M) > r_f$. Does this mean that in any period, the realized market return will always be higher than r_f? Is it possible to have a risky stock with expected return below that of the risk-free bond return rate r_f?
5. In a two-state world, suppose a stock A costs \$6 at $t = 0$ and pays at $t = 1$ either $q(\omega_1) = \$9$ in state 1 or $q(\omega_2) = \$6$ in state 2. The risk-free return over the period is 7/6. Find the state prices. Is there arbitrage opportunity in this economy? What is the price of an American put option on A with strike price at \$7, and maturing at $t = 1$?
6. Try solving the same problem in Sec. 6.2 but assuming a different utility function, $u(W) = \log W$.
7. Two stocks are described by the following factor structure.

   (a) $r_1 = a_1 + 3F_1 + 2F_2$
   (b) $r_2 = a_2 + 2F_1 + 4F_2$.

   A risk-free asset has return rate of 10%. If the expected returns of security 1 and 2 are 13% and 15% respectively, what are the values of $\lambda_0, \lambda_1, \lambda_2$?

Chapter 7

STOCHASTIC PROCESSES AND MARTINGALES

Diffusions, Markov processes, and martingales are separately different types of stochastic processes though they are closely linked. For example, a diffusion process is a Markov process that has continuous sample paths. There are Markov processes such as discrete-time processes that are not diffusions, and there are Poisson-jump processes that are Markov processes but not diffusions. Martingales, as studied in this chapter, arose out of the investigation of odds in infinite sequences of Bernoulli trials. Its root might have been linked in some ways to the type of betting strategy called a "martingale", which is also called the "doubling strategy", that was popular in 18th century France. Andrey N. Kolmogorov made many significant contributions to the axiomatic foundations of probability theory, while Paul P. Lévy introduced the concept of martingales in probability theory. Lastly, many of the early important results in martingales were developed by Joseph L. Doob.

This chapter contains some delicate treatment of the subject of investments over time. As such, the material in this chapter will be slightly longer and more detailed in its pedagogical exposition. This material forms the backbone of the key martingale method in modern finance theory with regard to derivative pricing under no-arbitrage condition, which is a major contribution by Harrison and Kreps (1979)[1] and Harrison and Pliska (1981),[2] among others.

7.1. Random Walk

A random walk is a discrete-time stochastic process that is analogous to continuous-time Brownian motion, a critically important class of stochastic processes that we shall discuss in Chap. 9. The random walk is an important workhorse that facilitates a more intuitive understanding of more complicated probabilistic concepts, and we shall use it as follows.

---

[1]Harrison, JM and DM Kreps (1979). Martingales and arbitrage in multiperiod securities markets. *Journal of Economic Theory*, 20, 381–408.

[2]Harrison, JM and S Pliska (1981). Martingales and stochastic integrals in the theory of continuous trading. *Stochastic Processes and their Applications*, 11(3), 215–260.

Consider independent Bernoulli RVs X_i, $\forall i$

$$X_i = \begin{cases} +1, & \text{with probability } p_i = \dfrac{1}{2} \\ -1, & \text{with probability } 1 - p_i = \dfrac{1}{2}. \end{cases}$$

Define $S_0 = 0$, and $S_n = \sum_{i=1}^{n} X_i$ for $n = 1, 2, \ldots,$ etc. The process S_n is a symmetric random walk process. If $p_i \neq 1/2$, it would be an asymmetric random walk. The symmetric random walk is a discrete martingale and also a Markov process.

A graphical illustration of a symmetric random walk is shown in Fig. 7.1. The next step in the random walk process is either up by $+1$ or down by -1 randomly with equal probability.

Let n be units of time. Suppose the random walk S_n first hits upper value or upper level 2 at time τ. Clearly, this first passage time of the random walk to level 2 is $\tau = 2$ as shown by the circle on the graph. The first passage time to the level -1 is $\tau = 9$.

In general, let τ_m be the first time the random walk after leaving 0 reaches level m. Then

$$\tau_m \triangleq \min\{n : S_n = m\}.$$

If at each time n, we have enough information to determine if the event $\{S_n = m \,|\, m\}$ has indeed occurred, the event is called a stopping time event, and n is called a stopping time when the event occurs. The RV τ_m is a stopping time, and is a first passage time to level m. Usually

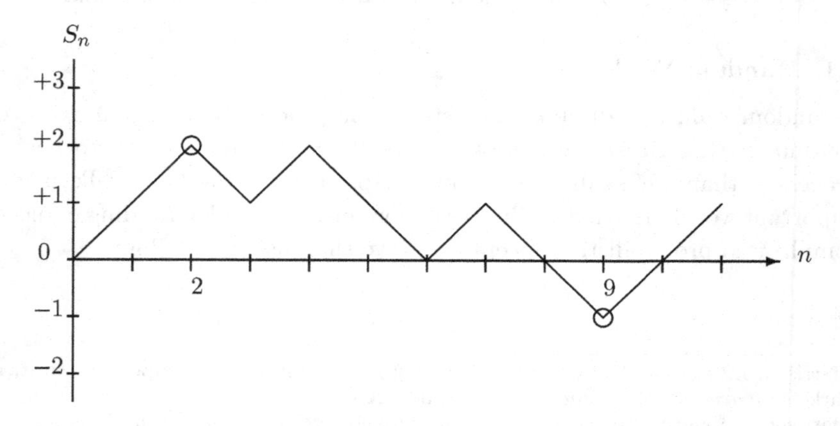

Fig. 7.1: A Random Walk Process

by definition, if the level m is never hit, then the stopping time is $+\infty$. A counting process N_t has a stopping time RV since at each t, we know whether event $\{N_t = m\}$ has occurred.

Suppose an urn contains an infinite number of balls marked with numbers $1, 2, 3, \ldots,$ etc. A person randomly draws a ball from the urn without replacement and then looks at the number on the ball. Let the number of the nth ball drawn be X_n. Define: $t_n = \min\{n : \text{rank}(X_n) = 3\}$, where rank 3 refers to the third-largest number of the drawn balls. Then t_n is a stopping time. Obviously, $t_n \geq 3$.

However, if the event is defined as $t_n = \min\{n : X_n < X_{n+1}\}$, then since the event can never be verified (as X_{n+1} does not belong to the filtration at n), t_n is not a stopping time RV in this case.

The hitting time or first passage time examples above are quite specific. Stopping time RV is a type of RV that maps events in $\mathcal{F}_t$ to a time index t for a domain that is time in $[0, \infty]$. (For technical reasons, in continuous-time processes, the mapping is formally to the set, time τ where $\tau \leq t$. But we should not worry about this here.) For discrete time processes, let $\tau = t$. Then what it means is that for time index $t = 1, 2, 3, \ldots, \infty$, for each $t = n$, we can visualize the mapping as follows (Table 7.1).

What the table means is that at time $t = n$, stopping events E_3, E_5, E_6 jointly occur with probability $P(n) = 1 - (p_1 + p_2 + p_4)$. When the stopping event does not occur, we put the stopping RV as ∞. When the stopping event occurs, the stopping RV takes the value n. The stopping events must be defined *a priori* in the context of the application, e.g., the first passage time of a stock price hitting barrier \$10 from below when its current price may be \$5. In Table 7.1, events E_i, $i = 1, 2, \ldots, 6$ belong to field or algebra $\mathcal{F}_n$. $\sum_{i=1}^{6} p_i = 1$. Thus, at time $t = n$ as seen, the probability of stopping, or a stopping event occurring, is $P(n) \in [0, 1]$.

If we perform the above analysis for every $t = 1, 2, 3, \ldots, \infty$, then we will obtain the following probability table (Table 7.2).

Table 7.1: Stopping Event Probabilities

Event	RV	Probability
E_1	∞	p_1
E_2	∞	p_2
E_3	n	p_3
E_4	∞	p_4
E_5	n	p_5
E_6	n	p_6

Table 7.2: Probability of Stopping Time

Time n	$P(n)$
1	$P(1)$
2	$P(2)$
3	$P(3)$
$\vdots$	$\vdots$
s	$P(s)$
$\vdots$	$\vdots$
∞	0

For stopping time RV, it takes a particular realized value with a particular probability, just as in any probability distribution of a RV except here the realized value is a time index n. Notice when stopping time equals ∞, it means that the stopping event did not occur, so the probability is 0. For a well-defined stopping-time RV, $\sum_{n=1}^{\infty} P(n) = 1$. Table 7.2, like any other RV probability distribution, is *ex ante*. As time progresses, suppose the stopping event such as the first passage time occurs at $n = 35$, then conditional on this information, Φ_{35} or $\mathcal{F}_{35}$, the probability of stopping RV taking values larger than 35 becomes zero. It is possible for $P(n)$ to be zero for some finite n. For example, in the first passage time to \$10 from \$5, suppose the stochastic price process is such that each time-step, the price moves can only be $\pm\$1$, then clearly it will require at least 5 time periods before there is first passage to \$10 from \$5, in which case, $P(n \leq 4) = 0$.

From the above, hopefully the concept of a simple stopping time RV τ is clearer. It is a RV because Table 7.2 shows that τ can take time index value n only with a chance or probability typically less than 1 (except in some special cases). The stopping RV realizes at the stopped (or stopping) event at time n, and so on. Once the event is stopped, the experiment in this setup is over.

Stopping time RVs are important in martingales (and some Markov) applications because RVs have the interesting property that many types of adapted stochastic processes are martingales at stopping times. An application which we will see in Chap. 10 is the use of lattice trees (binomial) to price American options, where implicitly the optimal stopping or exercise of American options does ensure the stopped price process is a martingale.

The set of events (Table 7.1) which forms the subset of $\mathcal{F}_n$ that yields the realization of a stopping event or stopping time $= n$ with probability

$P(n)$ is denoted as $\{\omega : E_i \ni \{\tau = n\}\}$, or in short-form as $\{\tau = n\}$. In other words, this set or $\{\tau = n\}$ is $\mathcal{F}_n$-measurable. As seen in Table 7.2, the stopping event RV is well-defined if $\{\tau = n\}$ is $\mathcal{F}_n$-measurable for every $n = 1, 2, 3, \ldots$ etc.

A stopping time RV τ is formally defined as follows. We start with a discrete-time process with a probability space and filtration $\{\mathcal{F}_n, n \geq 1\}$, for time $n = 1, 2, \ldots,$ etc. A stopping time is a random time according to the specified stopping event.

Definition 7.1 A RV $\tau \in [0, \infty]$ adapted to filtration $\mathcal{F}_n$ for $n = 1, 2, \ldots,$ is a stopping time (RV) if $\{\tau = n\} \in \mathcal{F}_n$ for all $n = 1, 2, \ldots$. ■

We shall move on to understanding martingales a bit more in order to reach our goal of explaining the martingale method of option pricing in this chapter. It is useful to more formally consider stochastic processes including the Markov property as follows, before we study martingales.

7.2. Stochastic Processes

We begin by recapitulating the ideas of algebras, filtrations, and adapted process in Chap. 2, and pay a short revisit to the binomial tree process in Chap. 3.

Generally, in a stochastic process (sequence of RVs over time), using the discrete binomial process (or lattice tree) in Fig. 7.2 as an example, the stock price S_t starting at \$1 at $t = 0$ is seen to follow the stochastic evolvement over time. At $t = 1$, S_1 takes value u with probability p or else value d with probability $q = 1 - p$. In each time period forward, the price level either increases by return u with probability p or decreases by return d with probability q. The lattice tree is constructed to show every possible sample path, so it is shown as a nonrecombining tree even if some nodes may have the same price values. Thus, at time $t = 3$, there are $2^3 = 8$ nodes. If this discrete stochastic process continues in time, then at $t = T$, there will be 2^T number of nodes or equivalently 2^T sample paths. Clearly the price level S_t is not stationary as its distribution expands over time to include more states of nature. However, the corresponding return process of this price process is stationary, as seen in Fig. 7.3.

In Figs. 7.2 and 7.3, we assume the stochastic process ends at $t = T = 3$. (This is only illustrative, as the concepts explained below extend to a very large T that may be infinite.) This is a discrete-space (discontinuous) process, so the values the RVs take are "jumpy" and not continuous

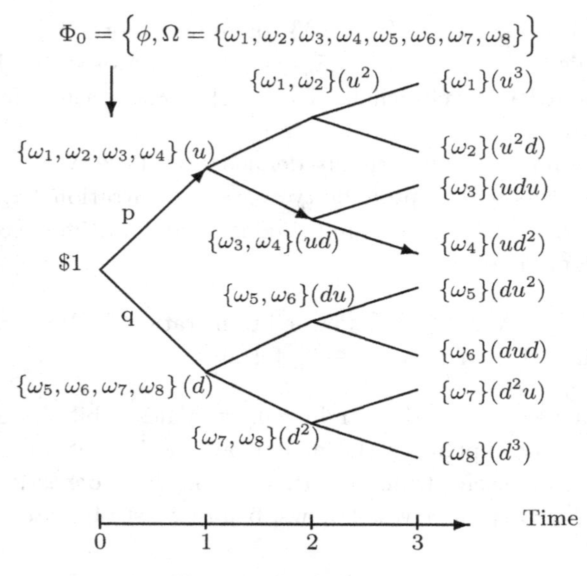

Fig. 7.2: Stock Price Process

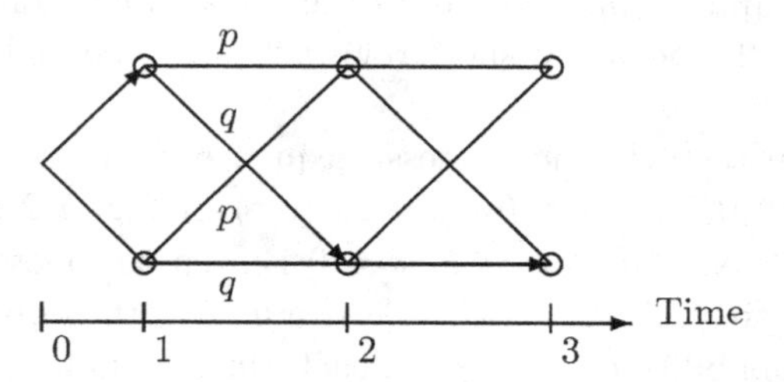

Fig. 7.3: Stock Return Process

(in space). For a continuous-time continuous(-space) process, particularly for most applications in finance, we assume the process is right-continuous with left limit (RCLL) or a càdlàg process. A càdlàg process includes both diffusion processes such as Brownian motion and also jump processes which may have some discontinuities in the values as time moves forward. Thus càdlàg processes are quite general, and the collection of such processes is known as a Skorokhod space. RCLL processes $X(t)$ allow jumps or discontinuities to be conveniently characterized as having taken place at a known time t', the jump being $X(t') - X(t'-)$.

We shall highlight some important ideas and definitions of stochastic processes as follows, using Figs. 7.2 and 7.3 for illustration.

(1a) The triple $(\Omega, \mathcal{F}, \mathcal{P})$ is a probability space where $\Omega = \{\omega_1, \omega_2, \ldots, \omega_8\}$, $\mathcal{F}$ or field is the collection or family of subsets of Ω e.g., $\{\omega_1, \omega_2, \omega_3, \omega_4\}$ which is called an event, and $\mathcal{P}$ is a probability measure on each event or element of $\mathcal{F}$.

We already saw in Chap. 2 how each possible set of events at time t, called $\mathcal{F}_t$ or information at time t, is mapped to a random variable $X(t)$ at t. Thus, each event or element in $\mathcal{F}_t$ takes a value under the RV when the particular event occurs or is realized. With suitable definition of the RV or the event, the RV can be a bijective function or a one-to-one correspondence between a measurable (i.e., with attached probability measure) event and a random value.

For each time t, the random variable $X(t)$ has therefore a probability measure attached for each of its realizations, and this constitutes a probability distribution of $X(t)$ at t, as seen in Chap. 1.

(1b) Suppose for each t, $X(t)$ is a well-defined RV taking values with a bijective relation to events in $\mathcal{F}_t$ (assuming $\mathcal{F}_t$ is defined for each t) so that $X(t)$ has a probability distribution $P(X(t))$, then the sequence $\{X(t)\}_{t\geq 0}$ is called a stochastic process of X.

(1c) The stochastic process X (sequence of RVs $X(t)$) can be expressed as a mapping

$$X:\ I \times \Omega \rightarrow \mathcal{R}$$

where $I = [0, T]$ in time domain, and $(I \times \Omega, t \times \mathcal{F})$, $t \in I$, is the measurable space. (This can be generalized to vector of RVs across time.) The stochastic process is measurable when there is a function f such that

$$f : X(t, \omega) \mapsto y \in \mathcal{R}$$

where $P(y) \in [0, 1]$ exists, and ω is defined as an element or event in $\mathcal{F}$.

For our purpose, we shall assume that the probability space of the stochastic process X exists, i.e., $(I \times \Omega, t \times \mathcal{F}, \mathcal{P})$ is well-defined.

(1d) $X(\cdot, \omega)$ is called the sample path of the stochastic process X, and is shown in Fig. 7.2 as the path traced out by the vectors (from state \$1 at $t = 0$ to state u at $t = 1$ to state ud at $t = 2$ and then to state ud^2 at $t = 3$.) In discrete time, the measurability of this sample path is easy to validate and can be found as probability $P\big(X(\cdot, \omega_4)\big) = pq^2$. This is equivalent to the vector path of the return process shown in Fig. 7.3, with the same path probability. The sum of probabilities of all measurable paths must equal to 1. In Fig. 7.2, at $t = T = 3$, there are $2^3 = 8$ paths (or nodes), and the sum of their probabilities from the top node to the bottom node is $p^3 + p^2q + pqp + pq^2 + qp^2 + qpq + q^2p + q^3 = 1$.

(1e) As seen in Chap. 2, we assume there is an evolution of information set Φ_t or in probability space, of field $\mathcal{F}_t$ which is a particular collection of subsets of Ω (it needs not be, and is usually smaller than the power set of 2^Ω that is the full-information set at the end of process time T). Field at t, $\mathcal{F}_t$ defines the events that (would) occur at time t, and which are measurable events, i.e., with attached probabilities such that they sum to 1. Let the events be $E_t^i \in \mathcal{F}_t$, for $i = 1, 2, \ldots, n$ so that $\sum_{i=1}^n P(E_i) = 1$. Then, the random variable at t, for each $t \in I$, $X(t, \cdot)$ is well-defined. In particular, we usually express the probability of event i at time t as $P(X(E_i)|\Phi_t)$ rather than the formal $P(X(t, E_i \in \mathcal{F}_t))$, though the latter should be borne in mind.

Now if at each t, field $\mathcal{F}_t$ is unchanged, then the set of events and their attached probabilities are constant and the corresponding RV $X(t)$ has the same probability distribution for every t. This is the special case of i.i.d. RV $X(t)$. However, probabilistic phenomena over time are rarely described by i.i.d. RVs alone. New information arises that change the field $\mathcal{F}_t$ or information set Φ_t. In financial economics, this means the information set becomes finer or more refined. Thus, events in fields later in time become less coarse. It means earlier fields are subsets of later fields, i.e.,

$$\mathcal{F}_t \subseteq \mathcal{F}_{t+1} \subseteq \mathcal{F}_{t+2} \subseteq \ldots \subseteq \mathcal{F}_T.$$

(1f) The family of fields over time with property

$$\mathcal{F}_t \subseteq \mathcal{F}_{t+1} \subseteq \ldots \subseteq \mathcal{F}_s \subseteq$$

for $t < s$, or an increasing family of fields or algebras, is called a filtration in the measurable space $(\Omega, \mathcal{F})$. A probability space $(\Omega, \mathcal{F}, \mathcal{P})$ with such an attached filtration $\mathcal{F}_t \subseteq \mathcal{F}$, $\forall t$, where $\mathcal{F}$ is the largest field generated by Ω, is called a filtered probability space and may be denoted as the quadruple $(\Omega, \mathcal{F}, (\mathcal{F}_t)_{t\geq 0}, \mathcal{P})$.

(1g) Any stochastic process $X(t, \omega)$ defined on the filtered probability space $(\Omega, \mathcal{F}, (\mathcal{F}_t)_{t\geq 0}, \mathcal{P})$ has its random variable $X(t)$ adapted to the filtration $(\mathcal{F}_t)_{t\geq 0}$, $\forall t$, or is $\mathcal{F}_t$-adapted or $\mathcal{F}_t$-measurable, which means that before time t, probabilities of events at t, $P(X(t, E_i \in \mathcal{F}_t))$, are well-defined; and at time t, the uncertainty is resolved so that $X(t)$ is realized or known at t, or takes a realized sample value $y \in \mathcal{R}$ at t.

It is in the above sense that in Chap. 2, for conditional probabilities and conditional expectations, we are able to construct *ex ante*, before time t, matrices $P(X(t)|\mathcal{F}_t)$ for elements $P(X(t,\omega)|E_i \in \mathcal{F}_t)$ and also vectors $E(X(t,\omega)|E_i \in \mathcal{F}_t)$. At time t, when the event is realized, these conditional probabilities of information at t, Φ_t, become 0 or 1.

For a stochastic process $X(t,\omega)$ (or short-hand $X(t)$) defined on the filtered probability space, the filtration generated by the sequence of increasing $X(t)$, $\sigma(X(s), 0 \leq s \leq t) = \sigma\big(\cup_{0 \leq s \leq t} \sigma(X(s))\big)$ forms a natural filtration $\mathcal{F}_t$ for the probability space. It is the smallest filtration such that X is adapted. Sometimes RV adapted $X(t)$ that is $\mathcal{F}_t$-measurable is called non-anticipative with respect to $\mathcal{F}_t$ or to some field generated by another RV such as a Brownian motion $\sigma(B(t), 0 \leq s \leq t)$ when $X(t)$ can be determined only by realized values of the RV $B(s)$ up to and including $B(t)$.

(1h) A stochastic process $Z(t,\omega)$ defined on the filtered probability space $\big(\Omega, \mathcal{F}, (\mathcal{F}_t)_{t\geq 0}, \mathcal{P}\big)$ is called predictable or previsible if for each t, $Z(t)$ is adapted to $\mathcal{F}_{t-1}$ or is $\mathcal{F}_{t-1}$-measurable. It means that at time $t-1$, the realized value $X(t)$ is known.

An example of a predictable process is the money (market) account process, $M(t)$, where $M(0) = \$1$ that is in a risk-free deposit, and in the next period, it grows by the risk-free rate r_0 already fixed and known at $t = 0$ to $M(1) = \$(1 + r_0)$ at $t = 1$. At $t = 2$, $M(2) = \$(1 + r_0)(1 + r_1)$. At $t = n$, $M(n) = \$\prod_{t=0}^{n-1}(1 + r_t)$. Since $r_{n-1}, r_{n-2}, \ldots, r_0$ clearly are random variables that are $\mathcal{P}$-measurable, i.e., have probabilities at $n-1$, $n-2$, etc., then $M(n)$ is $\mathcal{P}$-measurable w.r.t. $\mathcal{F}_{n-1} = \sigma(r_0, r_1, r_2, \ldots, r_{n-1})$, i.e., $M(n)$ is a predictable process. It would be known and is no longer a RV at $n-1$ already, even if the actual deposit amount $M(n)$ will be collectable only at $t = n$.

From Figs. 7.2 and 7.3, by (1c) the binomial stochastic process is well-defined since for each $t = 1, 2,$ or 3, $X : \{t\} \times \Omega$ maps to a particular node with an attached RV value. For example,

$$f : X\big(t = 2, \{\omega_1, \omega_2\} \in \mathcal{F}_{t=2}\big) \mapsto u^2.$$

From (1d), we already saw how the sample paths in Figs. 7.2 and 7.3 are well-defined and are $\mathcal{F}_t$-measurable. From (1b), every RV $X(t)$, $\forall t$ is well-defined. For example, at $t = 1$, $X(1) = u$ with probability p, and $X(1) = d$ with probability q. At $t = 2$, $X(2) = u^2$ with probability p^2, $X(2) = ud = du$ with probability $2pq$, and $X(2) = d^2$ with probability q^2. And so on.

For the RV $R(t)$ in Fig. 7.3 that is defined as the return of stock over time, i.e., $R(t+1) = X(t+1)/X(t)$, then R is stationary, and its probability at any time t, is $R(t) = u$ with probability p, and $R(t) = d$ with probability q.

In Fig. 7.2, the states (shown in brackets "$(\cdot)$" associated with each node at a time point) show the information structure of the economy as it moves forward in time. For example, at $t = 1$, in the up-state U (at node with value (u)), the realized event is (shown in $\{\cdot\}$) $\{\omega_1, \omega_2, \omega_3, \omega_4\}$. At $t = 1$, in the down-state D (at node with value (d)), the realized event is $\{\omega_5, \omega_6, \omega_7, \omega_8\}$. The associated information set at time $t = 1$ is

$$\Phi_1 = \mathcal{F}_1 = \{\phi, \Omega, E_1 = \{\omega_1, \omega_2, \omega_3, \omega_4\}, E_2 = \{\omega_5, \omega_6, \omega_7, \omega_8\}\}\,.$$

If we consider an "up" state U as happening with probability p and a "down" state D with probability q, then we may alternatively denote the sample points as $\omega_1 \equiv$ UUU, $\omega_2 \equiv$ UUD, $\omega_3 \equiv$ UDU, $\omega_4 \equiv$ UDD, $\omega_5 \equiv$ DUU, $\omega_6 \equiv$ DUD, $\omega_7 \equiv$ DDU, and $\omega_8 \equiv$ DDD. The order of the U's and D's matters, which explains why each sample point describes one sample path.

At $t = 0$, the information set is $\Phi_0 = \{\phi, \Omega\}$, i.e., either something happened — Ω; or nothing happened — ϕ, the empty set. They are of probabilities 1 or 0 respectively. These are somewhat trivial events, though technically they belong to each field.

At $t = 1$, the information set is

$$\Phi_1 = \{\phi, \Omega, \{UUU, UUD, UDU, UDD\}, \{DUU, DUD, DDU, DDD\}\}$$

At $t = 2$, the information set is

$$\begin{aligned}\Phi_2 = \Big\{&\phi, \Omega, \{UUU, UUD\}, \{UDU, UDD\}, \{DUU, DUD\}\\ &\{DDU, DDD\}, \{UUU, UUD, UDU, UDD\}\\ &\{DUU, DUD, DDU, DDD\}, \{UUU, UUD, DUU, DUD\}\\ &\{UDU, UDD, DDU, DDD\}, \{UUU, UUD, DDU, DDD\}\\ &\{UDU, UDD, DUU, DUD\}\\ &\{\{UUU, UUD\} \cup \{UDU, UDD, DUU, DUD\}\}\\ &\{\{\{UUU, UUD\} \cup \{UDU, UDD, DUU, DUD\}\}^c\}\\ &\text{etc.}\Big\}\end{aligned}$$

Note that in the above, the interesting minimal events (there are other events with the same probability because they include union with null events) with probabilities not zero or one are

$$\{\{UUU, UUD\}, \{UDU, UDD\}, \{DUU, DUD\}$$
$$\{DDU, DDD\}, \{UUU, UUD, UDU, UDD\}$$
$$\{DUU, DUD, DDU, DDD\}\}.$$

Trivial events such as $\{UDU, UDD, DUU, DUD\}$ at $t = 2$ has zero probability since this is not possible under the stochastic process. $\mathcal{F}_2$ or Φ_2 contains $2^4 = 16$ events.

At $t = 3$, when all paths are fully disclosed, the information set Φ_3 contains $2^8 = 64$ events measurable with respect to $\mathcal{F}_3$.

From the way the information sets are constructed, and how each set Φ_t is partitioned into events, it is seen that for any event $E_i \in \mathcal{F}_t$, the same event is also an event in $\mathcal{F}_{t+1}$ which contains finer partitions including unions of them. Thus, taking conditional probability of $X(t+1)$ based on $\mathcal{F}_u$ for any $u \leq t$ is well-defined. In Chap. 2, the more specialized situation of $P(X(T)|\Phi_s)$ and $E(X(T)|\Phi_s)$, $s \leq T$, is shown in an example. In general, we can compute $P(X(t)|\Phi_s)$, and $E(X(t)|\Phi_s)$, for any $s \leq t \leq T$.

Markov Process

With the filtered probability space $(\Omega, \mathcal{F}, \{\mathcal{F}_n\}, \mathcal{P})$, if for every function $f(X_{n+1})$, $n = 0, 1, 2 \ldots, T-1$, there is a function $g(X_n)$ such that

$$E_n\left[f(X_{n+1})\right] = g(X_n),$$

then X_0, X_1, $X_2, \ldots, X_T$ is called a Markov process. The expectation is taken w.r.t. the filtration $\mathcal{F}_n$, i.e., information set at time n.

As an example, suppose $X_0 = e_0$, $X_1 = 1 + e_1$, and for $n \geq 1$, $X_{n+1} = X_n + X_{n-1} + e_{n+1}$, where all e_i, $\forall i \geq 0$ are i.i.d. with mean zeros. Then, $E(X_2) = 1 + 0 + 0 = 1$, $E(X_3) = 1 + 1 + 0 = 2$, $E(X_4) = 2 + 1 + 0 = 3$, $E(X_5) = 3 + 2 + 0 = 5$, and so on, where the (unconditional) means form a Fibonacci series.

Clearly, for this process, the conditional mean

$$\begin{aligned} E_n[X_{n+1}] &= E[X_{n+1}|X_n, X_{n-1}, \ldots, X_0] \\ &= X_n + X_{n-1} + 0 \\ &\neq X_n \text{ and } \neq g(X_n) \end{aligned}$$

for any function g that depends only on X_n. Hence, this process is not a Markov process. In the next section of discussion on martingales, we shall see an example in the binomial process where it is a Markov process.

7.3. Martingales

Let $X_1, X_2, \ldots, X_t, \ldots, X_s, \ldots$, etc. be a timed sequence of real-valued adapted RVs with time $s > t > 0$. $E(|X_i|) < \infty$ so $X_i \in \mathcal{L}^1$. Assume there exists a filtration or increasing sequence of (σ-)algebras on a filtered probability space, including the natural filtration of X_t.

If $E(X_s|X_1, X_2, \ldots, X_t) = X_t$, a.s. for all s and $t < s$, the sequence $\{X_i\}_{i=1}^{\infty}$ is called a (discrete) martingale. (The original definition of a martingale to mean one time-step ahead conditioning, i.e., $s = t + 1$, can be extended to include conditioning many time-steps ahead, i.e., $s > t$, by using the law of iterated expectations.) If there is an initial condition where X_0 is a constant, then a martingale $\{X_t\}$ has a constant unconditional expectation equal to X_0.

If $E(X_s|X_1, X_2, \ldots, X_t) \geq X_t$, $\{X_i\}_{i=1}^{\infty}$ is called a submartingale. There is expectation for the sequence to increase over time.

If $E(X_s|X_1, X_2, \ldots, X_t) \leq X_t$, $\{X_i\}_{i=1}^{\infty}$ is called a supermartingale. There is expectation for the sequence to decrease over time.

An example of a martingale is the following. Suppose $X_1, X_2, \ldots$, are independent real-valued adapted RVs with $E(X_i) = 0$ for $i = 1, 2, \ldots$ etc.

Define partial sum $S_n = X_1 + X_2 + \cdots + X_n$. The conditional expectation or best forecast of S_{n+k}, given $S_1, S_2, \ldots, S_n$ is

$$
\begin{aligned}
&E[S_{n+k}|S_1, S_2, \ldots, S_n] \\
&= E[S_n + (X_{n+1} + X_{n+2} + \cdots + X_{n+k})|S_1, \ldots, S_n] \\
&= E[S_n|S_1, \ldots, S_n] + E[X_{n+1} + X_{n+2} + \cdots + X_{n+k}|S_1, \ldots, S_n] \\
&= S_n + \underbrace{E[X_{n+1} + X_{n+2} + \cdots + X_{n+k}]}_{=0} = S_n.
\end{aligned}
$$

Hence, the sequence of partial sums, $\{S_i\}_{i=1}^{n}$ is a martingale.

Note that when $f(X_{n+1}) = X_{n+1}$, and $E_n[X_{n+1}] = g(X_n)$ for a function g, and if another function $f'(X_{n+1})$ exists such that $E_n[f'(X_{n+1})] \neq g'(X_n)$ for any g', then $\{X_n\}$ is not a Markov process. Basically, a Markov process is also describable by the condition $P(X_{n+1}|X_0, X_1, X_2, \ldots, X_n) = P(X_{n+1}|X_n)$, i.e., of past history, only the most recent matters, there being no memory of the distant past. A Markov process is usually also a martingale when we put $f(X_{n+1}) = X_{n+1}$ and if a function $g(X_n) = X_n$ can be found. However, if such a function $g(X_n) = X_n$ cannot be found in a Markov process where $E_n[X_{n+1}] = g(X_n)$, then the process is Markov but not a martingale. A martingale is not necessarily a Markov process since $E_n[X_{n+1}] = X_n \not\Rightarrow E_n[f(X_{n+1})] = g(X_n)$ for any f and for each f such that $\exists$ a g.

Now, let us do some conditional expectations on the binomial stock price process shown in Fig. 7.2. Let the stock price at $t = 0$ be S_0, so that the outcomes in the binomial process are all multiplied by S_0, i.e., $\tilde{S}_n = S_0\tilde{X}_n$. Consider

$$\begin{aligned} P(S_3 = S_0u^3|S_2 = S_0u^2) &= \frac{P(\{\omega_1\} \cap \{\omega_1 \cup \omega_2\})}{P(\{\omega_1 \cup \omega_2\})} \\ &\equiv \frac{P(\{UUU\} \cap \{UUU, UUD\})}{P(\{UUU, UUD\})} \\ &= \frac{P(\{UUU\})}{P(\{UUU, UUD\})} \\ &= \frac{p^3}{p^3 + p^2q} = \frac{p}{p+q} = p. \end{aligned}$$

Similarly, we can show that

$$\begin{aligned} P(S_3 = S_0u^2d|S_2 = S_0u^2) &= \frac{P(\{\omega_2\} \cap \{\omega_1 \cup \omega_2\})}{P(\{\omega_1 \cup \omega_2\})} \\ &\equiv \frac{P(\{UUD\} \cap \{UUU, UUD\})}{P(\{UUU, UUD\})} \\ &= \frac{P(\{UUD\})}{P(\{UUU, UUD\})} \\ &= \frac{p^2q}{p^3 + p^2q} = \frac{q}{p+q} = q. \end{aligned}$$

Clearly, the above conditional probability is the same as the marginal probability of a Bernoulli RV with an "up" state U of probability p and a "down" state D of probability q, which we shall denote as RV Z_3 with $P(Z_3 = uS_2) = p$ and $P(Z_3 = dS_2) = q = 1 - p$, starting at the UU node at $t = 2$. Moreover, if you work through all the conditional probabilities, you will find that indeed for the binomial stock price process

$$P(S_{n+1}|S_n) \equiv P(Z_{n+1}),$$

where $Z_{n+1} = uS_n$ with probability p and $Z_{n+1} = dS_n$ with probability q.

In the binomial lattice example in Fig. 7.2, the stock price at period $n+1$ conditional on S_n is

$$S_{n+1} = \begin{cases} uS_n, & \text{with probability } p \\ dS_n, & \text{with probability } 1-p \end{cases}.$$

The probabilities are dependent only on S_n and not $S_{n-1}, S_{n-2}, \ldots$, etc.

Then, for any function $f(\cdot)$, $E_n[f(S_{n+1})] \equiv E[f(S_{n+1})|S_n, S_{n-1}, \ldots, S_0] = pf(uS_n) + (1-p)f(dS_n) = g(S_n)$. The second last term on the RHS is simply a function of S_n. Hence S_{n+1} here is a Markov process.

However, we saw in Chap. 5 that for any n, under risk neutrality and hence the risk-neutral probability measure Q:

$$S_0 = \frac{E_0^Q(S_n)}{R^n}, \quad \text{or more generally,} \quad S_n = \frac{E_n^Q(S_{n+k})}{R^k},$$

where $k = 1, 2, \ldots$, and R is the risk-free interest factor. Clearly, the stock price $E^Q[S_{n+1}|S_n] = R\, S_n \neq S_n$ when $R \neq 1$. Hence S_{n+1} is not a martingale, though a Markov process.

Although S_{n+1} is not a martingale, its discounted value $(S_{n+1})/(R^{n+1})$, however, is a martingale, under the risk-neutral probability measure. This is because at each time n, the conditional expectation

$$\frac{E_n^Q[S_{n+1}]}{R} = S_n.$$

So

$$\frac{E_n^Q[S_{n+1}]}{R^{n+1}} = \frac{S_n}{R^n}.$$

Let $S_n' = S_n/R^n$ be the discounted stock price, then clearly

$$E_n^Q[S_{n+1}'] = S_n',$$

and so the discounted stock price is indeed a martingale.

More generally, we allow the risk-free interest factor to be a RV $\tilde{R}$ over time instead of a constant, i.e., R_0 is the interest factor at $t = 0$ over period $[0,1]$, R_1 is the interest factor at $t = 1$ over period $[1,2]$, $R_1 \neq R_0$, and so on.

Let M_n be a money account or deposit accumulating interest, starting from $M_0 = 1$, $M_1 = R_0$, $M_2 = R_0R_1, \ldots$, to $M_n = \prod_{t=0}^{n-1} R_t$. M_n is a predictable or previsible process in which its value is known at $n-1$. M_n is the money account value at time $t = n$.

Then

$$S_n = \frac{E_n^Q(S_{n+1})}{R_n} = E_n^Q\left(\frac{S_{n+1}}{R_n}\right).$$

Since at time $n-1$, $R_0, R_1, R_2, \ldots, R_{n-1}$ are known, therefore dividing both sides by $\prod_{t=0}^{n-1} R_t$ yields

$$\frac{S_n}{\prod_{t=0}^{n-1} R_t} = E_n^Q\left(\frac{S_{n+1}}{\prod_{t=0}^{n} R_t}\right), \quad \text{or}$$

$$\frac{S_n}{M_n} = E_n^Q\left(\frac{S_{n+1}}{M_{n+1}}\right). \tag{7.1}$$

Hence, the discounted stock price under stochastic risk-free interest rate is also a martingale.

In Eq. (7.1), there are really two prices happening at each time n. One is the stock price S_n, the other is the money account price or value M_n, and both are expressed in the same currency, e.g., \$. However, when we divide one by the other, as in this case, we obtain quotient $(S_n)/(M_n)$ which is really the amount of stock value expressed as number of units of the money account. Recall that one unit of money account starts with an initial deposit of \$1 at time $t = 0$. Hence, if $S_n = \$5$, and $M_n = \$1.25$, then $S_n/M_n = 4$ indicates that one stock at time $t = n$ is worth 4 units of the money account. Notice that we have shifted from expressing value of stock in dollars to its value in terms of a new numéraire, unit of money account. M_n could have been some other entities as long as these are predictable, i.e., known one period earlier. Thus, we see that although S_{n+1} is not a martingale when expressed in \$, by expressing it in a suitable numéraire, it can become a martingale.

If we define a risk-free discount bond with n period to maturity as having price B_n. Then, $B_n = 1/M_n$. At maturity, the discount bond is redeemed and the investor is paid the par value of \$1. Then, Eq. (7.1) can also be expressed as

$$B_n S_n = E_n^Q\left(B_{n+1} S_{n+1}\right).$$

Here, $Y_{n+1} \triangleq B_{n+1} S_{n+1}$ is a martingale.

Some Martingale Results

(2a) If X_t is a martingale, so $E(X_{t+1}|\mathcal{F}_t) = X_t$, $\forall t \leq T-1$, then $E(\triangle X_{t+1}|\mathcal{F}_t) = 0$, where $\triangle X_{t+1} = X_{t+1} - X_t$.

(2b) Suppose RV W_t is adapted to filtration $\{\mathcal{F}_t\}$, $E(W_t) = 0$, and $W_{t+1} - W_t$ is independent of $\mathcal{F}_t$. Then, $E(W_{t+1} - W_t|\mathcal{F}_t) = E(W_{t+1} - W_t) = 0$. Thus, $E(W_{t+1}|\mathcal{F}_t) = W_t$. Hence, W_t is a martingale.

(2c) If X_{t+1}, with X_0 arbitrarily fixed at 0, is a martingale adapted to $\{\mathcal{F}_{t+1}\}$, where filtration $\{\mathcal{F}_{t+1}\} = \sigma(X_{t+1}, X_t, \ldots)$, and ϕ_{t+1} is a bounded predictable process adapted to $\{\mathcal{F}_t\}$, $\forall 0 \leq t \leq T-1$, and suppose we define

$$Z_{t+1} \triangleq \sum_{i=1}^{t+1} \phi_i \triangle X_i = (\phi_1 \triangle X_1 + \phi_2 \triangle X_2 + \cdots + \phi_{t+1} \triangle X_{t+1}),$$

then $\triangle Z_{t+1} = Z_{t+1} - Z_t = \phi_{t+1} \triangle X_{t+1}$. Thus,

$$E(\triangle Z_{t+1} | \mathcal{F}_t) = E(\phi_{t+1} \triangle X_{t+1} | \mathcal{F}_t) = \phi_{t+1} E(\triangle X_{t+1} | \mathcal{F}_t) = 0,$$

for $0 \leq t \leq T$. The last term reflects the result in (2a). Hence, Z_{t+1} is a martingale.

Note that ϕ_{t+1} is known at t, whereas X_{t+1} will only be known at $t+1$. Z_{t+1} is called the martingale transform of X by ϕ. The continuous-time version is $dZ = \phi dX$ or integrating, $Z = \int \phi dX$, hence $\phi = dZ/dX$. This is a stability property showing that the transform of a martingale X under a predictable or previsible process ϕ preserves Z_{t+1} as a martingale.

Under continuous time, if X_t is a Wiener process or Brownian motion — also a martingale — and Z_t is any martingale w.r.t. filtration $\sigma(X_t)$, then Z_t can always be represented as

$$Z_t = Z_0 + \int_0^t \phi_u dX_u$$

where ϕ_u is a predictable process adapted to the same filtration $\sigma(X_t)$. This is called the Martingale Representation Theorem.

(2d) If process X is such that $E(X_{t+1} | \mathcal{F}_t) \leq (\geq) X_t \quad \forall t \leq T-1$, then X is called a supermartingale (submartingale).

Theorem 7.1 (Doob-Meyer Decomposition): *Any submartingale (supermartingale) $(X_t, \mathcal{F}_t)$, i.e., submartingale (supermartingale) process X_t adapted to $\mathcal{F}_t$, can be uniquely decomposed into the sum of a martingale $(M_t, \mathcal{F}_t)$, and an increasing (a decreasing) predictable process $(A_t, \mathcal{F}_{t-1})$, so*

$$X_t = M_t + A_t \quad \text{for } t \in [0, T]$$

with $A_0 = 0$ and $X_0 = M_0$. ■

Proof. Define adapted process $(M_t, \mathcal{F}_t) \ni$

$$M_{t+1} \triangleq M_t + X_{t+1} - E(X_{t+1} | \mathcal{F}_t),$$

and predictable process $(A_t, \mathcal{F}_{t-1}) \ni$

$$A_{t+1} \triangleq A_t - X_t + E(X_{t+1}|\mathcal{F}_t).$$

Then

$$E(M_{t+1}|\mathcal{F}_t) = E(M_t|\mathcal{F}_t) + E(X_{t+1}|\mathcal{F}_t) - E(X_{t+1}|\mathcal{F}_t) = M_t,$$

and M_t is thus a martingale. A_{t+1} is a predictable process since its components: A_t, X_t, and $E(X_{t+1}|\mathcal{F}_t)$ are all $\mathcal{F}_t$ measurable. Now, we have

$$\begin{aligned} M_{t+1} + A_{t+1} &= [M_t + X_{t+1} - E(X_{t+1}|\mathcal{F}_t)] + [A_t - X_t + E(X_{t+1}|\mathcal{F}_t)] \\ &= X_{t+1} + (M_t + A_t - X_t). \end{aligned}$$

Now, $M_0 + A_0 - X_0 = 0$, hence $M_1 + A_1 = X_1$ or $M_1 + A_1 - X_1 = 0$. In turn, $M_2 + A_2 = X_2$, and so on.

To show that the decomposition of X_t is unique, suppose another two processes satisfy

$$M'_{t+1} \triangleq M'_t + X_{t+1} - E(X_{t+1}|\mathcal{F}_t)$$

and

$$A'_{t+1} \triangleq A'_t - X_t + E(X_{t+1}|\mathcal{F}_t).$$

Similarly, as derived earlier, M'_t is a martingale, and

$$M'_{t+1} + A'_{t+1} = X_{t+1} = M_{t+1} + A_{t+1},$$

for $t \geq 0$.

From $M'_{t+1} + A'_{t+1} = M_{t+1} + A_{t+1}, \quad t \geq 0$, by taking conditional expectation on $\mathcal{F}_t$,

$$M'_t + E(A'_{t+1}|\mathcal{F}_t) = M_t + E(A_{t+1}|\mathcal{F}_t).$$

Starting at $t = 0$, since $M'_0 = X_0 = M_0$,

$$E(A'_{t+1}|\mathcal{F}_t) = A'_{t+1} = A_{t+1} = E(A_{t+1}|\mathcal{F}_t),$$

hence $A'_1 = A_1$. In turn, $M'_1 = M_1$, then $A'_2 = A_2$, and so on. Hence uniqueness.

Now $M_{t+1} + A_{t+1} = X_{t+1}$, hence

$$M_t + E(A_{t+1}|\mathcal{F}_t) = E(X_{t+1}|\mathcal{F}_t).$$

Thus

$$M_t + E(A_{t+1}|\mathcal{F}_t) - (M_t + A_t) = E(X_{t+1}|\mathcal{F}_t) - X_t,$$

or

$$E(A_{t+1}|\mathcal{F}_t) - A_t = E(X_{t+1}|\mathcal{F}_t) - X_t.$$

The LHS is $E(A_{t+1}|\mathcal{F}_t) - A_t = A_{t+1} - A_t \geq (\leq) 0$ which is an increasing (a decreasing) predictable process. Therefore, the RHS $E(X_{t+1}|\mathcal{F}_t) \geq (\leq) X_t$. Hence, X_t is a submartingale (supermartingale). ■

A local martingale is a stochastic process X_t adapted to filtered space $(\Omega, \mathcal{F}, \{\mathcal{F}_t\}, \mathcal{P})$ satisfying a localized version of the martingale property in the sense that there is a sequence of increasing stopping times τ approaching $+\infty$ such that $X_{t \wedge \tau}$ is a martingale.

Every martingale is a local martingale, but the converse is generally not true. There are continuous local martingales that are uniformly integrable, but are not martingales. However, a bounded local martingale is a martingale.

A continuous semimartingale is a continuous stochastic process that can be expressed as $X_t = X_0 + M_t + A_t$ where M_t is a continuous local martingale and A_t is a continuous adapted process of finite variation. Thus, almost all the no-arbitrage results we shall see later in the chapter in terms of martingales carry over as well in terms of semimartingales. Semimartingales is about the most general class of Itô stochastic processes that satisfy property (2c) preserving martingale transforms.

Just as in martingales, Markov stochastic processes that are stopped are called strong Markov processes and are restrictions to (or special cases of) the Markov processes. In this case, strong Markov processes are Markov, but Markov processes may not be strong Markov processes.

7.4. Multi-period Trading Strategies and Arbitrage

In Chap. 3, we showed how taking the binomial asset pricing to convergence and taking the expectation on a convergent lognormal distribution leads to the Black–Scholes option formula. A critical ingredient in those processes is the assumption of risk neutrality. In reality, investors are typically risk averse and would not be indifferent between a certainty value of \$1M and a gamble returning \$2M with 50% probability and \$0 with the other 50% probability. Why is it justifiable to use the risk neutrality assumption there and under more general processes such as a diffusion process? There it is seen that the empirical probability measures do not feature in the pricing of securities (it does feature if we had instead modeled differently via an equilibrium risk-premia approach like in CAPM). But via the no-arbitrage equilibrium approach, existence of a risk-neutral probability measure is assured in pricing securities as if investors are risk neutral. The no-arbitrage result in Chap. 5 was obtained in a one-period setup. Now, we look at the more general multi-period setting.

Let S_t be a vector of security or asset prices at time t, and X_{t+1} be a vector of the number of shares held in each security in period $(t, t+1]$. At $t = 0$, security prices S_0 are known, and then in the next instance the investor decides his trading strategy (selects his portfolio of securities) X_1 (strictly speaking, this should be X_{0+}, but dealing with discrete time models has this drawback of indexing by $t = 1$ to denote that the event happens after $t = 0$, and more importantly to denote a decision with given information at $t = 0$). Hence, it should be clear henceforth that in the discrete framework, asset prices S_t are observable at time point t, whereas numbers of shares bought (sold if it is a negative number) on the assets, X_{t+1} is a decision and action taken an instance[3] after obtaining information on S_t, and thus becoming observable an instance after time t, but before S_{t+1} is observed. Our model assumes actions X_{t+1} do not affect the formation of next period asset prices S_{t+1}. Implicitly, asset prices are determined by some exogenous processes.

The investor's initial portfolio market value is $V_0 = X_1 \cdot S_0$, which we assume is the investor's initial endowment, where "$\cdot$" refers to a dot product or inner product in vector algebra or $X_1^T S_0$ in matrix format. The investor then holds the portfolio till $t = 1$ when S_1 is realized. Then, his/her portfolio value becomes $V_1(X_1) = X_1 \cdot S_1$. In the next instance, the investor then readjusts his/her portfolio to X_2 and holds it to $t = 2$.

Hence, the price process S_t is a stochastic process that is adapted to a filtration, in the sense that S_t is $\mathcal{F}_t$-measurable, for $t = 0, 1, 2, \ldots$, etc. where $\mathcal{F}_t$ is σ-generated by RV S_t, i.e., $\mathcal{F}_t = \sigma(S_t)$. Since the trading strategy or vector stochastic process $\{X_{t+1} : t \geq 0\}$ is a process dependent on given information in the price process $\{S_t : t \geq 0\}$, that is adapted

[3]Some other textbooks use instead the description that X_{t+1} is determined before $t+1$, after t, and the portfolio must be held until after the announcement of S_{t+1}, without being explicit on the interval between the point of decision and the next price revelation. Of course, in continuous time, when the time interval shrinks to zero, the definitions are the same. However, in discrete time, deciding X_{t+1} at t_+ ensures portfolio purchased at t_+ with cost $X_{t+1}S_t$ (inner product) bears interest r_t till $t+1$ but also collects dividends in $(t, t+1]$.

to $\mathcal{F}_t$, then for each t, the way we define the processes of price revelation followed instantly next by portfolio choice or rebalancing, X_{t+1} is adapted to $\mathcal{F}_t \equiv \sigma(S_t)$ (assuming X_{t+1} is not a random decision, but one which is a proper function of given S_t). Hence, by the usual definition, X_{t+1} is predictable or previsible[4] based on $\sigma(S_t)$. The diagram below illustrates the decision and revelation sequence.

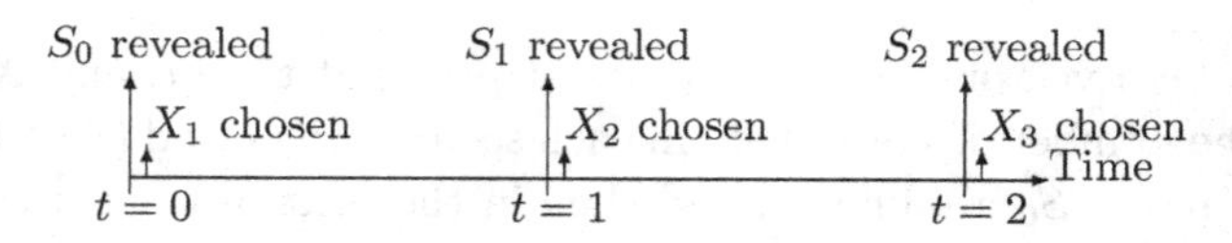

To be more precise in our analyses, let there be $N + 1$ securities each with traded market prices s_t^i where superscript i refers to the ith security and subscript t refers to the time in which it is traded. The 0th security is a risk-free bond with $s_0^0 = 1$ and per period interest (factor) R_t, so $s_1^0 = R_0$, $s_t^0 = \prod_{j=0}^{t-1} R_j$.

An investor's portfolio at time $(t-1)_+$ or more clearly within time $(t-1, t]$, is given by vector of $(N+1) \times 1$ security shares: $X_t = (x_t^0, x_t^1, x_t^2, \ldots, x_t^N)$ where again the superscript denotes association with the ith security, and not power per se.

The security price vector at t is an $(N+1) \times 1$ vector $S_t = (s_t^0, s_t^1, s_t^2, \ldots, s_t^N)$. The value of the portfolio at time t is

$$V_t(X_t) = X_t \cdot S_t = \sum_{i=0}^{N} x_t^i s_t^i, \quad \text{for } t \geq 1.$$

For a concrete numerical example, suppose there are only two securities A and B. At $t = 0$, $s_0^A = \$10$ and $s_0^B = \$5$. An investor holds 10 A shares and 20 B shares, so his/her initial endowment is $V_0 = 10 \times \$10 + 20 \times \$5 = \$200$ (during $(0, 1)$).

At $t = 1$, the prices become $s_1^A = \$5$ and $s_1^B = \$10$. His/her portfolio value at instant $t = 1$ becomes

$$V_1 = 10 \times \$5 + 20 \times \$10 = \$250.$$

His/her capital thus increased from V_0 of \$200 to V_1 of \$250 by a value gain of \$50.

At $t = 1_+$, he/she adjusts his/her portfolio to c shares in A and d shares in B. In order that it is a self-financing trading strategy, there must

[4]As in the numéraire case seen in the previous section, $X_{t+1} E_t(S_{t+1}) = E_t(X_{t+1} S_{t+1})$ since X_{t+1} is known at t.

be no cash withdrawal or fresh cash input, so he/she must use the value V_1 of \$250 to redistribute his/her share holdings. Thus

$$c \times \$5 + d \times \$10 = \$250.$$

Suppose $c = 6$ and $d = 22$. Then, his/her trading portfolio has changed from $X_1 = (10, 20)$ to $X_2 = (6, 22)$ adapted to σ-algebra generated by price information $S_0 = (\$10, \$5)$ and $S_1 = (\$5, \$10)$, respectively.

Thus, a self-financing trading strategy or stochastic process $\{X_t\}_{t=1,2,\cdots}$ has value changes in the portfolio coming only from net gains or losses realized on the investments with no cash withdrawal or fresh cash input, and is characterized by

$$X_{t+1} \cdot S_t = X_t \cdot S_t, \quad \text{for } t = 1, 2, \ldots \text{ etc}$$

or

$$(X_{t+1} - X_t) \cdot S_t = \triangle X_{t+1} \cdot S_t = 0, \quad \text{for } t = 1, 2, \ldots \text{ etc} \tag{7.2}$$

Self-financing is characterized by orthogonality of vector of portfolio adjustment to prior price vector. In addition, value change process is

$$\begin{aligned} \triangle V_{t+1} &= X_{t+1} \cdot S_{t+1} - X_t \cdot S_t \\ &= X_{t+1} \cdot S_{t+1} - X_{t+1} \cdot S_t \\ &= X_{t+1} \cdot \triangle S_{t+1} \end{aligned} \tag{7.3}$$

where we apply self-financing in the process.

A self-financing trading strategy $\Leftrightarrow$ Eq. (7.2) $\Leftrightarrow$ Eq. (7.3). Now

$$\begin{aligned} V_{t+1} = {} & (V_{t+1} - V_t) + (V_t - V_{t-1}) + (V_{t-1} - V_{t-2}) \\ & + \cdots + (V_2 - V_1) + (V_1 - V_0) + V_0 \end{aligned}$$

where we may interpret V_0 as a given initial endowment at $t = 0$. $V_0 = X_1 \cdot S_0$.

Since the value gain at $t+1$ is $\triangle V_{t+1}$, the cumulative gains at $t+1$ is defined as

$$G_{t+1} = \triangle V_{t+1} + \triangle V_t + \cdots + \triangle V_2 + \triangle V_1 = V_{t+1} - V_0. \tag{7.4}$$

Under self-financing constraint (7.2), this cumulative gain can be expressed as

$$G_{t+1} = \sum_{s=1}^{t+1} X_s \cdot \triangle S_s \tag{7.5}$$

via (7.3).

It is interesting to note that according to property (2c) of the martingale results, G_{t+1} is a martingale if S_{t+1} is a martingale, since X_{t+1} is predictable. Further, even if S_{t+1} itself is not a martingale, from the numéraire result earlier, it is possible to transform it for a class of stochastic processes $\{S_{t+1}\}$, into a martingale by expressing it suitably in another numéraire such as a discounted process $S^*_{t+1} = S_{t+1}/M_{t+1}$.

Then, from Eqs. (7.3) and (7.5), we obtain $\triangle V^*_{t+1} = X_{t+1} \cdot \triangle S^*_{t+1}$, and $G^*_{t+1} = \sum_{i=1}^{t+1} X_i \cdot \triangle S^*_i$, where V^*_{t+1} and G^*_{t+1} are martingales.

Suppose the numéraire (unit measure of transaction) is not \$ but a good (e.g., a gram of salt) that has a traded price \$ Z_t per unit of the numéraire good at time t. (Non-traded numéraires are of no use here, e.g., a personal effect.) Then, a share value \$ S_t is now worth $(S_t)/(Z_t)$ number of units of the numéraire good. We shall just say its value is now $(S_t)/(Z_t)$ in terms of the new numéraire. If $Z_t = 1$ for all t, then we get back the \$ as numéraire. Note that the "numéraire process" $Z_t > 0$ is a single-dimension RV. Any numéraire will not affect the self-financing characterization in (7.2), viz.

$$\triangle X_{t+1} \cdot \left(\frac{S_t}{Z_t}\right) = \frac{1}{Z_t}(\triangle X_{t+1} \cdot S_t) = \frac{1}{Z_t} \times 0 = 0,$$

iff self-financing applies to the old numéraire. Hence, the self-financing trading strategy or the decisions in X_t is invariant to (does not change with) any numéraire.

The above can be stated as a lemma:

Lemma 7.1 *Let Z_t be a numéraire (dollar price of a numéraire good). A trading strategy $\tilde{X}$ is self-financing w.r.t. S_t iff $\tilde{X}$ is self-financing w.r.t. $(S_t)/(Z_t)$.* ■

In finance modeling, it is usually convenient to use the economy-wide risk-free bond as the numéraire where this bond is characterized by prices $s^0_0 = 1$, $s^0_1 = R_0$, $s^0_2 = s^0_1 R_1$, and so on. It is equivalent to a money account or deposit process of putting a dollar $M_0 = 1$ in a risk-free deposit, and having the account grow to $M_1 = R_0$, then $M_2 = M_1 R_1$, and so on. Hence, the bond price under the new numéraire is always $s^0_t/s^0_t = 1$, which makes for slightly easier presentation with one variable less in R_t, and share prices are s^i_t/s^0_t for $i = 1, 2, \ldots, N$. Hence, instead of using $S_t = (s^0_t, s^1_t, \ldots, s^N_t)$, we can use

$$S_t^* = (1, s_t^{1*}, s_t^{2*}, \ldots, s_t^{N*}),$$

where $s_t^{j*} = s_t^j / s_t^0$ is a discounted price process.

Similarly, $V_t^* = X_t \cdot S_t^*$ is a discounted value process, and $G_t^* = V_t^* - V_0^*$ is a discounted gains process. These discounted processes are in terms of the bond or money account as the numéraire. As a corollary to Lemma 7.1, we have the following.

Corollary 7.1 *A trading strategy $\tilde{X}$ is self-financing w.r.t. S_t iff $\tilde{X}$ is self-financing w.r.t. $\frac{S_t}{M_t} \equiv S_t^*$, where M_t is the money account at t.* ■

It is worthwhile to emphasize the meaning of self-financing: that once an initial trading position is started, there is neither any fresh injection of new capital nor withdrawal of profit if any, till the end of the multi-period trading at time T.

In the literature, the context in which the prices and trading concepts in the multi-period model in this section is constructed is called the market model. Specifically, the market model is described by a filtered probability space $(\Omega, \mathcal{F}_t, \{\mathcal{F}_t\}, \mathcal{P})$, and a time sequence in which trading or portfolio decisions are made, $\mathcal{T}$, e.g., $0 \leq t \leq T$, and the definition of a sequence of RV's S_t, e.g., asset prices, that are adapted to the filtration $\{\mathcal{F}_t\} = \sigma(S_t, S_{t-1}, \ldots, S_0)$. Thus, the market model can be indicated in short form as a space $(\Omega, \mathcal{F}_t, \{\mathcal{F}_t\}, \mathcal{P}, \mathcal{T}, \{S_t\})$. Sometimes $\mathcal{T} \times \{S_t(\omega)\}$ is called the marketed space, or collection of points $S(t, \omega)$ in $\mathcal{T} \times \{S_t(\omega)\}$.

Definition 7.2 An arbitrage opportunity is defined as some self-financing trading strategy in an admissible class Θ where trading is up to finite time $T \geq t > 0$, and $V_t \geq B$ for some lower bound $-\infty < B \leq 0$, and

(3a) $V_0 = 0$
(3b) $V_T \geq 0$
(3c) $P(V_T > 0) > 0$. ■

This definition of admission or not of an arbitrage opportunity is similar to the arbitrage condition (1a) of Chap. 5. In the multi-period model here, when there is self-financing constraint, it is like a one-period model with initial condition $V_0 = 0$ and the final condition $V_T > 0$ with at least one state having strictly positive payoff.

Condition (1b) of Chap. 5 is also similar in this context here by considering that if initial portfolio cost is $V_0 < 0$, i.e., being paid to own the portfolio, then one could use the paid money to buy some securities, whereby future portfolio value $V_T > 0$ with non-zero probability. Hence, we shall only deal with arbitrage as represented by trading strategies in Θ, i.e., satisfying equation (7.2), and which in addition, have properties $V_0 = 0$, $V_T \geq 0$, and $P(V_T > 0) > 0$. The properties (3b) and (3c) $\Rightarrow E(V_T) > 0$. On the other hand, (3b) and $E(V_T) > 0 \Rightarrow$ (3c). Hence, we may replace (3b) and (3c) with (3b) and (3d): $E(V_T) > 0$.

Since $V_0^* = V_0/s_0^0 = 0$, $V_T^* = V_T/s_T^0 \geq 0$ (as $s_T^0 > 0$), and $E(V_T^*) = E(V_T/s_T^0) > 0$ (as $s_T^0 > 0$), then some self-financing strategy Θ is an arbitrage opportunity iff they satisfy

(4a) $V_0^* = 0$;
(4b) $V_T^* \geq 0$; and
(4c) $E(V_T^*) > 0$.

This uses Lemma 7.1 where self-financing applies similarly to the discounted price and value processes. As long as the trading does not end in bankruptcy, self-financing makes the intermediate condition of $V_t \geq B$ irrelevant in the final outcome at T.

An alternative definition of an arbitrage opportunity is:

Definition 7.3 An arbitrage opportunity is defined as some self-financing trading strategy Θ such that:

(5a) $V_0^* = 0$
(5b) $G_T^* \geq 0$
(5c) $E(G_T^*) > 0$. ■

The above definition is equivalent by using the definition $G_T^* = V_T^* - V_0^*$. It can be seen that for a given $V_0^* = 0$, then $G_T^* = V_T^*$, (4b) implies (5b), and (4c) implies (5c).

Now, we define what is a martingale measure, and explain what are equivalent martingale measures.

Definition 7.4 A martingale measure is a probability measure Q on a stochastic process W_t such that $(\Omega, \mathcal{F}_t, \{\mathcal{F}_t\}, Q)$ is a filtered probability space, and W_t is a martingale under Q, i.e., $E_Q(W_{t+1}|\mathcal{F}_t) = W_t$. ■

For notation, Q is used, instead of the empirical or physical measure $\mathcal{P}$, for the martingale measure. Q is sometimes also called the risk-neutral measure, because when W_t is the discounted security price S_t/M_t shown earlier, then being a martingale means $E^Q(S_{t+1}/M_{t+1}|\mathcal{F}_t) = S_t/M_t$, where superscript Q denotes integration or expectation operator is taken w.r.t. probability measure Q. Since money account is predictable or previsible, $E^Q(S_{t+1}|\mathcal{F}_t) = (M_{t+1}/M_t)S_t = (1 + r_t)S_t$ which indicates risk neutrality.

More specifically in our market model context, we have:

Definition 7.5 A martingale measure in the market model

$$(\Omega, \mathcal{F}_t, \{\mathcal{F}_t\}, \mathcal{P}, \mathcal{T}, \{S_t\})$$

is a probability measure Q on the discounted security prices $S^*_{t+1} \triangleq S_{t+1}/M_{t+1}$ (where M_t is the money account) such that S^*_{t+1} is a martingale under Q, i.e., $E_Q(S^*_{t+1}|\mathcal{F}_t) = S^*_t$. Sometimes, S^*_{t+1} or the discounted security price process is called the Q-martingale. ■

Definition 7.6 A probability measure Q is absolutely continuous w.r.t. another probability measure P if for each P-measurable event $A \in \mathcal{F}_t$, $P_P(A) = 0$ implies $P_Q(A) = 0$, where $P(\cdot)$ denotes probability. Then measure Q is said to be dominated by measure P, and we write $Q \ll P$. If both $Q \ll P$ and $P \ll Q$, then probability measure Q is said to be equivalent to probability measure P. ■

As a simple example, if RV X takes values $2, 4, -1$, and under a probability (measure) P, $P_P(2) = 0$, $P_P(4) = 0.8$, and $P_P(-1) = 0.2$, while under probability Q, $P_Q(2) = 0.5$, $P_Q(4) = 0.5$, and $P_Q(-1) = 0$, then Q is not absolutely continuous w.r.t. P. Neither is P absolutely continuous w.r.t. Q. However, if $P_Q(2) = 0$, $P_Q(4) = 1$, and $P_Q(-1) = 0$, then Q is absolutely continuous w.r.t. P, but the converse is not true. If $P_Q(2) = 0$, $P_Q(4) = 0.6$, and $P_Q(-1) = 0.4$, then P and Q-measures are equivalent. In terms of notation, sometimes, $P_P(A)$ and $P_Q(A)$ are written as $P(A)$ and $Q(A)$, respectively where P, Q now denote the respective probability measures. We use the latter notations below.

Any event A is typically described w.r.t. a particular $\mathcal{F}_t$ at time t, but in what follows we drop the subscripts to the field where it is understood that the event is adapted to the field at that time. It is also assumed that there exists a Q-measure that is equivalent to the physical or empirical measure P almost everywhere as in Definition 7.6. Therefore, the probability $P(V_T > 0) > 0$ in Definition 7.2, and expectation $E(G^*_T) > 0$ in Definition 7.3 can be defined w.r.t. probability measure Q as well. Henceforth, we treat all the entities as measurable w.r.t. Q.

Theorem 7.2 *For a probability space $(\Omega, \mathcal{F}, P)$, if there exists a strictly positive RV Z (>0 a.s.) with $E_P(Z) = 1$, we can define a new probability measure Q by*

$$Q(A) = \int_A Z(\omega)\, dP(\omega) \quad \textit{for all events } A \in \mathcal{F}.$$

Then, ratio $\frac{dQ}{dP}(\omega) = Z(\omega) > 0$ is called the Radon-Nikodým derivative of Q w.r.t. P.
Moreover, if X is another RV, then

$$E_Q(X) \equiv \int_{\omega\in\Omega} X\, dQ = \int_{\omega\in\Omega} X\, Z\, dP = E_P(XZ).$$

Since $\frac{dQ}{dP}(\omega) = Z(\omega) > 0$ a.s., then $\exists\ \frac{dP}{dQ} = \frac{1}{Z}$ for all $\omega \in \Omega$, and

$$E_P(X) \equiv \int_{\omega\in\Omega} X\, dP = \int_{\omega\in\Omega} \frac{X}{Z}\, dQ = E_Q\left(\frac{X}{Z}\right).$$

Moreover, $Q \ll P$ and $P \ll Q$, so the measures Q and P are equivalent.

■

The converse of the above, when P and Q are equivalent probability measures, is that there exists a RV $Z > 0$ a.s. such that $E_p(Z) = 1$, and $Q(A) = \int_A Z(\omega)\ dP(\omega)$ for all events $A \in \mathcal{F}$. This is called the Radon-Nikodým theorem. The proof is technically more difficult, so it is left out here. ■

In the above theorem, condition $E_P(Z) = 1$ is required so that $Q(\Omega) = \int_\Omega Z(\omega) dP(\omega) = E_P(Z) = 1$ defines a proper probability measure Q that adds to one. Recall from Chap. 6 Section 1 that in discrete states of the world in a single period, the state-price kernel or deflator or density has the same form as the Radon-Nikodým derivative, i.e., $\frac{Q(A)}{P(A)} = Z(A) > 0$.

Now, we state what is known as the first fundamental theorem of asset pricing.

Theorem 7.3 **(First fundamental theorem of asset pricing):** *There is no arbitrage opportunity ($\forall t = 1, 2, \ldots, T$) in a market model $(\Omega, \mathcal{F}_t, \{\mathcal{F}_t\}, \mathcal{P}, \mathcal{T}, \{S_t\})$ iff there exists an equivalent*[5] *martingale probability measure.* ■

[5]The origination of this term "equivalent martingale measure" in this context refers to the equivalence with the physical measure P as expressed in Shiryaev, AN (1999). *Essentials of Stochastic Finance*. Singapore: World Scientific.

Proof. If there exists a market martingale probability measure, then by Definition 7.5, let the martingale measure be Q on the discounted security prices $S^*_{t+1} \triangleq S_{t+1}/M_{t+1}$ s.t. S^*_{t+1} is a martingale, i.e., $E_Q(S^*_{t+1}|\mathcal{F}_t) = S^*_t$. (If there are other equivalent measures that satisfy the condition that S^*_{t+1} is a martingale, then these are non-unique equivalent martingale measures.) From Eq. (7.3) (self-financing is assumed), using the discounted version, then $\triangle V^*_{t+1} = X_{t+1} \cdot \triangle S^*_{t+1}$. Thus, V^*_{t+1} is a martingale.

Employing $V^*_{t+1} - V^*_0 = G^*_{t+1}$, $\forall t = 0, 1, 2, \ldots, T-1$, we have

$$E_Q(G^*_T|\mathcal{F}_0) = E_Q(V^*_T - V^*_0|\mathcal{F}_0) = V^*_0 - V^*_0 = 0.$$

By Definition 7.3, since $E_Q(G^*_T) \not> 0$, given $V^*_0 = 0$, there is no arbitrage opportunity. This applies $\forall t = 1, 2, \ldots, T$.

For the converse, by Definition 7.3, if there is no arbitrage opportunity $\forall t = 1, 2, \ldots, T$, then for $V^*_0 = 0$, $E_Q(G^*_t) = 0$ for every $t = 1, 2, \ldots, T$. Since $G^*_{t+1} = \triangle V^*_{t+1} + \triangle V^*_t + \ldots + \triangle V^*_1$, then $E_Q(\triangle V^*_{t+1}) = 0$, $\forall t$. By the self-financing equation (7.2) and thus (7.3), therefore $E_Q\left(X_{t+1} \cdot \triangle S^*_{t+1}\right) = 0$, $\forall t$. Hence, there exists the Q-measure according to Definition 7.5. ■

Theorem 7.3 can also be restated as: *There is no arbitrage opportunity in a market model iff the discounted security price process is a martingale under a probability measure Q.* ■

Now, suppose there is a contingent claim (or multi-period derivative such as a European option) which has terminal value at T of $\tilde{C}_T(\omega \in \Omega) \geq 0$, with $P(\tilde{C}_T > 0) > 0$. In other words, $P(\{\omega : C_T(\omega) > 0\}) > 0$. In terms of the money account numéraire, $C_T/M_T = C^*_T \geq 0$ and has $P(\tilde{C}^*_T > 0) > 0$.

This contingent claim is said to be attainable if there is a self-financing trading strategy X_t on the (underlying) securities, starting with $V^*_0 = C^*_0$ that is not necessarily zero, that generates or produces $V^*_T = C^*_T$, i.e., $V^*_T(\omega) = C^*_T(\omega)$ for all $\omega \in \Omega$. The trading strategy $X_{t+1} \in \mathcal{F}_t$ is also called a replicating or generating strategy for $\tilde{C}^*_T$. The binomial tree example in Chap. 3 shows it is possible to replicate a derivative security exactly by rebalancing a stock and a bond to attain the final payoff outcomes at T.

Lemma 7.2 *Generating or replicating self-financing strategies are unique for a given attainable $C^*_T(S^*_T)$. In particular, the cost of starting the self-financing strategy to end with RV C^*_T is unique and the value process V^*_t is unique.* ■

First, note that here we are considering only one given derivative, a RV C_T at T. Second, we can always express in terms of the numéraire M_t, and thus consider it as a given RV $C_T^* = C_T/M_T$, as stated in Lemma 7.2. Henceforth, we shall prove using the values under the money account numéraire. Third, Lemma 7.2 does not consider whether or not there is arbitrage opportunity per se.

At this point, it is useful to recapitulate the trading process diagrammatically as follows. The bold prints denote where either the price or the decision shares or the value is instantly updated at that time point on the row.

Time u	Security prices S_u^*	Number of shares X_u	Portfolio value $V_u^* = X_u S_u^*$	Value gain $G_u^* = V_u^* - V_{u-1}^*$
t	$\mathbf{S_t^*}$	X_t	$\mathbf{X_t \cdot S_t^*}$	...
t_+	S_t^*	$\mathbf{X_{t+1}}$	$X_{t+1} \cdot S_t^*$	0
$t+1$	$\mathbf{S_{t+1}^*}$	X_{t+1}	$\mathbf{X_{t+1} \cdot S_{t+1}^*}$	$\mathbf{V_{t+1}^* - V_t^*}$
$(t+1)_+$	S_{t+1}^*	$\mathbf{X_{t+2}}$	$X_{t+2} \cdot S_{t+1}^*$	0
$t+2$	$\mathbf{S_{t+2}^*}$	X_{t+2}	$\mathbf{X_{t+2} \cdot S_{t+2}^*}$	$\mathbf{V_{t+2}^* - V_{t+1}^*}$
$(t+2)_+$	S_{t+2}^*	$\mathbf{X_{t+3}}$	$X_{t+3} \cdot S_{t+2}^*$	0
$t+3$	$\mathbf{S_{t+3}^*}$	X_{t+3}	$\mathbf{X_{t+3} \cdot S_{t+3}^*}$	$\mathbf{V_{t+3}^* - V_{t+2}^*}$
$\vdots$	$\vdots$	$\vdots$	$\vdots$	$\vdots$
$\vdots$	$\vdots$	$\vdots$	$\vdots$	$\vdots$
$T-1$	$\mathbf{S_{T-1}^*}$	X_{T-1}	$\mathbf{X_{T-1} \cdot S_{T-1}^*}$	$\mathbf{V_{T-1}^* - V_{T-2}^*}$
$(T-1)_+$	S_{T-1}^*	$\mathbf{X_T}$	$X_T \cdot S_{T-1}^*$	0
T	$\mathbf{S_T^*}$	X_T	$\mathbf{X_T \cdot S_T^*}$	$\mathbf{V_T^* - V_{T-1}^*}$

The proof of Lemma 7.2 is as follows.

Proof. The proof applies similarly for whatever numéraire. The numéraire w.r.t. the money account is applied here. Suppose replicating self-financing strategies X and $Y \in \Theta$ with starting values V_0^X and V_0^Y respectively attain $V_T^X = V_T^Y = C_T^*$. Working backward from time T

$$X_T \cdot S_T^* = V_T^X = V_T^Y = Y_T \cdot S_T^*.$$

Hence, $X_T = Y_T$, since this applies for arbitrary positive S_T^* that is strictly positive in at least one state, $\omega \in \Omega$.

From the self-financing equation (7.2), for strategies X and Y

$$X_T \cdot S_{T-1}^* = X_{T-1} \cdot S_{T-1}^* \tag{7.6}$$

$$Y_T \cdot S_{T-1}^* = Y_{T-1} \cdot S_{T-1}^*. \tag{7.7}$$

Use $X_T = Y_T$ on the LHS of (7.6) and (7.7), then

$$Y_{T-1} \cdot S^*_{T-1} = X_{T-1} \cdot S^*_{T-1}.$$

Hence, $X_{T-1} = Y_{T-1}$.

In the same way working backward, we can show that for all $t = 1, 2, \ldots, T$, $X_t = Y_t$.

Now, at the start when $t = 0_+$, the starting value of strategy X is $X_1 \cdot S^*_0 = V^X_0$, while that of strategy Y is $Y_1 \cdot S^*_0 = V^Y_0$. Since $X_1 = Y_1$, then $V^X_0 = V^Y_0$. Thus, we show that indeed the generating strategy for C^*_T is unique and has a unique starting cost. Since for every t, $X_t = Y_t$, so $V^X_t = X_t \cdot S^*_t = Y_t \cdot S^*_t = V^Y_t$, then the value process $\{V_t\}$ is also unique.

■

For a discrete state market with N states $i = 1, 2, \ldots, N$, and N uniquely priced state claims, suppose the ith claim has terminal payoff $C_T(S_T(\omega))$ bounded, $j > 0$ for $j = i$ and 0 otherwise, for every $\omega \in \Omega$, then the market model is said to be complete. Thus, every derivative security $\tilde{C}_T$ is attainable, or can be perfectly hedged without risk. If an issuer sells such a derivative, then at maturity time T, the issuer is exposed to claim $\tilde{C}_T$ against him or her. To perfectly hedge this exposure means that the issuer can employ a self-financing trading strategy in Θ such that the final payoff is $\tilde{C}_T$ in order to nullify the exposure, and according to Lemma 7.2, this strategy is unique.

Theorem 7.4 (Second fundamental theorem of asset pricing): *If a market model* $(\Omega, \mathcal{F}_t, \{\mathcal{F}_t\}, \mathcal{P}, \mathcal{T}, \{S_t\})$ *has a martingale probability measure, then the market is complete iff the martingale probability measure is unique.* ■

Proof. Given that the market model has a martingale measure, say, Q. Then, by Definition 7.5, there is a self-financing strategy such that $C^*_T = X_T \cdot S^*_T$, $C^*_0 = X_0 \cdot S^*_0$, and

$$E_Q(X_T \cdot S^*_T | \mathcal{F}_0) = X_0 S^*_0 \equiv V^*_0 = C^*_0.$$

By Lemma 7.2, this self-financing strategy that replicates $X_T \cdot S^*_T$ is unique, and $C^*_0 > 0$ is also unique. Thus for N states of the world,

$$\sum_{i=1}^{N} \big(X_T \cdot S^*_T\big)(\omega_i)\ P_Q(\omega_i) = C^*_0.$$

If the market is complete, it means there are N state contingent claims at T each with a terminal payoff $C^*_T(\omega_j) = \big(X_T \cdot S^*_T\big)(\omega_j) > 0$, and 0 otherwise for $\omega \neq \omega_j$. Then,

$$C_T^*(\omega_j)\ P_Q(\omega_j) = C_0^*(\omega_j),$$

for the jth claim. The RHS is unique, and thus equivalent martingale probability measure (EMM) $P_Q(\omega_j)$ is unique for all j.

Conversely, if Q is unique which means $P_Q(\omega_j)$ is unique for all $j = 1, 2, \ldots, N$, then any j^{th} state claim at T with terminal payoff $C_T^*(\omega_j) > 0$, and 0 otherwise for $\omega \neq \omega_j$ can be priced as

$$C_T^*(\omega_j)\ P_Q(\omega_j) = C_0^*(\omega_j).$$

There is a unique self-financing strategy to replicate $C_T^*(\omega_j)$ with price $C_0^*(\omega_j)$. Thus all state claim prices $C_0^*(\omega_j)$ are unique, and the market is complete. ∎

So far our treatment of securities including stocks do not explicitly consider the situation where stocks do issue dividends. Suppose the security price P_t^* at time t is ex-dividend price and the dividend D_t^* is also issued at time t. In terms of existing notation, then the cum-dividend value of the security as revealed at t is

$$S_t^* = P_t^* + D_t^*.$$

Note that we work in terms of prices that are denominated with the money account as numéraire.

If at $(t-1)_+$, numbers of securities X_t are purchased, then value of portfolio at t (cum dividend) is

$$V_t^* = X_t \cdot (P_t^* + D_t^*).$$

In terms of self-financing, adjustments in portfolio at t_+ is

$$X_{t+1} \cdot P_t^* = X_t \cdot \left(P_t^* + D_t^*\right). \tag{7.8}$$

In other words, monies from the dividends are reinvested in the securities at their ex-dividend prices (usually ex-dividend prices are lesser than cum-dividend prices by the amount of the dividends if there is no tax effect).

The (portfolio) value change at $t+1$ is

$$\begin{aligned}\triangle V_{t+1}^* &= X_{t+1} \cdot (P_{t+1}^* + D_{t+1}^*) - X_t \cdot (P_t^* + D_t^*)\\ &= X_{t+1} \cdot (P_{t+1}^* + D_{t+1}^*) - X_{t+1} \cdot P_t^*\\ &\qquad \text{(the last term from Eq. (7.8))}\\ &= X_{t+1} \cdot (\triangle P_{t+1}^* + D_{t+1}^*).\end{aligned}$$

Therefore, the value change process is: $\triangle V_1^* = X_1 \cdot (\triangle P_1^* + D_1^*)$, $\triangle V_2^* = X_2 \cdot (\triangle P_2^* + D_2^*)$, $\triangle V_3^* = X_3 \cdot (\triangle P_3^* + D_3^*), \ldots, \triangle V_t^* = X_t \cdot (\triangle P_t^* + D_t^*)$, $\triangle V_{t+1}^* = X_{t+1} \cdot (\triangle P_{t+1}^* + D_{t+1}^*)$, ... and so on.

In the no-dividend case, Eq. (7.3) shows V_{t+1}^* is a martingale when S_{t+1}^* is a martingale, and then when V_{t+1}^* is a martingale, G_{t+1}^* is a martingale, which leads to the equivalence of no-arbitrage and existence of Q-martingale measure in Theorem 7.3.

In the dividend case here, define $D_0^* = 0$, and suppose $P_{t+1}^* + \sum_{i=0}^{t+1} D_i^*$ is a martingale under Q-measure. By the martingale definition, therefore, for all $0 \leq t \leq T-1$

$$\begin{aligned}&E_Q\left(P_{t+1}^* + \sum_{i=0}^{t+1} D_i^* \middle| \mathcal{F}_t\right) = P_t^* + \sum_{i=0}^{t} D_i^* \Rightarrow\\ &E_Q(\triangle P_{t+1}^* + D_{t+1}^* | \mathcal{F}_t) = 0. \end{aligned} \tag{7.9}$$

Since X_{t+1} is previsible and adapted to $\mathcal{F}_t$, then for all $0 \leq t \leq T-1$

$$\begin{aligned}E_Q(\triangle V_{t+1}^* | \mathcal{F}_t) &= E_Q\big(X_{t+1} \cdot (\triangle P_{t+1}^* + D_{t+1}^*) | \mathcal{F}_t\big)\\ &= X_{t+1} \cdot E_Q\big(\triangle P_{t+1}^* + D_{t+1}^* | \mathcal{F}_t\big) = 0.\end{aligned}$$

Applying iterated expectations, therefore for all $0 \leq t \leq T-1$,

$$E_Q(\triangle V_{t+1}^* | \mathcal{F}_0) = 0.$$

Thus, for all $0 \leq t \leq T-1$,

$$E_Q(G_{t+1}^* | \mathcal{F}_0) = E_Q\left(\sum_{i=1}^{t+1} \triangle V_i^* | \mathcal{F}_0\right) = 0.$$

By the same argument as in the proof of Theorem 7.3, there is no arbitrage opportunity. Likewise, if there is no arbitrage, then there exists $P_{t+1}^* + \sum_{i=0}^{t+1} D_i^*$ as a Q-martingale.

From Eq. (7.9), we can obtain, $\forall 0 \le t \le T-1$,

$$E_0^Q\left(P_{t+1}^* + \sum_{i=1}^{t+1} D_i^*\right) = P_0^* + D_0^* = P_0^*,$$

or

$$E_0^Q\left(\frac{P_{t+1}}{M_{t+1}} + \sum_{i=1}^{t+1} \frac{D_i}{M_i}\right) = P_0. \tag{7.10}$$

Equation (7.10) is the present value formula for security price at $t = 0$ of its future stream of dividends plus an end-value price P_{t+1} discounted by risk-free return M_{t+1} since it is under the Q-measure.

From Eq. (7.9), we can also obtain

$$\frac{1}{P_t^*} \cdot E_Q(\triangle P_{t+1}^* + D_{t+1}^* | \mathcal{F}_t) = E_Q\left(\frac{1}{P_t^*} \cdot (\triangle P_{t+1}^* + D_{t+1}^*) \middle| \mathcal{F}_t\right) = 0.$$

Or re-arranging,

$$E_Q\left(\frac{1}{P_t} \cdot (P_{t+1} + D_{t+1}) \middle| \mathcal{F}_t\right) = \frac{M_{t+1}}{M_t} = R_t$$

or the fact that the Q-measure expected return of any security is always equal to the risk-free return over the same period.

Theorem 7.3 may also be expressed as follows.

Corollary 7.2 *There is no arbitrage opportunity in a market model iff the ratio of security return to risk-free return is a martingale under a Q-measure.* ■

In this case, the Q-conditional expectation of this ratio is always 1. This is a case of risk-free rate being a numéraire for the random returns.

So far in the treatment of arbitrage, we consider only self-financing trading strategies. This is actually quite general in the following sense. At any time t, suppose value of a portfolio is $X_t \cdot S_t = \sum_{k=0}^N x_t^k s_t^k$. The investor can always borrow W_0 more by short-selling bonds (security "0"), i.e., choose $x_{t+1}^0 < 0$ and invest the proceeds on other stocks to start a second self-financing investment process with initial value W_0. Thus, self-financing is not about borrowing or not borrowing, but that no money leaves the market, and no new personal wealth or endowment adds to the original portfolio.

Moreover, to avoid some rare self-financing trading strategies that technically (but not realistically) can produce arbitrage profits, such as a doubling strategy, some additional restrictions are usually put on the self-financing trading strategies. These restrictions limit the strategies to "admissible" class or "admissible trading strategies". The reason why there is such a strong prior or an *a priori* reason to develop theories of no arbitrage than theories of arbitrage is that in the real world, if arbitrage were rampant, then there would be no equilibrium prices and markets would break down. Thus, no-arbitrage theories, and in this case, via martingales, are the order of business for financial theoretical developments. This of course does not mean there is no such a thing as arbitrage, and indeed those who become rich are the ones who secretly found recipes to prove these theories wrong. Or perhaps they were lucky with a few big bets and got away. But still, no-arbitrage theories explain the majority of market events, and going by them for purposes of risk management and thinking about markets is a far safer bet than attempting some "arbitrage" strategies that may prove to become, say, Nick Leeson's downfall.

In the doubling strategy, say in a casino setting (though doing it in a stock market setting is also technically feasible), start by borrowing a dollar and betting on roulette red or black (assuming equal probability of win or loss). If a win occurs, quit with a dollar gain. If a loss occurs, borrow two more dollars to bet. Next, if a win occurs, quit with a dollar gain. If a loss occurs again, borrow 2^2 dollars, and so on. If after n straight losses, the total borrowing owed is $1 + 2 + 2^2 + 2^3 + \ldots + 2^{n-1} = 2^n - 1$, then borrow another 2^n to bet. If a win occurs, quit with a dollar gain. If a loss occurs, go on to borrow 2^{n+1}. The probability of an eventual win of \$1 is $2^{-1}+2^{-2}+2^{-3}+\ldots+2^{-n}+\cdots = 1$. This self-financing strategy technically produces positive arbitrage profit, but may take an infinitely long time to do so, or may incur an infinitely large amount of debt to do so.

The latter is not realistic in actual markets. Thus, our admissibility is a restriction to self-financing trading strategies in a class with a finite time $T < \infty$, and also (implicitly) a lower bound B such that $V_t \geq B$, where $B < 0$. The whole idea of restriction of the admissible strategies is to ensure the prices, values, and gains, are $\in \mathcal{L}^1$ in some numéraire, and are thus martingales.

So far in this discussion of the multiperiod securities market and trading strategies, we have not imposed any specific distributional assumptions on the security prices except that their discounted values follow a martingale. The story on the change of probability measure as we see above in the Radon Nikodým derivative is not quite complete and has not been developed into its full potential without applying the Girsanov theorem, shifting drift to make a diffusion process become an equivalent martingale measure. This will be the central theme when it involves specific distributional assumptions and stochastic processes such as a diffusion process in the security price. An exact specification of some stochastic processes such as the diffusion class is especially tractable with the tools of Itô calculus, and are useful because we can then find analytical formulas or sharper assessments of contingent claim prices under no-arbitrage conditions. However, as this involves more continuous-time apparatus such as Itô's calculus, we shall reserve this for Chap. 9. Nevertheless, the results in this chapter on the close relationships between martingales and no-arbitrage under some mild admissible restriction to class Θ, is a major finance theory breakthrough and thus were named "Fundamental Asset Pricing Theorems" by Dybvig and Ross (1987).[6] No-arbitrage equilibrium, from an economic angle, is easy to justify as otherwise prices will not settle at all. Once there is no-arbitrage, the fundamental theorem gives the powerful result that one can employ expectations on the Q-martingale measure to find the current price of any security and derivative without having to construct an equilibrium theory of risk premium.

7.5. Problem Set 7

1. Show that the symmetric random walk S_{n+1} described at the beginning of this chapter is a martingale and also a Markov process.
2. If $\{X_{t+1}\}$ is an i.i.d. stochastic process that is adapted to filtration $\{\mathcal{F}_{t+1}\}$, and if $\{Y_{t+1}\}$ is an i.i.d. predictable process w.r.t. to filtration $\{\mathcal{F}_{t+1}\}$, and moreover X_t and Y_t are equal in distribution (i.e., have the same probability distribution) for every t, what is the ratio of the variance of $E(X_{t+1}|\mathcal{F}_t) - E(X_t|\mathcal{F}_t)$ to $E(Y_{t+1}|\mathcal{F}_t) - E(Y_t|\mathcal{F}_t)$. Does $E(X_{t+1}Y_{t+1}|\mathcal{F}_t)$ exist?
3. If $\{M_s\}_{s=1,2,...}$ is a martingale, show that $E(M_s) = M_t$ for every $s > t$.
4. If S_n is a stock price that follows a binomial price process with the stock price evolution as shown in the graph and where the risk-neutral probabilities of an "up" state is 0.5 and of a "down" state is 0.5. Does this stock price follow a Markov process? (Note that $S_0 = \$6$.)

[6]Dybvig, PH and SA Ross (1987). Arbitrage. In *The New Palgrave Dictionary of Economics*, J Eatwell, M Milgate and P Newman (eds.). London: Macmillan.

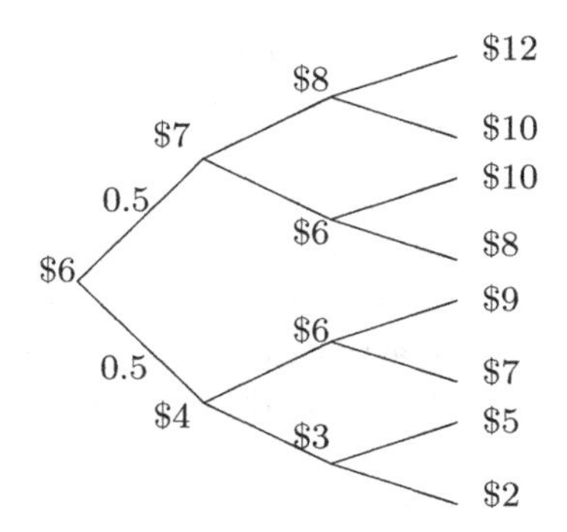

5. In the binomial stock price process in Fig. 7.3, if $u = 10/9$, $d = 9/10$, and risk-free interest rate factor $R = 1$, $p = 9/19$, $q = 10/19$, a maximum lookback call option with a strike price of \$0.5 and 3 periods to go, has maturity value C_3 equal to $(\max\{S_0, S_1, S_2, S_3\} - 0.5, 0)$. (a) Is the stock price S_n a martingale? (b) Is the maximum lookback call option price C_n a martingale?
6. Let $\{X_n\}$ be a stochastic process with RV's $X_0, X_1, X_2, \ldots, X_n, \ldots$. Define $Y_t \triangleq E(X_n|\mathcal{F}_t)$, where $\mathcal{F}_t$ is the σ-field at time $t \leq n$. Thus, Y_t is the best mean-square forecast of X_n given information at time t. Assume that $\{\mathcal{F}_t\}$ is a natural filtration, and thus $\mathcal{F}_t$ increases with t (in the set-sense). Show if $\{Y_t\}$ is a martingale.
7. Consider a factory that produces a widget with a probability q of defect and $1 - q$ of satisfying the quality standard. Each production is independent of the other, i.e., not dependent on the machine. Let N_t be the number of defects after t widgets are produced. Let $Z_t = N_t - qt$, and $\mathcal{F}_t$ be the algebra corresponding to the observations of the t widgets. Is Z a martingale? Show.
8. If $C_T^*(S_T^*)$ is the terminal value of a contingent claim in a market with no arbitrage opportunity (a "viable market"), show that the value processes of all generating strategies for C_T^* are the same.
9. What is the expected payoff of the self-financing doubling strategy, starting with a bet of \$1?
10. Suppose the underlying stock process is a binomial lattice. The market has only one such stock and a bond with constant per period interest rate. After T periods, the number of possible stock prices at T is 2^T, each associated with a state $\omega \in \Omega$. The market is complete if every state is attainable in the sense that there is a state contingent claim that pays \$1 iff state ω occurs, or else nothing. Is it true therefore

that in this case the market is not complete since there are no state contingent claims or other derivatives trading but only the stock and bond are trading?

11. In the market model, if X_{t+1} is a previsible process adapted to $\mathcal{F}_t$ in a filtration $\{\mathcal{F}_s\}$, $s = 1, 2, \ldots, T$, and security price S_t^*, $0 < t \leq T$, is a martingale, show that under self-financing trading, $E_0(X_{t+1}S_{t+1}^*) = X_1 S_0^*$. Is $X_{t+1}S_{t+1}^*$ a martingale? (Note that subscript "0" to the expectation operator denotes conditioning w.r.t. $\mathcal{F}_0$.)

Chapter 8

DYNAMIC PROGRAMMING AND MULTI-PERIOD ASSET PRICING

Herbert Simon once said, "Human knowledge has been changing from the word go and people in certain respects behave more rationally than they did when they didn't have it. They spend less time doing rain dances and more time seeding clouds." Paul Samuelson also said, "Funeral by funeral, theory advances." The key themes are that humans make mistakes, including economists in their models or the oversimplifying assumptions that are embedded; they learn, and they advance for the better. The Ramsey growth model[1] was a pioneering work that built on earlier works of Ludwig von Mises and also contemporaries such as Friedrich von Hayek and others, and led to later generations of economic growth models in macroeconomics as well as offshoots into stochastic models of dynamic consumption and investments in finance. In this chapter, we look at the issue of social planning and neo-classical economic concepts (rational choice theory and optimization analyses of households and producers resulting in aggregate outputs, incomes, and market prices) of Pareto optimality, leading to multi-period optimal consumption and investment models with implications for rational asset pricing in finance.

8.1. Log-Optimal Strategy

The log utility was seen as a special case of the CRRA class of utilities. It carries another interesting result.

Suppose W_0 is the original wealth. There are two securities A and B yielding returns per period of R_t^A and R_t^B respectively at time t. In other words, if an investor puts \$ W_0 on security A at $t = 0$, then after one period, at $t = 1$, the new wealth is $W_0 \times R_1^A$. Similarly, for investment on B, the new wealth would be $W_0 \times R_1^B$. The return accrues over time $(0, 1]$. If the returns on A and B are stationary ("no change to the underlying

[1]Ramsey, FP (1928). A mathematical theory of saving. *Economic Journal*, 38(152), 543–559.

probabilities of the returns over time"), wealth at T of investment in A is $W_T = R_T^A R_{T-1}^A R_{T-2}^A \ldots R_3^A R_2^A R_1^A W_0$. Wealth at T of investment in B is $W_T = R_T^B R_{T-1}^B R_{T-2}^B \ldots R_3^B R_2^B R_1^B W_0$.

Taking natural logarithm of both sides, for $K = A$ or B

$$\ln W_T = \ln W_0 + \sum_{t=1}^{T} \ln R_t^K.$$

Then

$$\ln \left(\frac{W_T}{W_0}\right)^{1/T} = \frac{1}{T}\sum_{t=1}^{T} \ln R_t^K \longrightarrow g(K) \quad \text{as } T \uparrow \infty, \tag{8.1}$$

where $g(K) = E[\ln R_t^K]$ is the population mean of RV $\ln R_t^K$, and varies from security to security, i.e., varies across K. The convergence to $g(K)$ follows the Law of Large Numbers. Hence, as $T \uparrow \infty$

$$\left(\frac{W_T}{W_0}\right)^{1/T} \longrightarrow e^{g(K)}.$$

Then

$$W_T \longrightarrow W_0\, e^{T\, g(K)}.$$

Thus, for large T, wealth grows approximately (because we are still in finite time) exponentially with time T at a rate of $g(K)$. Therefore, to have a highest rate of wealth growth over a long time period by repeatedly investing in a security, one should choose the security with the highest $g(K)$. That is, choose security with the highest expected log return, as seen in (8.1), which is $\max_K E(\ln R_t^K)$, or the log-optimal strategy. Note that it is not to choose security with the highest expected return, i.e., not $\max_K E(R_t^K)$. The latter may lead to short-run maximal gains but to long-run suboptimal gains.

Suppose the investor has log utility, and there is a large set of N securities to invest in. Each period, he or she maximizes expected utility as

$$\max\ E[\ln(W_{t+1})] \equiv \ln(W_t) + \max\ E[\ln R_{t+1}^P]$$

for portfolio P. In other words, he/she selects the portfolio that maximizes expected log returns. Thus, if the investor has log utility, the log-optimal strategy is also a short-run strategy for maximal satisfaction since he or she maximizes expected log return each period. In general, however, a different utility would imply that a period-by-period optimization is suboptimal in the long run.

Applications

Suppose in blackjack, a wager of $1 either pays off $2 in a win or nothing in a loss. The return in a win is 100% and is −100% in a loss. Suppose a player starts with capital W_0 and is trying to decide what fraction of it he/she will wager in each game. Let this fraction be θ. If he/she wins, his/her capital increases by a return factor $1+\theta$. If he/she loses, his/her capital is reduced by factor $1-\theta$. The probability of a win is p and of a loss is $1-p$ where $p \in (0,1)$.

To maximize long-run growth, he/she uses the log-optimal strategy and maximizes the expected log return by choosing the appropriate optimal θ:

$$\max_\theta E[\ln R] \equiv \max_\theta p\ln(1+\theta) + (1-p)\ln(1-\theta).$$

The FOC is

$$\frac{p}{1+\theta} - \frac{1-p}{1-\theta} = 0$$

or $\theta = 2p - 1$.

The above is called Kelly's betting rule (based on the idea of log-optimal strategy). If $p = 0.5075$ with professional blackjack players, $\theta = 0.015$ or 1.5% of capital in each wager. The growth rate $g(K) = E(\ln R) = 0.5075\ln(1.015) + 0.4925\ln(0.985) = 0.000113$ or 0.0113% gain each wager. Notice how small the % bet (1.5%) is, as this is a long-run strategy.

Next, consider an interesting investment situation involving a high-risk high-return stock A and a zero-risk bank deposit B. See payoff Table 8.1.

Table 8.1: Stock Returns

Prob.	A's Return rate	B's Return rate
0.4	100%	0%
0.6	−50%	0%

Each month, suppose θ fraction of capital is invested in A and $1-\theta$ in B:

$$\begin{aligned}\max_\theta \quad & 0.4\ln[\theta \times 2 + (1-\theta)\times 1] + 0.6\ln\left[\theta \times \frac{1}{2} + (1-\theta)\times 1\right] \\ &= 0.4\ln[1+\theta] + 0.6\ln\left[1 - \frac{1}{2}\theta\right].\end{aligned}$$

Optimal θ equals 0.2 with expected log return 0.97% per month or growth $g(K)$ of 0.97% per month.

Notice this strategy brings about some rebalancing. In a month when the stock market moves up and A achieves a return rate of 100%, the capital allocation of fixed 20% to A means a smaller number of A shares since its price has gone up. On the contrary, when A's price goes down by 50%, a fixed 20% allocation of capital to A means more of the cheaper A shares will be purchased next period. Thus, such fixed log-optimal strategy does have a volatility-pumping effect of improving capital if the market is randomly moving up and down.

Consider an investor who goes for short-run gain of maximizing expected return: $\max_\theta \ 0.4[1+\theta] + 0.6[1-\frac{1}{2}\theta] = 1 + 0.1\ \theta$. In this case, optimal $\theta = 1$. But two successive downturns could reduce capital to 1/4 rather quickly.

8.2. Pareto Efficiency

Italian economist Vilfredo Pareto (1848–1923) developed the concept of efficiency in exchange of goods between two parties. If consumption or utility goods are distributed to individuals in an allocation, the allocation is Pareto efficient (or Pareto optimal or having Pareto improvement to the point where no more improvement is possible) if the goods cannot be redistributed or reallocated to make at least one individual better off without making someone else worse off.

An illustration of the Pareto efficiency concept is shown using the Edgeworth diagram in Fig. 8.1.

Two individuals or firms A and B are each allocated a portfolio of two resources X and Y. The total amount of the X and Y resources in the system is fixed at $x_A + x_B = 12$ and $y_A + y_B = 7$, respectively. The circled

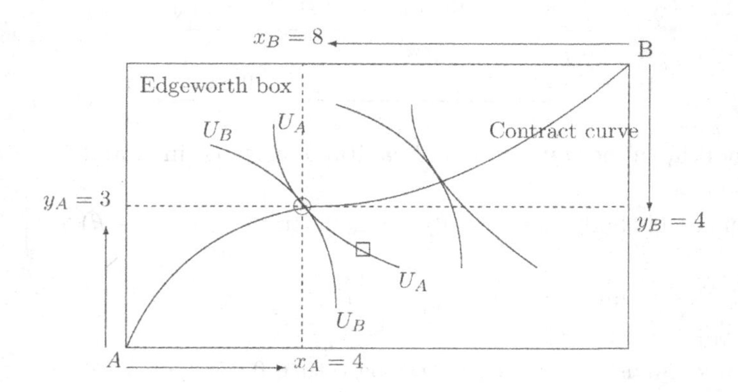

Fig. 8.1: Illustration of Pareto Efficiency

point shown is an allocation whereby A gets $(x_A = 4, y_A = 3)$ and B gets $(x_B = 8, y_B = 4)$. Note the axes for A are east for X-axis and north for Y-axis originating at the south-west corner of the box, whereas the axes for B are west for X-axis and south for Y-axis originating at the north-east corner of the box. All points in the box represent feasible allocations to A and B, i.e., allocations pertaining to the total of 12 units of X and 7 units of Y.

The indifference curves for A, shown as $U_A - U_A$ on the diagram are strictly convex to the origin of A at the south-west corner. The indifference curves for B, shown as $U_B - U_B$ on the diagram are strictly concave to the origin of A at the south-west corner. They represent diminishing preference to scale of each resource as more is being obtained by trading off the other resource. An initial allocation at the square point shown is not in equilibrium because A is indifferent between this square point and the circled point, but B is definitely better off at the circled point than at the square point. Thus, B has incentive to trade or barter with A, the outcome being that eventually both will be better off if A exchanges some of its X for more of Y with B. Eventually, the equilibrium will be reached at points along the shown contract curve. The contract curve links all points which are on the tangents of both A's and B's indifference curves, i.e., at a point where both A's and B's curves just touches. For example, at the circled point, there is no incentive for either A or B to do any further exchange, as any movement away from an allocation at the circled point means either less preference for A or else less preference for B though the other party would benefit. (We assume the economic axiom of nonsatiability and that people are neither sacrificial nor altruistic.)

Thus, all initial allocations would end up on a point along the contract curve which is a Pareto efficient allocation. There is a bit of philosophy involved in the concept of the efficiency or optimality. The origination of the allocation is not a theoretical issue. Pareto optimality, however, does not address issues relating to the socially desirable distribution of resources. Pareto optimality just says that given such an initial allocation, the welfare of both parties can be improved if they exchange or trade so they move to a Pareto optimal point. Thus, if for some historical reason, the initial allocation is such that 1% of the population enjoys 99% of all consumable resources while the other 99% of the population starve in deprivation with only 1% of the resources, it is still Pareto optimal since it is not Pareto optimal to take something away from the insatiable super-rich 1% to give to the utterly poor 99%.

This is capitalism at its heart, or the market economy — do not fight the fact one is poor or one is rich, but use the market as best as possible to improve one's well-being. A communist or Marxist view would be more radical — that it is ideologically not right for one to be born with or stuck with an inferior initial allocation, that there should be state control (or else a revolt) to change the initial allocations to something more egalitarian or fair, something which, however, can be quite elusive since all human reallocation mechanisms can be flawed.

The market economy works, of course, by the price mechanism. A and B can declare their prices for X versus Y, and eventually agree on an equilibrium price with which to bring their transactions or exchange onto the contract curve so that both are better off.

Economic research, spilling into financial economics since the 1930s, has a large bearing on these (Pareto) efficient allocation and resulting equilibrium issues. These ideas were fascinating in those times, and over the years have taken on more sophisticated applications in ideas in industrial economics about the strategic behavior of firms, the structure of markets and how firms in certain market structures would behave, considering transaction costs, information, regulations, and competitiveness of entry and exit. In markets that do not trade in every uncertain state and are imperfect with transaction costs and information asymmetries, competitive equilibrium is in general not Pareto optimal.

Ignoring the first issue about the fairness of initial allocations and whether philosophically or practically one should readjust that using regulations or fiscal transfers (tax the rich and subsidize the poor), Pareto improvement — moving to a Pareto optimal point — is an appealing and intuitively correct concept from the point of view of improving everyone's welfare without having to cause anyone to suffer for it. Of course, the exchange or trading mechanism is assumed to operate under a perfect market situation. It may be thought of as the "win-win" proposition in any two-party negotiation for an outcome or contract which both will find it beneficial to sign. Of course, the win-win may be more sophisticated involving not just exchange, but also increased production as well for both.

Before we leave these ideas for more theoretical grounds, consider another pertinent example to illustrate Pareto optimality. Consider a monopolist and a perfectly competitive market on the demand and supply of a good. The two regimes are shown below on the marginal analysis (analysis of marginal revenues versus marginal costs) diagrams in Figs. 8.2(a) and 8.2(b).

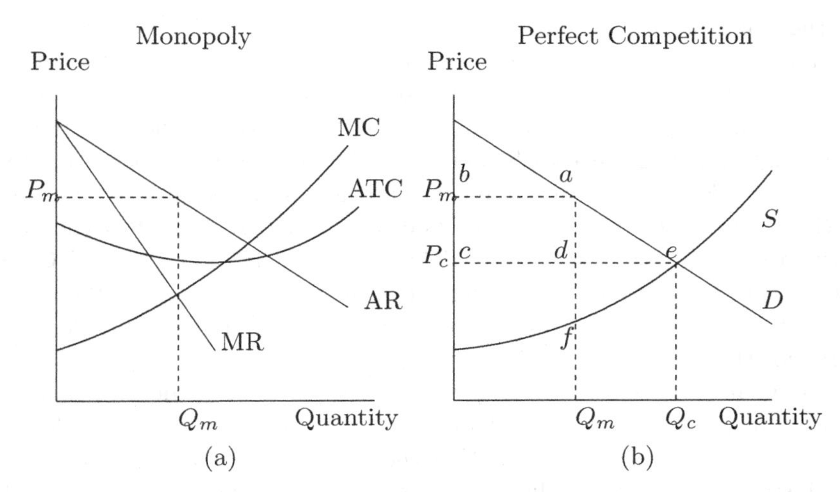

Fig. 8.2: Marginal Analyses in a Monopoly versus a Competitive Market

In Fig. 8.2(a), average revenue AR = (Price $P\times$ Quantity $Q)/Q = P$. Hence, the AR curve is the market demand curve which is function $D = Q(P)$. The monopolist's supply curve is in fact a vertical line cutting at Q_m with price P_m since the monopolist controls both price and output. Total cost TC = fixed cost + variable cost. The marginal cost curve MC cuts the average variable cost, AVC, curve and the average total cost, ATC, curve at their minimum points from below. For the monopolist, he/she produces and supplies at Q_m where MC = MR in order to maximize his/her profit, which is $P_m \times Q_m - ATC(Q) \times Q_m > 0$. The TC already incorporates "rent" or premium for entrepreneurship, risk taking, and expertise, so the positive profit is a pure monopoly profit over and above ownership and entrepreneurial rent.

Suppose there is antitrust law, and the market is "contestable" (i.e., other firms are assumed to be able to enter and exit the competition without additional costs other than the usual production costs), then other firms will enter and increase the aggregate market supply of goods. As seen in Fig. 8.2(b), this supply curve then shifts right from point (Q_m, P_m) to an upward sloping schedule S, and a new competitive price is established at $D = S$ with $P_c < P_m$ and $Q_c > Q_m$.

In Fig. 8.2(a), suppose there is the monopolist A, and there are 99 other competiting firms shut out, making zero business or profit, for whatever reasons (think of this as the initial condition). If somehow, this is moved to a situation of perfect (market) competition in Fig. 8.2(b), where now all the other 99 firms could participate in offering supplies, assuming all are now identical firms, and given that demand remains the same. Then the increased supply pushes to a contestable market perfectly competitive equilibrium at (Q_c, P_c). In this perfect market competition, the AR curve is now a flat line at P_c since each firm cannot influence price by adjusting its own output. Each firm's MC is upward sloping and its output is determined at the condition $P_c = MC$. Hence, its own supply curve is in fact the MC curve above the point when its AVC is zero, i.e., it must cover AVC to want to produce, given that fixed cost is a sunk cost. Adding all 100 firms' upward sloping MCs thus produces the upward sloping aggregate supply curve for the market.

Now, all 100 firms share the producer surplus — area "def" (increased revenue above MC which is not obtained when they produce at a lower level Q_m), but forego area "abcd" if they had colluded and produced at Q_m instead. Thus, net producer surplus, or the $ equivalent improvement in well-being to producers, is area "def-abcd".

On the other hand, by bringing the market price from P_m down to P_c, the consumer surplus, or $ equivalent improvement in well-being to consumers, is saving of area "abcd" for those who had demanded even when price is P_m, plus area "aed" which is what marginal consumers at P_m would enjoy if they can consume now at price P_c. Thus, net consumer surplus is area "abcd + ade".

By moving from a pure monopoly to a perfect competitive market, the net benefit to society (assuming this comprises the total of all producers and consumers) is net producer surplus + net consumer surplus = area "def-abcd + abcd + aed" = area "aed + def" = area "aef". This is a net increase in benefit to society, which is otherwise lost due to monopoly. The latter loss is called a deadweight loss.

Is the move Pareto improving? The other 99 firms and the consumers are better off, but not the monopolist. Hence, strange as it seems, it is not Pareto improving. Thus, it is not a capitalist idea to dethrone a monopolist. However, modest socialism, including government intervention, could introduce, for example, a proposal to the monopolist to be "paid" from an imposed consumer tax (in effect not allowing consumers to enjoy the new consumer surplus) so it is not worse off, and the rest of the 99 firms continue to be better off now that they can earn at least entrepreneurial and ownership rent rather than zero income before. Sometimes, such a solution may not be feasible. But when it is feasible, it is a socially meaningful proposition that improves everyone's welfare and is thus Pareto improving.

The Pareto optimality is an exchange efficiency concept. If A can unilaterally do to B without B's agreement (or agreed exchange) and improve its own welfare without regard to B's choice, it may still not be Pareto optimal even if B's welfare is improved as a result because it may not be B's optimal under an exchange agreement. But then this benign outcome, is Pareto improving, though it is a rare situation. Maximization of total societal welfare is more of a general efficiency concept than Pareto efficiency concept, as it includes the case of Pareto efficiency, the benign case above, and the monopolist case without government intervention earlier. To emphasize again, Pareto efficiency is only concerned with exchange efficiency — that is, how to bring individuals to their best preferences given their initial endowments. It does not concern itself with how welfare is to be distributed or with equity issues of initial endowments.

In the Edgeworth box, there is an infinite number of Pareto optimal points along the contract curve, all of which are exchange efficient and each results from some initial endowment point in the box. But, there is possibly only one maximum point if we think of a total social welfare or societal welfare function that is some function of the utilities of all individuals. However, arriving at this maximum point may not be Pareto optimal as some may actually suffer for it.

Pareto efficiency implies net increase in social welfare, but the converse is not necessarily true. Nevertheless, financial economics looks at characterizations of competitive equilibrium situations in asset pricing and it would be nice if the equilibrium also satisfies conditions of Pareto efficiency. This would give more weight to achieving such equilibrium results. Indeed, it can be shown that generally, a competitive equilibrium in a market without friction or transaction costs (a perfect competitive market) and where each agent can go to the market and trade freely (under Walrasian price equilibrium — i.e., until excess demand or excess supply equals zero at an equilibrium price under conditions where each agent fully utilizes his/her budget for consumption and investment) is also Pareto efficient as the agent maximizes his or her own preference given the market price to trade on.[2]

In addition, Pareto efficient allocations (where any reallocation of consumption cannot make all better off without making some worse off)

[2]This is called the first fundamental theorem of welfare economics. The first theorem is informally explained as Adam Smith's "invisible hand" hypothesis where markets under price competition lead to an efficient allocation of resources.

under concave continuous utility functions of agents can be attained in a competitive equilibrium. An equilibrium is sustainable where there is a Walrasian equilibrium market price and all individuals have expended their budgets and have no more incentive to trade on the market price.[3] A competitive equilibrium associated with a specific positive equilibrium market price and initial endowments, e.g., a point on the contract curve, is attainable by any initial endowment point and a price line from that endowment point to the Pareto optimal point.

In a larger picture, the idea of efficient social exchange (as in interaction between society agents) leads to advocates of free market competition and free international trades. Liberalization of international trades is usually argued as a way to improving overall countries' welfare, though usually it is not Pareto efficient in the sense that some parties will be adversely affected, such as the farmers under previously protected prices or workers being retrenched because of cheaper labor elsewhere. Thus, the social issues are a lot more complex. The converse is easier to argue against, i.e., imposing more regulations or restrictions to existing free market mechanisms or free trades, are certainly moving away from Pareto efficiency if there was one, since certainly some parties will be adversely affected by the new regulations and restrictions. In the mathematics of optimization covered in earlier chapters, constrained optimization is always suboptimal to unconstrained optimization.

However, free markets can fail, as demonstrated by the recent domino chain reaction in the financial markets as a result of the 2008 subprime crisis. Market failure in the form of Pareto inefficiency could be due not just to a monopoly, but also to the inadequate disclosure of information (e.g., consumers or investors not pricing their goods correctly and thus failing to enjoy the anticipated welfare *ex post*). Inefficiency could also be due to negative externalities e.g., consumers buying more TVs when the factories producing TVs are dumping toxic wastes down the river to the backyards of the consumers, who unwittingly suffer while presumably enjoying more TVs. It could also be due to an inadequate supply of public goods that all consumers enjoy because no private firms would produce them, being of negative profitability since revenue cannot be collected (e.g., producing

[3]This is called the second fundamental theorem of welfare economics, that any efficient allocation can be achieved or sustained by a competitive equilibrium.

more oxygen by building more trees). Public goods provision is, of course, a positive externality.

8.3. Competitive Exchange Market

Continuing from the concept of Pareto efficiency above, we now show a key result in comparing the economic outcomes of social planning (central planning) versus competitive market mechanism.

First, we consider competitive market equilibrium in a complete market setting over multiple periods where every time-state t, s is spanned and has a traded state security that pays 1 unit of a consumption good when state ts occurs and 0 otherwise, and its traded market price is ϕ_{ts}.

Assume each individual k's lifetime utility is time-additive and state independent, i.e., $\sum_t U^k(C^k_{ts})$. In other words, t and s do not affect the functional form of U except enter into its argument via C^k_{ts}. Assume consumption is on a single good, and that $U' > 0$ and $U'' < 0$. Assume also that all the individuals have homogeneous (identical) *ex ante* beliefs of probabilities of time-states π_{ts}.

We assume competitive market conditions where there are many atomistic individuals, none of whom can affect the market price. As such, the price is taken as given. Each individual k solves his or her expected utility maximization as follows. Once this is solved, subject to the feasibility consumption constraint the individual faces, there is no incentive for the individual to exchange or trade anymore, hence there is Pareto efficiency under competitive market exchange or trade.

At $t = 0$, the individual k chooses the dollar amount of current consumption C^k_0 and also the amount of state ts securities, C^k_{ts} to purchase. As the states become realized at each future period, the individual then consumes \$ C^k_{ts} amount of goods at realized time-state ts. The individual solves the following Lagrangian

$$\max_{C^k_0, C^k_{ts}, \lambda^k} L = U^k_0(C^k_0) + \sum_t \sum_s \pi_{ts} U^k(C^k_{ts}) + \lambda^k \left[W^k_0 - C^k_0 - \sum_t \sum_s \phi_{ts} C^k_{ts} \right], \tag{8.2}$$

where k's initial endowment wealth is W^k_0, and state ts securities are priced at ϕ_{ts} each. Throughout the rest of this section, we assume that there are N finite number of states, $S = 1, 2, \ldots, N$, and T finite number of periods

in a lifetime, $t = 1, 2, \ldots, T$. λ^k is the Lagrange multiplier. Notice that in a complete market setting above, the individual just needs to solve the problem L once and for all at time $t = 0$.
FOC:

$$\frac{\partial L}{\partial C_0^k} = U_0^{k'} - \lambda^k = 0 \Rightarrow \lambda^k = U_0^{k'}.$$

$$\frac{\partial L}{\partial C_{ts}^k} = \pi_{ts} U^{k'}(C_{ts}) - \lambda^k \phi_{ts} = 0 \Rightarrow \phi_{ts} = \frac{\pi_{ts} U_{ts}^{k'}}{U_0^{k'}}, \forall t, s.$$

Competitive equilibrium as well as Pareto efficiency imply

$$\frac{\phi_{ts}}{\pi_{ts}} = \frac{U_{ts}^{k'}}{U_0^{k'}} \quad \text{and} \tag{8.3}$$

$$\frac{\phi_{ts'}}{\pi_{ts'}} = \frac{U_{ts'}^{k'}}{U_0^{k'}}, \quad \text{or} \quad \frac{\phi_{ts}/\pi_{ts}}{\phi_{ts'}/\pi_{ts'}} = \frac{U_{ts}^{k'}}{U_{ts'}^{k'}}$$

for any two different states $s, s' \in S_t$, and every individual k.

Equation (8.3) implies that

$$\frac{U_{ts}^{k'}}{U_{ts'}^{k'}} = \frac{U_{ts}^{j'}}{U_{ts'}^{j'}}, \tag{8.4}$$

for any two individuals k and j. For a full solution, the market-clearing conditions of $S_0 = \sum_k C_0^k$ and $S_{ts} = \sum_k C_{ts}^k$, where S_0, S_{ts} are aggregate supplies at $t = 0$ and time-states ts, respectively, must also apply.

Suppose $C_{ts}^k > C_{ts'}^k$, then since $U' > 0$ and $U'' < 0$, (8.4) implies

$$\frac{U_{ts}^{k'}}{U_{ts'}^{k'}} = \frac{U_{ts}^{j'}}{U_{ts'}^{j'}} < 1,$$

and thus, $C_{ts}^j > C_{ts'}^j$.
Aggregating over all individuals, then

$$\sum_i C_{ts}^i > \sum_i C_{ts'}^i.$$

If this is market equilibrium, then the aggregate market supplies S_{ts} and $S_{ts'}$ of consumption units of the good in state s and s' are different, i.e., $\sum_i C_{ts}^i = S_{ts}$, and $\sum_i C_{ts'}^i = S_{ts'}$, where $S_{ts} > S_{ts'}$. If aggregate supply (or equivalently aggregate consumption under equilibrium) in some

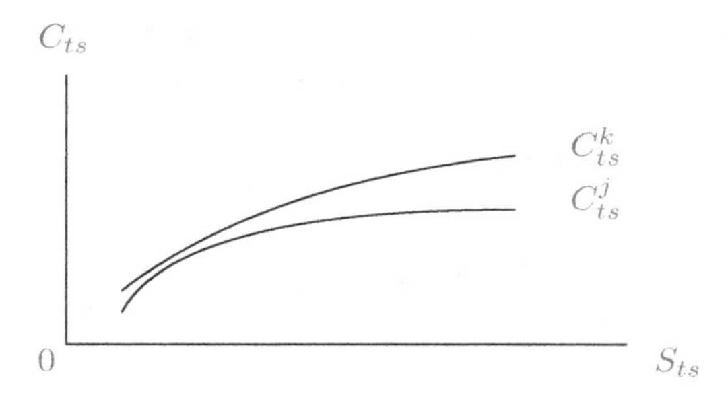

Fig. 8.3: Consumption in Different States

of the states e are the same, then clearly C^k_{te} is constant for k across those states e.

From the above, it is seen that for any individual k, his or her consumption in state ts, C^k_{ts}, increases with the aggregate supply of consumption goods in state ts, S_{ts}. We can depict this using a graph at each t, across states $s \in S_t = \{1, 2, 3, \ldots\}$ where S_{ts} increases with $s \in S_t$. This is shown in Fig. 8.3. Note that individuals j and k need not have identical utility functions.

Hence, we may express for each individual k, his or her time-state consumption as $C^k_{ts} = f^k(S_{ts}, t)$ where for each t, $f^k(\cdot)$ is a monotone function in S_{ts}. Note that $f^k(\cdot)$ is state independent. In addition, the functions $f^k(\cdot)$ must satisfy $S_{ts} = \sum_k f^k(S_{ts}, t), \forall t, s$. In other words, given supply S_{ts} at t and in state s, individual k gets to share amount $f^k(S_{ts}, t)$ of the consumption good. The function $f^k(S_{ts}, t)$ allocating amount to individual $k, \forall k$, is called the (Pareto) optimal sharing rule. We shall next show how social planning could also achieve Pareto optimality with some strong assumptions. The flipside is of course that if a central or social planner can achieve a Pareto optimal allocation, then a freely competitive market system can also perform the same job. This latter idea is in fact the ideology of free market systems (typically under democracies), and was a particularly strong incentive for economic research during the cold war and before the collapse (or diminution) of large central-planning communist systems.

Assume a social or central planner knows the utility functions of all consumers in the economy (this is a strong assumption) and attempts to maximize some linearly weighted combination of the individuals' expected utility functions using $a^k s$ as the weights, the objective being a social welfare function. The social planner decides how much consumption goods at each time-state to allocate to each individual k in order to maximize the social welfare function. Assume that the market is complete as before.

$$\max_{C_0^k, C_{ts}^k, \lambda_0, \lambda_{ts}} L \equiv \sum_k a^k U^k(C_0^k) + \lambda_0\left(S_0 - \sum_k C_0^k\right) + \sum_k a^k \sum_t \sum_s \pi_{ts} U^k(C_{ts}^k) + \sum_t \sum_s \lambda_{ts}\left[S_{ts} - \sum_k C_{ts}^k\right], \tag{8.5}$$

where λ_0 is the Lagrange multiplier for total supply S_0 at $t = 0$, and λ_{ts} is the Lagrange multiplier for each time-state aggregate supply of goods constraint.

FOC:

$$\frac{\partial L}{\partial C_0^k} = a^k U_0^{k'} - \lambda_0 = 0$$

$$\Rightarrow a^k U_0^{k'} = \lambda_0, \quad \forall k.$$

$$\frac{\partial L}{\partial C_{ts}^k} = a^k \pi_{ts} U_{ts}^{k'} - \lambda_{ts} = 0$$

$$\Rightarrow a^k \pi_{ts} U_{ts}^{k'} = a^j \pi_{ts} U_{ts}^{j'} = \lambda_{ts}, \quad \forall t, s, k, j.$$

$$\sum_k C_{ts}^k = S_{ts}, \quad \forall t, s, \quad \text{and} \quad \sum_k C_0^k = S_0.$$

As the utility functions $U^k(C_{ts}^k)$ for each individual at ts are additively separable across time and states, and the aggregate consumption constraint for each time-state, S_{ts} are independent across t and s, then the maximization in (8.5) can be performed alternatively $\forall t, s$, as

$$\max_{C_0^k, C_{ts}^k, \lambda_0, \lambda_{ts}} L' \equiv \sum_k a^k U^k(C_0^k) + \lambda_0\left(S_0 - \sum_k C_0^k\right) + \pi_{ts} \sum_k a_k U_t^k(C_{ts}^k) + \lambda_{ts}\left[S_{ts} - \sum_k C_{ts}^k\right]. \tag{8.6}$$

Thus, FOC:

$$\frac{\partial L'}{\partial C_0^k} = a^k U_0^{k'} - \lambda_0 = 0$$

$$\Rightarrow a^k U_0^{k'} = \lambda_0, \quad \forall k. \tag{8.7}$$

$$\frac{\partial L'}{\partial C_{ts}^k} = a^k \pi_{ts} U_{ts}^{k'} - \lambda_{ts} = 0 \tag{8.8}$$

$$\Rightarrow a^k \pi_{ts} U_{ts}^{k'} = a^j \pi_{ts} U_{ts}^{j'} = \lambda_{ts} \quad \forall k, j, t, s.$$

$$\sum_k C_{ts}^k = S_{ts}, \quad \forall t, s, \quad \text{and} \quad \sum_k C_0^k = S_0.$$

Hence solutions to programs (8.5) and (8.6) are identical. The optimal allocations, $C_0^k, C_{ts}^k, \forall t, s, k$ depend on the *ex ante* beliefs $\{\pi_{ts}\}$, the aggregate economy supply S_0 at $t = 0$ and $\{S_{ts}\}$ at each time-state, the individual utility functions U_{ts}^k, and the social planner's weights a^k.

Equations (8.7) and (8.8) imply that

$$\frac{\pi_{ts} U_{ts}^{k'}}{U_0^{k'}} = \frac{\lambda_{ts}}{\lambda_0}. \tag{8.9}$$

Comparing this with equation (8.3), where

$$\frac{\pi_{ts} U_{ts}^{k'}}{U_0^{k'}} = \phi_{ts},$$

then the solution to the social planner's problem and the competitive market equilibrium problem are identical if the market price $\phi_{ts} = \frac{\lambda_{ts}}{\lambda_0}$, and the social planner chooses, from (8.7):

$$a^k = \frac{\lambda_0}{U_0^{k'}} \quad \forall k.$$

From Eq. (8.8)

$$\frac{U_{ts}^{k'}}{U_{ts'}^{k'}} = \frac{U_{ts}^{j'}}{U_{ts'}^{j'}},$$

for any two individuals k and j, which is the same as in the competitive market situation in Eq. (8.4). Hence, the same optimal sharing rule $f^k(S_{ts}, t)$ can be obtained by the social planning problem here.

Optimal Linear Sharing Rule

The monotone optimal sharing rule we characterized earlier can be made more specific if we assume certain utility functions. One class is the HARA utility functions for individual i where the marginal utility is characterized by $U^{i'}(\tilde{c}_i) = d_i(A_i + B\tilde{c}_i)^{-1/B}$. It is also a linear tolerance class of utility functions with property

$$-\frac{U^{i'}(\tilde{c}_i)}{U^{i''}(\tilde{c}_i)} = A_i + B\tilde{c}_i,$$

where the LHS is a tolerance coefficient, the inverse of Arrow–Pratt risk-aversion coefficient, and B is called a cautiousness coefficient. B is assumed to be a constant across all individuals i.

From Eqs. (8.3) and (8.4) of a competitive equilibrium, we obtain

$$\frac{U^{i'}(\tilde{c}_i)}{U^{j'}(\tilde{c}_j)} = \frac{d_i(A_i + B\tilde{c}_i)^{-1/B}}{d_j(A_j + B\tilde{c}_j)^{-1/B}} = \frac{\lambda_i}{\lambda_j},$$

where λ_i, λ_j are the Lagrange multipliers on individual i and j's wealth constraints respectively. Moreover, $\tilde{c}_k$ can represent current consumption c_0^k or state ts consumption c_{ts}^k as in Eq. (8.2).

Therefore

$$\frac{\lambda_i^B/d_i^B}{\lambda_j^B/d_j^B}(A_i + Bc_i) = A_j + Bc_j.$$

Let $\mu_i = \left(\frac{\lambda_i}{d_i}\right)^B, \forall i$. Then

$$\frac{\mu_i}{\mu_j}(A_i + Bc_i) = (A_j + Bc_j).$$

Summing over all individuals j

$$\left(\sum_j \frac{1}{\mu_j}\right)(\mu_i(A_i + Bc_i)) = \sum_j A_i + B\sum_j c_j.$$

Let aggregate consumption by all individuals $\tilde{C}^T = \sum_j \tilde{c}_j$, and $A^T = \sum_j A_j$. Then

$$\mu_i(A_i + B\tilde{c}_i)\left[\sum_j \frac{1}{\mu_j}\right] = A^T + B\tilde{C}^T.$$

Rearranging,

$$\tilde{c}_i = \frac{A^T + B\tilde{C}^T}{B\mu_i \sum_j \frac{1}{\mu_j}} - \frac{A_i}{B}. \tag{8.10}$$

Hence, each individual i's consumption c_i is a linear sharing rule on aggregate consumption $\tilde{C}^T$.

Representative Agent and Aggregation Property

From Eq. (8.10), first note that $\sum_i^M \tilde{c}_i = \tilde{C}^T$, where M is the number of individuals in the economy. Summing over all individuals i, and dividing both sides by M, we obtain

$$\hat{c} \triangleq \frac{\tilde{C}^M}{M} = \frac{1}{M}\sum_i \frac{1}{\mu_i}\left[\frac{A^T + B\tilde{C}^T}{B\sum_j \frac{1}{\mu_j}}\right] - \frac{A_r}{B},$$

where $A_r = A^T/M$. Let $\frac{1}{\mu_r} = \frac{1}{M}\sum_i \frac{1}{\mu_i}$. Therefore,

$$\hat{c} = \frac{A^T + B\tilde{C}^T}{B\mu_r \sum_j \frac{1}{\mu_j}} - \frac{A_r}{B}. \tag{8.11}$$

Comparing Eq. (8.11) with Eq. (8.10), (8.11) is the linear sharing rule of an rth individual with consumption $\tilde{c}_r = \hat{c}$, and HARA tolerance coefficient $A_r + B\hat{c}$. Also, this rth individual has marginal HARA utility characterized by $d_r(A_r+B\hat{c})^{-1/B}$ and shadow price of wealth constraint $\lambda_r = d_r\mu_r^{1/B}$ with individual wealth now $W_0^r = \frac{1}{M}\Sigma_i W_0^i$. If all individuals in this economy are identical, then the aggregation of consumption and sharing implies that this identical individual is the rth individual.

In this case, the economy produces the same equilibrium prices and consumption sharing as if all individuals are identical, like the rth individual. (This is analogous to the idea of risk-averse agents behaving as if their preferences are risk neutral in a no-arbitrage situation as expounded in Chap. 7.) The rth individual is called the representative agent of the economy, and one can work out the full equilibrium results with just the representative individual's optimal consumption-investment solution since he/she represents everyone in the economy. Utility functions such as HARA that allows such an aggregation into a representative agent is said to possess the aggregation property. Representative agents, such as the above should be distinguished from assumption of homogeneous or identical agents where there may not be the market aggregation property.

Besides the first key idea that a competitive market economy under a free price mechanism is Pareto optimal, a second key idea is that when the market moves from being incomplete to being complete, then constrained Pareto optimality in the former moves to improved total welfare and unconstrained Pareto optimality in the latter. Thus, there is a strong advocacy for market completion, of introducing more financial products to span hitherto unreached states of uncertainty.

This can be intuitively explained as follows. Suppose individuals' utility functions are now functions of multiple goods A, B, and C. A and C are traded but there is no market for B, so B is not traded. Thus, there is no market price for B. Individual X may be well-endowed with lots of units of B and few units of A, so he/she would prefer to sell B to buy A and increase his/her welfare if indeed he/she could sell B. But he/she cannot, since there is no market for B. On the other hand, another individual Y could be endowed with lots of A and few units of B, and would prefer to sell A for more B. Under market equilibrium, their original endowed amounts of B remain unchanged while X and Y can only trade A and C. This equilibrium can be Pareto-improved by introducing an added market for the trading of good B.

Markets such as stock securities exist to enable risk sharing by investors (each individual choosing risk-return tradeoffs to maximize their own preferences with given traded market prices under general equilibrium). In more primitive times when stock or securities markets were not readily accessible, societies and associations existed for mutual help and risk sharing. Risk sharing is also like insurance. Individuals who do not like the high risk of state s occurring (when he or she would be poorer) can buy more of state s securities, and sell in turn more of state r securities to another who does not like the high risk of state r occurring.

In general, opening up more stocks or securities is said to make the markets more complete and thus improve risk sharing and enable Pareto improvement. (Note that Pareto improvement does not imply that the old allocation was or was not ex-ante Pareto optimal. It just means an increase in total welfare under the new Pareto optimality.)

While this is an underlying dogma of increasing the number of securities to increase society welfare, think of why the huge increase of derivatives and structured products may have caused Warren Buffett to suggest that derivatives are financial weapons of mass destruction (this certainly appeared to be the case in the global financial crisis of 2008). The key reason in a one-line answer is that there is no contradiction about the good of completing a market, but in so doing, if hedge funds or banks leveraged so excessively to bet or speculate on one side of the market using these derivatives (where trading exposures were not transparent), then any loss by a large entity that was deemed too big to fail would jeopardize the entire global market. Thus, it is seen also that the persuasion of the Pareto optimality doctrine does not consider future possible dynamics, but is only content with the current allocation.

8.4. Multi-period Optimal Consumption and Investment

We first consider what is a complete market setting where every future uncertain state of nature is spanned or attainable by existing securities. In other words, there exists time-state contingent (or Arrow–Debreu certificates) claims: each time-state security in time-state $t, s \in S$ pays \$1 iff when future state s occurs at t, and 0 otherwise. Such a ts-claim costs ϕ_{ts} at current time $t = 0$. We assume logarithmic utility of the CRRA class, but add a time-preference discount $0 < \delta < 1$ so that more distant utilities are smaller by this discount factor, i.e., less utility enjoyment of 1 unit of consumption in future than now. This also has the effect of making the utility function time-dependent in a simple way, viz. $U(C,t) = \delta^t \ln(C_t)$. Since it is log utility which belongs to the HARA class, we can assume the presence of a representative individual investor who lives till T and whose objective function is to maximize lifetime preference as follows:

$$\max_{C_0, C_{ts}, \lambda} L \equiv \ln(C_0) + \sum_t^T \sum_s^S \pi_{ts} \delta^t \ln(C_{ts})$$
$$+ \lambda \left((W_0 - C_0) - \sum_t^T \sum_s^S \phi_{ts} C_{ts} \right), \tag{8.12}$$

where W_0 is initial wealth at $t = 0$, and C_0 is consumption at $t = 0$. After the consumption at $t = 0$, the investor then allocates his/her investment in a portfolio of state securities with C_{ts} number of time-state ts-securities. This ensures the individual will enjoy consumption units C_{ts} in time-state ts when that occurs. We may, without loss of generality, think of the consumption units as dollars.

Note that the $t+1$ wealth or consumption is indeed \$ $C_{t+1,s}$ given that state s occurs at $t+1$. As usual, $\lambda \geq 0$ is the Lagrange multiplier. It is noted that unless there is new information with respect to *ex ante* probability beliefs about the occurrences of the states, there will not be any more trading once the market clears at time $t = 0$. This is a unique situation with complete markets where all future time-state simple securities can be purchased at time $t = 0$.

The FOC's are

$$\frac{\partial L}{\partial C_0} : 1/C_0 = \lambda$$

$$\frac{\partial L}{\partial C_{ts}} : \frac{\pi_{ts}}{C_{ts}} \delta^t - \lambda \phi_{ts} = 0 \Rightarrow \phi_{ts} = \frac{\pi_{ts} \delta^t}{C_{ts} \lambda}, \quad \text{for every } t, s$$

Therefore

$$\phi_{ts} = \pi_{ts}\delta^t \frac{C_0}{C_{ts}} \Rightarrow C_{ts} = \frac{\pi_{ts}}{\phi_{ts}}\delta^t C_0.$$

Or

$$\sum_t \sum_s \phi_{ts} C_{ts} = C_0 \sum_t \sum_s \pi_{ts}\delta^t = W_0 - C_0$$

$$\Rightarrow C_0 \left[1 + \sum_t \sum_s \pi_{ts}\delta^t\right] = W_0$$

$$\Rightarrow C_0 = \frac{W_0}{[1 + \sum_t \sum_s \pi_{ts}\delta^t]} = \frac{W_0}{1 + \sum_t \delta^t}.$$

Thus

$$C_{ts} = \delta^t \frac{\pi_{ts}/\phi_{ts}}{1 + \sum_t \delta^t} W_0. \tag{8.13}$$

From Eq. (8.13), consumption in time-state ts is proportional to π_{ts}/ϕ_{ts}, and is some fraction of W_0. Rearranging terms,

$$\phi_{ts} C_{ts} = \pi_{ts} \frac{\delta^t}{1 + \sum_t \delta^t} W_0.$$

Then, $\sum_s \phi_{ts} C_{ts} = \frac{\delta^t}{1+\sum_t \delta^t} W_0$. And so,

$$\frac{\phi_{ts} C_{ts}}{\sum_s \phi_{ts} C_{ts}} = \pi_{ts}. \tag{8.14}$$

Equation (8.14) says that the fraction of time-t consumption spent on state s claims at time t, for consumption in that time-state, is equal to the probability of state s at time t. This fraction is independent of wealth. Indeed, from Eq. (8.13), the numbers of units of contingent claims ts that are purchased are increasing in the *ex ante* probability of making the claims in those states, i.e., π_{ts}, and decreasing in the claim's price ϕ_{ts}.

So far, the multi-period consumption-investment problem of the representative agent in Eq. (8.2) is similar to that in Eq. (8.12) except for introducing a time-preference discount. So, the outcomes of the optimization in this section also attains the properties of Pareto optimality under competitive equilibrium, given an initial endowment W_0 and the returns distribution.

In a complete market such as this where every time-state contingent claim is traded, consumption and investment decisions need only be made at the beginning time $t = 0$. In effect, the market opens for trading only once at $t = 0$, and all insurance contracts on future time-state consumptions are bought. This is tantamount to a single-period model where each time-state is basically serving as a single-period state and the single-period utility is a time-aggregated utility.

In a less trivial model where the market opens each period for trading in all the time-state securities, then the heterogeneous individual (whose beliefs of *ex ante* probabilities of future states may change over time) can trade each period and can adjust his or her position, depending on the state of the world up to that time. In such a situation, it is clearly Pareto improving for the heterogeneous individuals. The worst the individual can do is to choose to trade once at the beginning and then decide not to trade later. Hence, being able to trade later will improve preferences. It is like optimization without the constraint of no subsequent trading and thus leads to increase in welfare. However, another possible trade system is where in each period, the individuals can trade only state securities pertaining to the next period and not future time-state securities. In this case, as the future unravels, there is more information for the individuals to make better state security allocations. Thus, it is Pareto improving.

We now consider a more realistic multi-period optimal consumption and investment problem where trading occurs in every period, such that the market needs not be complete, and show that in general, an optimization technique called dynamic programming is required.

In terms of nomenclature, we stick to each period being an interval $[0, 1)$ for period 1, $[1, 2)$ for period 2, and so on. There is a decision or event time-point at $t = 0$ which is the beginning of period 1, at time-point $t = 1$ which is the beginning of period 2, and so on. This avoids some confusion in the literature where sometimes a "period" is treated as a time-point.

An individual lives from $t = 0$ till $t = \mathrm{T}$. He/she maximizes lifetime utility as follows:

$$\max_{\{c_t, w_t\}} E_0 \left[\sum_{t=0}^{T} \delta^t \ln(C_t) \right] \equiv J(W_0, S_0, t = 0),$$

subject to initial wealth W_0, exogeneous security prices S_0, and no new injection of capital or wealth. $J(\cdot)$ is called the value function. Any withdrawal is only in the form of consumption in each period. (Some finance models do not impose withdrawal but are just concerned about optimal portfolio and final investment or wealth outcomes. Those tend to be models trying to establish optimal investment rules. The models with consumption withdrawal tend to be involved in equilibrium outcomes and may use aggregate consumption as a factor in asset pricing.)

There are N securities (which may include a risk-free bond) for investments, and the weight or fraction of wealth allocated to the particular security i is w_t^i. w_t is a vector $\left(w_t^1, w_t^2, w_t^3, \ldots, w_t^N\right)$. The weights add to one, or $w_t^T I = 1$, where I is a N by 1 vector of ones as elements.

Using the dynamic programming method, we first assume that all consumption of a single good $(c_0, c_1, \ldots, c_{T-2})$ and portfolio investment decisions $(w_0, w_1, \ldots, w_{T-2})$ up to and including time $T-2$ are optimal. Then, at $T-1$, or the beginning of $[T-1, T)$, given S_{T-1}, W_{T-1}, the program is to solve:

$$J(W_{T-1}, S_{T-1}, T-1) = \max_{c_{T-1}, w_{T-1}} \ln(c_{T-1}) + E_{T-1}[\delta \ln(c_T)]$$

where S_t is the vector of N security prices at time t, $\forall t$.

Thus, at $T-1$, the program becomes

$$\begin{aligned} J(W_{T-1}, S_{T-1}, T-1) \equiv \max_{c_{T-1}, w_{T-1}} & \ln(c_{T-1}) \\ & + E_{T-1}[\delta \ln([W_{T-1} - c_{T-1}] r_T)] \end{aligned}$$

s.t. $r_T = w_{T-1}^T R_T$, where R_T is the vector of security returns over $(T-1, T]$, and r_T is the scalar portfolio return over $(T-1, T]$. Or

$$\begin{aligned} J(W_{T-1}, S_{T-1}, T-1) \equiv \max_{c_{T-1}, w_{T-1}} & \ln(c_{T-1}) + \delta[\ln(W_{T-1} - c_{T-1})] \\ & + \delta E_{T-1}\left[\ln\left(w_{T-1}^T R_T\right)\right] \end{aligned} \tag{8.15}$$

s.t. $I^T w_{T-1} = 1$.
FOC:

$$\begin{aligned} \frac{\partial}{\partial c_{T-1}} : \frac{1}{c_{T-1}} - \frac{\delta}{W_{T-1} - c_{T-1}} &= 0 \\ \Rightarrow W_{T-1} - c_{T-1} &= \delta c_{T-1} \\ \Rightarrow c_{T-1} &= \frac{W_{T-1}}{1+\delta}. \end{aligned}$$

We see that under log utility, consumption is a constant $1/(1+\delta)$ proportion of wealth and independent of past wealth or other state variables. This is called a myopic behavior. It is also independent of the investment opportunities R_T. Investment and consumption decisions are independent of each other in this special case of log utility.

The objective function in Eq. (8.15) can also be written as a Lagrangian:

$$\max_{c_{T-1}} \ln(c_{T-1}) + \delta[\ln(W_{T-1} - c_{T-1})]$$
$$+ \delta \max_{w_{T-1}} E_{T-1}\left[\ln\left(w_{T-1}^T R_T\right)\right] + \lambda_{T-1}\left(1 - w_{T-1}^T I\right). \tag{8.16}$$

The associated optimization problem is the portfolio return optimization at $T-1$:

$$\max_{w_{T-1}} \sum_s \pi_s \ln\left(\sum_i w_{T-1}^i R_s^i\right) + \lambda_{T-1}\left(1 - \sum_i w_{T-1}^i\right) \tag{8.17}$$

where R_s^i is security i's return in state s. Suppose the number of states is N. Then, $\sum_s^N \pi_s = 1$.
For FOC:

$$\frac{\partial L}{\partial w_{T-1}^i} : \sum_s \pi_s \frac{R_s^i}{R_s^p} - \lambda_{T-1} = 0, \quad \forall i$$

where $R_s^p = \Sigma_i w_{T-1}^i R_s^i$. This implies

$$E_{T-1}\left(\frac{\tilde{R}_i}{\tilde{R}_p}\right) = \lambda_{T-1},$$

where $\tilde{R}_i$ is the ith element in R_T and $\tilde{R}_P$ is r_T.

Existence of a risk-free asset with return R_f implies

$$E_{T-1}\left(\frac{\tilde{R}_i - R_f}{\tilde{R}_p}\right) = 0, \quad \forall i.$$

Hence

$$E(\tilde{r}_i - r_f) = \frac{\operatorname{cov}(\tilde{r}_i, -1/\tilde{R}_p)}{E(1/\tilde{R}_p)} \tag{8.18}$$

where $r_i = R_i - 1$ and $r_f = R_f - 1$. Here, we see that higher covariance between r_i and R_p leads to higher risk premium on the LHS.

At $T-1$, the value function (remaining value of the optimal objective function) is

$$J(W_{T-1}, S_{T-1}, T-1) = \ln\left[\frac{W_{T-1}}{1+\delta}\right] + \delta\left[\ln\left(W_{T-1} - \frac{W_{T-1}}{1+\delta}\right)\right] + \delta\zeta_{T-1}$$

where $\zeta_{T-1} = E_{T-1}\left[\ln\left(w_{T-1}^T R_T\right)\right]$. Or

$$J(W_{T-1}, S_{T-1}, T-1) = \ln[W_{T-1}] + \delta\ln[W_{T-1}] + \delta\zeta_{T-1} + g_{T-1}(\delta)$$
$$= (1+\delta)\ln[W_{T-1}] + \delta\zeta_{T-1} + g_{T-1}(\delta),$$

where the last term contains expressions in δ that are collected together. In this case, $g_{T-1}(\delta) = \delta \ln(\delta) - (1+\delta)\ln(1+\delta)$.

Now, we roll time backwards and optimize at $T - 2$, assuming all consumption and portfolio investment decisions up to and including time $T-3$ are optimal, and making use of the value function $J(W_{T-1}, S_{T-1}, T-1)$ that now represents the final remaining optimized value ahead. The existence and use of this function is a cornerstone of the backward dynamic programming approach. The value function essentially allows the multi-period problem to be simplified to a "two-period" problem, with the remaining future all lumped into one value function as if it were the final state. This is the principle of optimality allowing the recurrence "2-period" equation to be solved at each time t backwards. The equation is also called the Bellman equation.

Similar to the step in (8.15), at $T - 2$

$$
\begin{aligned}
J(W_{T-2}, S_{T-2}, T-2) &= \max_{c_{T-2}, w_{T-2}} \ln(c_{T-2}) \\
&\quad + \delta E_{T-2}[J(W_{T-1}, S_{T-1}, T-1)] \\
&= \max_{c_{T-2}, w_{T-2}} \ln(c_{T-2}) + \delta E_{T-2}[(1+\delta)\ln[W_{T-1}] \\
&\quad + \delta \zeta_{T-1} + g_{T-1}(\delta)] \\
&= \max_{c_{T-2}, w_{T-2}} \ln(c_{T-2}) \\
&\quad + \delta(1+\delta) E_{T-2}[\ln[W_{T-2} - C_{T-2}] r_{T-1}] \\
&\quad + \delta^2 E_{T-2}[\zeta_{T-1}] + \delta g_{T-1}(\delta)
\end{aligned}
$$

$\ni r_{T-1} = w_{T-2}^T R_{T-1}$, and $I^T w_{T-2} = 1$. In the same way using FOC, we solve

$$
c_{T-2} = \frac{W_{T-2}}{1+\delta+\delta^2},
$$

and the optimal portfolio weight problem using the approach in Eq. (8.17), obtaining similar risk premium characterization in Eq. (8.18). This is continued backward till finally $J(W_0, S_0, 0)$ is obtained.

Complete versus Incomplete Markets

The idea of Pareto optimality as a weak but perhaps minimal sense of exchange or trading efficiency was discussed. For the related but dissimilar idea of valuation or asset pricing, the same framework of rational consumption-investment problem is developed. We now provide further in-depth observations. We employ the concept of a representative agent. (We assume such an agent exists under suitable utility function classes, and thus need not be concerned with heterogeneous agent preferences under the utility class.)

The complete markets framework as developed in (8.2) and (8.3) shows that at time t, a state s contingent claim (paying \$1 iff state s occurs) is priced as follows. It could also be construed as a one-period model as all purchases of time-state securities are done altogether at $t = 0$. Suppose we add a time discount factor $0 < \delta \leq 1$ for the next period (or end-of-period) utility function, to reflect a weakly stronger preference for current consumption of \$1 to a future consumption of the same \$1, *ceteris paribus.* Then, state s security price is

$$\phi_s = \pi_s \delta \frac{U'(C_s)}{U'(C_0)}. \tag{8.19}$$

Equation (8.19) under complete markets holds for every state s. Thus, a state price ϕ_s is available in the market for every state s. Equation (8.19) cannot be obtained in an incomplete market. For example, suppose state s' does not have a traded state s' security. If $U(C_{s'}) > 0$ and $\pi_{s'} > 0$, still there does not exist a traded price $\phi_{s'} = \pi_{s'} \delta \frac{U'(C_{s'})}{U'(C_0)}$.

Now, consider a canonical problem in incomplete markets, for an equivalent one-period model of a representative agent.

$$\max_{C_0, w_1, \ldots, w_N} \quad U(C_0) + E_0[\delta U(C_1)]$$

$$\text{s.t.} \quad C_1 = (W_0 - C_0) \left(\sum_{i=1}^{N} w_i \frac{P_{1i}}{P_{0i}} \right)$$

$$\text{and} \quad \sum_{i=1}^{N} w_i = 1,$$

where P_{ti} is the asset i's price at time t, and W_0 is initial wealth. Note that all end of period wealth W_1 is consumed, i.e., $C_1 = W_1$. It can be re-expressed as a Lagrangian L:

$$\max_{C_0, w_1, \ldots, w_N, \lambda} L \equiv U(C_0) + E_0 \left[\delta U \left((W_0 - C_0) \left(\sum_{i=1}^{N} w_i \frac{P_{1i}}{P_{0i}} \right) \right) \right] + \lambda \left(1 - \sum_{i=1}^{N} w_i \right).$$

The FOCs are:

$$U'(C_0) - \delta E_0 \left[U'(C_1) \left(\sum_i w_i \frac{P_{1i}}{P_{0i}} \right) \right] = 0 \quad (8.20)$$

$$\delta(W_0 - C_0) E_0 \left[U'(C_1) \frac{P_{1i}}{P_{0i}} \right] = \lambda, \forall_i \quad (8.21)$$

and

$$\sum_i w_i = 1.$$

Equations (8.20) and (8.21) together imply

$$E_0 \left[\delta \frac{U'(C_1)}{U'(C_0)} R_{1i} \right] = 1, \quad (8.22)$$

where return to any traded security i at $t = 1$ is $R_{1i} = \frac{P_{1i}}{P_{0i}}$. Or in terms of prices, we can also obtain

$$P_{0i} = E_0 \left[\delta \frac{U'(C_1)}{U'(C_0)} P_{1i} \right]. \quad (8.23)$$

Comparing the equilibrium asset pricing outcomes of complete markets in Eq. (8.19) and of incomplete markets in Eq. (8.23), it is seen that the complete markets situation provides more information. We can price a security such as stock i with payoff $P_{1i}(s)$ in different states s, using (8.19) (for the simple state securities of a complete market):

$$\begin{aligned} P_{0i} &= \sum_s \pi_s \left[\delta \frac{U'(C_s)}{U'(C_0)} P_{1i}(s) \right] \\ &= \sum_s \phi_s P_{1i}(s). \end{aligned}$$

However, we cannot use (8.23) to price all state securities, certainly not those states s' that are not spanned or traded.

For asset pricing, in most of these models that are derived by marginal analyses, i.e., price of a share today is equal in value to foregone consumption today for a better tomorrow by investing the price in the share that will yield some return tomorrow, there is a unifying theme. Whether the model is complete or incomplete markets, as soon as the returns distributions are assumed to be multivariate normal, then all the pricing models will yield a version of the Sharpe-Lintner CAPM. Thus, the central idea of positive risk premium, or return compensation to investor, being some kind of systematic undiversifiable risk, is very powerful indeed, as first discussed in Chap. 6. We show below how all the models we have come across so far can be related to the SL-CAPM.

(1a) In the incomplete markets model above, via Eq. (8.22), let $M_1 = \delta U'(W_1)/U'(C_0)$ which is called the stochastic discount factor or price kernel, so

$$E_0[(1 + r_{1i})M_1] = 1,$$

where we denote return rate $r_{1i} = R_{1i} - 1$. M_1 is called a stochastic discount factor because $E_0[P_{1i}M_1] = P_{0i}$, so M_1 acts like a discount factor.

In general, we shall use

$$E_t[(1 + r_{t+1})M_{t+1}] = 1,$$

where the expectation is taken a period earlier in a two time-point framework, and we do not show the i subscript and shall treat the return rate as of any tradeable security or portfolio in the market. Then, we can obtain

$$E(1 + r_{t+1}) = \frac{1}{E(M_{t+1})} - \frac{\text{cov}(r_{t+1}, M_{t+1})}{E(M_{t+1})}.$$

This is also true for a risk-free asset, assuming it exists (e.g., treasury bonds). Then

$$E(1 + r_f) = \frac{1}{E(M_{t+1})} - \frac{\text{cov}(r_f, M_{t+1})}{E(M_{t+1})} = \frac{1}{E(M_{t+1})},$$

since r_f, the risk-free rate in this period is treated as a constant. (Over more periods, it is possible for the risk-free rate to vary.) Taking the difference in the last two equations, therefore

$$E(r_{t+1} - r_f) = -\frac{\text{cov}(r_{t+1}, M_{t+1})}{E(M_{t+1})}.$$

This can also be expressed using $M_{t+1} = \delta U'(W_{t+1})/U'(C_t)$ as

$$E(r_{t+1} - r_f) = -\frac{\text{cov}\big(r_{t+1}, U'([W_t - C_t]R^p_{t+1})\big)}{E\big(U'([W_t - C_t]R^p_{t+1})\big)} \tag{8.24}$$

where R^p_{t+1} is the market portfolio return (factor).

(1b) In Chap. 6, Eq. (6.20), we also saw the incomplete market state preference model, where RVd_{t+1} is the state price density, and

$$E_t\left(R_{t+1}\frac{d_{t+1}}{1+r_f}\right) = 1$$

$$\Rightarrow \text{cov}\left(R_{t+1}, \frac{U'\left([W_t - C_t]R^p_{t+1}\right)}{E_t\left(U'\left([W_t - C_t]R^p_{t+1}\right)\right)}\right) + E(R_{t+1}) = 1 + r_f,$$

or

$$E(r_{t+1} - r_f) = -\text{cov}\left(R_{t+1}, \frac{U'([W_t - C_t]R^p_{t+1})}{E_t(U'([W_t - C_t]R^p_{t+1}))}\right), \tag{8.25}$$

since $M_{t+1} = d_{t+1}/(1 + r_f)$ and $E_t(d_{t+1}) = 1$.

Equations (8.24) and (8.25) are identical. Hence, these models are different ways of saying the same thing. Assuming multivariate normality $\Rightarrow r^p_{t+1}$ and r^j_{t+1} for any security j are bivariate normal, then we can apply a result as follows. For bivariate normal distribution of general RVs Y and Z, for differentiable function $f, \text{cov}(Y, f(Z)) = E(f'(Z))\text{cov}(Y, Z)$ for bounded $f'(Z)$. This is also called Stein's lemma. We also assume utility function is twice differentiable.

RHS of Eq. (8.24) becomes:

$$-\frac{\text{cov}\left(r_{t+1}, U'\left(W'_t R^p_{t+1}\right)\right)}{E\left(U'\left(W'_t R^p_{t+1}\right)\right)} = -\frac{\text{cov}\left(r_{t+1}, W'_t R^p_{t+1}\right) E_t\left[U''\left(W'_t R^p_{t+1}\right)\right]}{E\left(U'\left(W'_t R^p_{t+1}\right)\right)}.$$

Hence

$$E\left(r^j_{t+1} - r_f\right) = -W'_t \frac{\text{cov}\left(r^j_{t+1}, R^M_{t+1}\right) E\left[U''\left(W'_t R^M_{t+1}\right)\right]}{E\left(U'\left(W'_t R^M_{t+1}\right)\right)},$$

putting the representative agent's optimal portfolio as the market portfolio M. Thus also,

$$E\left(r^M_{t+1} - r_f\right) = -W'_t \frac{\text{cov}\left(r^M_{t+1}, R^M_{t+1}\right) E_t\left[U''\left(W'_t R^M_{t+1}\right)\right]}{E\left(U'\left(W'_t R^M_{t+1}\right)\right)}.$$

where $W'_t = W_t - C_t$.

Dividing one by the other of the above two equations, we obtain the SL CAPM.

For Eq. (8.18) in the multi-period framework under log utility where the investment problem becomes separable, again we can show it is CAPM if the distribution is normal. This is left as an exercise. The CAPM results can also be obtained when we assume in single-period models that investors have quadratic utility function as discussed in Chap. 6.

8.5. Modern Asset Pricing Framework

In general, analytical solutions to a dynamic multi-period consumption-investment problem may not be available. Analytical solutions such as the above are possible with simplifying conditions such as additive utility functions over time, and state-independent utilities. The log case is especially myopic in the sense that the consumption optimization and the investment optimization are separable, and security returns that are i.i.d. over time are assumed to simplify the period-by-period backward optimization. It is Markov in the sense it is conditional only on the last period observed security prices S_t for the consumption and investment decisions at t. This makes the solution a lot easier and tractable.

Empirical efforts to validate models and to provide clues to the pricing anomalies and provide guidances for valuation of securities,[4] and related theoretic macroeconomic decision-making (since aggregate consumption is involved) meanwhile press for some standard framework of asset pricing that is not mathematically too cumbersome. Simplicity and parsimony means that it does not require a full and complete solution to a multi-period problem for $t = 0, 1, 2, \ldots, T$, and that it does not make too many simplifying assumptions so as to be implausible right from the start.

A couple of important papers, coupled with a path-breaking econometric technique (generalized method of moments[5]) introduced the idea of testing asset pricing models under rationality (or rational optimization

---

[4]Excellent books covering these subjects ranging from more classical 1960s, 1970s, and 1980s treatments to present-day issues in asset pricing encompassing some new behavioral finance ideas, include, not exhaustively: Ingersoll, JE Jr., (1987). *Theory of Financial Decision-Making.* Rowman and Littlefield; Huang, CF and RH Litzenberger (1988). *Foundations for Financial Economics.* North-Holland; Campbell, JY, AW Lo and AC MacKinlay (1997). *The Econometrics of Financial Markets.* Princeton University Press; Cochrane, JH (2001). *Asset Pricing.* Princeton University Press; Bossaerts, P (2002). *The Paradox of Asset Pricing.* Princeton University Press; Campbell, JY and LM Viceira (2002). *Strategic Asset Allocation.* Oxford University Press; and for more macroeconomics flavor, Stokey, N, RE Lucas, Jr. and EC Prescott (1989). *Recursive Methods in Economic Dynamics.* Harvard University Press; Korn, R (1997). *Optimal Portfolios.* World Scientific.

[5]Hansen, (1982). Large sample properties of generalized method of moments. *Econometrica* 50; and Hansen and Singleton (1982). Generalized instrumental variables estimation of nonlinear rational expectations models. *Econometrica* 50.

and market equilibrium conditions) using only the necessary condition of optimality, sometimes called the Euler equation.[6]

This is shown as follows. The representative individual's problem is

$$\max_{\{C_t, w_t\}} E_0 \left[\sum_{t=0}^{\infty} \delta^t U(C_t) \right] \text{ s.t. } W_{t+1} = (W_t - C_t) \sum_i w_i R_i, \sum_i w_i = 1, \ \forall t$$

where we assume stationary return distribution of all stock returns R_i and W_t is wealth at time t, with initial condition as W_0. $w_t = (w_1, w_2, \ldots, w_N)$ were N is the number of securities. Moreover, returns are independent of consumption decisions.

Assuming there is an optimal solution to this dynamic optimization problem, but which we do not wish to add more assumptions to derive the solution explicitly, then at each t, there exists a value function $V_t(W_t)$ (we can think of this as an indirect utility function) such that the individual maximizes:

$$V_t(W_t) = \max_{C_t, w_t} U(C_t) + \delta E_t V_{t+1}(W_{t+1}) \tag{8.26}$$

$$\text{s.t. } W_{t+1} = (W_t - C_t) \sum_i w_i R_i \quad \text{and} \quad \sum_i w_i = 1. \tag{8.27}$$

The above Eq. (8.26) is a recursive equation for each t, and is the Bellman equation. The value function at time t is equal to the utility of optimal consumption at time t plus the expected value of the discounted optimal value function at time $t+1$. If we know the functional form of $V(\cdot)$, then the agent's optimization is reduced to a single-period (two time-point decisions) problem.

FOC is:

$$\frac{\partial}{\partial C_t} : U'(C_t) - \delta E_t \left[V'_{t+1}(W_{t+1}) \sum_i w_i R_i \right] = 0$$

$$\Rightarrow U'(C_t) = \delta E_t \left[V'_{t+1}(W_{t+1}) \sum_i w_i R_i \right]. \tag{8.28}$$

[6]This nomenclature may also be called the Euler–Lagrange equation in calculus of variation and is basically a differential equation, discrete or continuous, in which the solutions are the functions for which a given functional is stationary. The property of being a differentiable functional which is stationary at its local maxima and minima is useful for solving optimization problems, i.e., given some functional, one seeks the function minimizing or maximizing it. For example, $V(\cdot)$ is a function of policy function $C(W_t)$.

$$\frac{\partial}{\partial w_i} : \delta E_t\left[V'_{t+1}(W_{t+1})(W_t - C_t)R_i\right] - \lambda = 0$$

$$\Rightarrow \delta E_t\left[V'_{t+1}(W_{t+1})R_i\right] = \frac{\lambda}{W_t - C_t}, \tag{8.29}$$

where λ is the Lagrange multiplier over the portfolio weight constraint.

If we sum the last equation over all weights w_i in Eq. (8.29):

$$\sum_i w_i(\delta E_t[V'_{t+1}(W_{t+1})R_i]) = \delta E_t\left[V'_{t+1}(W_{t+1})\sum_i w_i R_i\right]$$
$$= \frac{\lambda}{W_t - C_t}.$$

From (8.28), LHS is $U'(C_t)$. So, $\frac{\lambda}{W_t - C_t} = U'(C_t)$. Using Eq. (8.29), we obtain

$$U'(C_t) = \delta E_t\left[V'_{t+1}(W_{t+1})R_i\right]. \tag{8.30}$$

Now, we still have two different quantities, viz. U' on the LHS and V' on the RHS of Eq. (8.30).

Consider the Bellman Eq. (8.26). On its LHS, function $V_t(W_t)$ is actually an optimized form of $V_t\big(W^*_{t+1}(C_t), W_t\big)$ in which W^*_{t+1} is maximized given W_t via control on C_t and portfolio weights w_i.

We can make use of the Envelope theorem: $\max f(x,a) = \max f(x(a),a)$, where $x(a)$ is evaluated at optimal f given a, i.e., $\frac{\partial f}{\partial x}|_a = 0$ yields $x^* = g(a)$. Then, for maximum f, FOC yields:

$$df/da = \frac{\partial f}{\partial x}(dx/da) + \frac{\partial f}{\partial a} = 0,$$

or,

$$\frac{df(g(a),a)}{da} = 0 + \frac{\partial f}{\partial a}|_{x^*=g(a)}$$

for optimality. Above, C_t is like x and W_t is like parameter a.

We can therefore take the derivative w.r.t. W_t on the LHS and also a partial derivative on the RHS of Eq. (8.26) to obtain

$$V'_t(W_t) = \delta E_t\left[V'_{t+1}(W_{t+1})\frac{\partial W_{t+1}}{\partial W_t}\right], \tag{8.31}$$

where the RHS is evaluated at optimal C_t and w_t.

Employing budget constraint (8.27):

$$V_t'(W_t) = \delta E_t\left[V_{t+1}'(W_{t+1})\sum_i w_i R_i\right]. \tag{8.32}$$

Then, Eqs. (8.28) and (8.32) imply $U'(C_t) = V_t'(W_t)$. If $U(C_t)$ is a stationary policy, so is $V(W_t)$ and we drop the subscripts t. So also $V'(W_{t+1}) = U'(C_{t+1})$, where it means the marginal utility of consuming one real dollar less at time t or marginal cost is equated to the marginal benefit or expected marginal utility from investing one real dollar and consuming at $t+1$.

Substitute $U'(C_{t+1})$ for $V'(W_{t+1})$ in Eq. (8.30) to obtain

$$U'(C_t) = \delta E_t[U'(C_{t+1})R_i]. \tag{8.33}$$

Putting $R_i = \frac{P_{t+1}^i}{P_t^i}$ where P_t^i is stock i's price at time t, then

$$P_t^i U'(C_t) = \delta E_t\left[U'(C_{t+1})P_{t+1}^i\right].$$

The last equation is called the stochastic Euler equation, and is a necessary but not sufficient condition for any solution to a rational asset pricing equilibrium. This necessary condition is a heavily tested condition in much of empirical asset pricing literature.

We may write the stochastic Euler Eq. (8.33) as

$$E[M_{t+1}R_{t+1}] = 1, \tag{8.34}$$

where $M_{t+1} = \delta\frac{U'(C_{t+1})}{U'(C_t)}$ is the price kernel ("kernel" is used as it refers to the integrand or integral kernel in calculus, being a term within the expectation operator) or stochastic discount factor (as it discounts the future stock price to the current price), and letting R_{t+1} denote the return to any traded security at time $t+1$. $R_{t+1} = P_{t+1}/P_t$. Unconditional expectation has also been taken over the conditional expectation in (8.33) where the Law of Iterated Expectations applies.

The Euler Eq. (8.34) then implies

$$E(R_{t+1})E(M_{t+1}) + \text{cov}(M_{t+1}, R_{t+1}) = 1.$$

As the risk-free asset with return R_f also satisfies (8.34), then $R_f E(M_{t+1}) = 1$, or $E(M_{t+1}) = 1/R_f$. Hence

$$E(R_{t+1}) - R_f = -\text{cov}(M_{t+1}, R_{t+1})/E(M_{t+1}) \tag{8.35}$$

or

$$E(R_{t+1}) = R_f + \frac{\text{cov}(-M_{t+1}, R_{t+1})}{\text{var}(M_{t+1})} \times \frac{\text{var}(M_{t+1})}{E(M_{t+1})}. \quad (8.36)$$

If we define $\beta_C = \frac{\text{cov}(-M_{t+1}, R_{t+1})}{\text{var}(M_{t+1})}$, and $\lambda_M = \frac{\text{var}(M_{t+1})}{E(M_{t+1})}$ in (8.36), then λ_M behaves as a market premium (> 0) or the price of common risk to all assets, and β_C is a consumption beta specific to the asset with return R_{t+1}.

The intuition of the consumption beta, β_C above is as follows. Suppose the asset's $\beta_C > 0$, and thus its return R_{t+1} correlates positively with $-M_{t+1}$. In turn, this implies a positive correlation between R_{t+1} and C_{t+1} since $U''(C_{t+1}) < 0$. Holding the asset in a portfolio thus adds to the consumption volatility. Since consumption life-cycle theory suggests consumption smoothing (less consumption volatility) as desirable for a risk-averse individual over his/her lifetime, adding to consumption volatility would require risk compensation in the form of higher expected returns for the asset. This is indeed the case for (8.36) when the market premium is strictly positive.

Equation (8.35) can be re-written as

$$E(R_{t+1}) - R_f = -\rho_{MR}\sigma_M\sigma_R/E(M_{t+1}),$$

where ρ_{MR} is the correlation coefficient between M_{t+1} and R_{t+1}, and σ_M, σ_R denote the respective standard deviations. Then

$$\frac{E(R_{t+1} - R_f)}{\sigma_R} = -\rho_{MR}\frac{\sigma_M}{E(M_{t+1})} \leq \frac{\sigma_M}{E(M_{t+1})}. \quad (8.37)$$

In Eq. (8.37), the LHS is the Sharpe ratio or performance of the stock showing its equity premium (or excess return over risk-free rate) divided by its return volatility. The RHS is called the Hansen–Jagannathan bound, which provides a theoretical upper bound to the Sharpe ratio or to the equity premium. (Recall that M here refers to the price kernel or stochastic discount factor, and should not be confused with the market return that also typically uses this notation.) The RHS is also $R_f \times \sigma_M$. One interesting research agenda in finance is to account for why post-WWII US equity premiums have been observed to be too high and have exceeded empirical measures of volatility of the stochastic discount factor.

8.6. Problem Set 8

1. In the stock A and safe bank deposit B situation in Table 8.1, suppose the investor has negative exponential utility function $-e^{-aW_0}$. What

is the investor's preferred allocation of wealth W_0 to investment in A when $aW_0 = 3.8$? and when $aW_0 = 7.6$? (a is usually interpreted as the constant absolute risk aversion coefficient in the theory.) Does the percentage optimal investment in the risky asset A decrease with increase in wealth W_0 if a is constant?

2. Suppose there is a representative agent in a single period complete market world where at $t = 0$, he/she maximizes preference $U(C_0) + \sum_s^S U(C_s)$ subject to state s security price ϕ_s and initial wealth W_0. At the end of period, the state s of the world is realized, and the agent receives C_s units of goods for consumption. Let each unit be worth \$1. There is no bequest. His/her *ex ante* belief of the probability of occurrence of state s is π_s.
   (a) Show that $\phi_s = \frac{\pi_s U'(C_s)}{U'(C_0)}$.
   (b) Suppose a complex security or structured product j is sold with terminal or end-of-period payoffs in units of consumption goods as follows: x_s or x units in state s, $\forall s \in S$. Show that its current price is
$$p_j = \sum_s^S \pi_s \left[\frac{U'(C_s)}{U'(C_0)} x_s \right].$$
   (c) Use the above result in (b) to show that the one-period CAPM attains
$$E(\tilde{r}_j - r_f) = \frac{\text{cov}(\tilde{r}_j, \tilde{r}_m)}{\text{var}(\tilde{r}_m)} E(\tilde{r}_m - r_f)$$
   where $1 + \tilde{r}_j = x_s/p_j$ for security j with state payoff x_s, and market portfolio return $\tilde{r}_m$ is such that end-of-period wealth $M = (W_0 - C_0)$ $(1 + \tilde{r}_m)$. (Hint: For bivariate normal distribution of general RVs Y and Z, for twice differentiable function U, $\text{cov}(Y, U'(Z)) = E(U''(Z))\text{cov}(Y, Z)$.)

3. In the dynamic model under log utility, suppose asset valuation is recursively obtained as, over each period t
$$E(\tilde{r}_i - r_f) = \frac{\text{cov}(\tilde{r}_i, -1/\tilde{R}_M)}{E(1/\tilde{R}_M)}$$
   where R_M is the next period market portfolio return. Show under bivariate normality of r_i and $r_M = R_M - 1$, that Sharpe–Lintner CAPM holds. Apply the same hint as in the previous problem.

4. Show how the power utility function $\frac{C^{1-\gamma}}{1-\gamma}$ and also the log utility are both members of the HARA class with the property $U'(C) = d(A + BC)^{-1/B}$. (The power utility including the log utility has the nice feature that when maximizing the utility function, it separates out the original wealth W_0 from the return factor to be maximized. Thus, the maximizing does not depend on the level of wealth. This gives rise to the power utility having a constant relative risk aversion coefficient. HARA becomes the power utility when $A = 0$.)
5. A representative individual lives for 2 periods $[0, 2]$. At $t = 0$, he/she solves: $\max_{C_0, C_s, C_{ss}} \ln(C_0) + [0.5\ln(C_u) + 0.5\ln(C_d)] + [0.25\ln(C_{uu}) + 0.25\ln(C_{ud}) + 0.25\ln(C_{du}) + 0.25\ln(C_{dd})] + \lambda(W_0 - C_0 - \phi_u C_u - \phi_d C_d - \phi_{uu}C_{uu} - \phi_{ud}C_{ud} - \phi_{du}C_{du} - \phi_{dd}C_{dd})$, where the uncertainty evolves as in the two-period binomial tree with up-state U and down-state D each period of probability 0.5 each state. Original endowment is \$ $W_0 = \$100$. C_s is the number of state s contingent claims bought and hence the number of \$ he/she receives if state s at time $t = 1$ occurs, and C_{ss} is the number of \$ he/she receives if state ss at time $t = 2$ occurs. If the market opens only once at $t = 0$, what is his/her optimal consumption C_0 at $t = 0$? Assume state prices $\phi_u = \$0.11$ and $\phi_d = \$0.8$, what are his/her optimal consumptions in state u and in state d?
6. We assume power utility of the CRRA class, $U(C_{ts}) = C_{ts}^{\gamma}$ where $\gamma < 1$, and C_{ts} is the individual's consumption in state ts, i.e., at time t state s. Empirical probability of state s at t as seen at $t = 0$ is π_{ts}. Trading of all state-securities takes place once at $t = 0$. The single individual investor's problem at $t = 0$ is to

$$\max_{\{C_{ts}\}} C_0^{\gamma} + \sum_t^T \sum_s^S \pi_{ts}\delta^t C_{ts}^{\gamma} + \lambda\left(W_0 - C_0 - \sum_t \sum_s \phi_{ts}C_{ts}\right)$$

where δ is the per time-period preference discount, i.e., tomorrow's preference of \$1 is a little less than today's preference of the same \$1 by factor δ — this is more of a specification of utility functions across time. Hence, the investor is in fact maximizing expected time-additive utility function $\sum_t^T U(t, C_{ts}) = \sum_t^T \delta^t \tilde{C}_t^{\gamma}$.

(a) Show that $\phi_{ts} = \frac{\pi_{ts}\delta^t C_{ts}^{\gamma-1}}{C_0^{\gamma-1}}$ for all time-states ts.

(b) Show how C_0 and C_{ts} are explicit functions of W_0.

7. Eq. (8.34) shows the Euler condition with the stochastic discount factor M_{t+1}. Considering a single-period model, show that CAPM is equivalent to the model $M = A + Br^M$ where the stochastic discount factor is affine (linear translation) in market portfolio return r^M. (This shows that the marginal utility of wealth expressed in M is linear in final wealth expressed in $W_0(1 + r^M)$ and does not depend on other state variables in the CAPM. This is similar to saying that the CAPM returns depend only on r^M and not on other state variables.)

Chapter 9

CONTINUOUS-TIME OPTION PRICING

In 1826–1827, the botanist Robert Brown observed the irregular motions of tiny pollen particles suspended in water, that the path of a particle has no tangent at any point (i.e., no slope!), and each movement appeared independent of the previous movement (in 3D, or in 2D if projected on a flat surface). Such movements came to be called Brownian motion (BM). A French mathematician named Louis Bachelier wrote a PhD thesis[1] in 1900 that achieved some success in modeling mathematically similar movements observed in stock prices, and produced the earliest known analytical forms of option prices for standard calls and puts. The Bachelier Finance Society, formed in 1996 by a group of prominent finance mathematicians, was named in his honor.

Albert Einstein had also attempted to model BM mathematically at some point. Norbert Wiener, a mathematician, who once taught at MIT and had discussed electromagnetism with Einstein on a train to Switzerland,[2] developed the generalized process of continuous-time random walk with drift, and such a mathematical process was named after him as the Wiener process. A Wiener process without drift is essentially the BM. In this way, BM spawned an enormous library of works and studies because of its importance and central place in the modeling of stochastic processes.

Financial economists such as Case Sprenkle and Paul Samuelson worked on option pricing again in the early 1960s. Soon after that, Fischer Black and Myron Scholes[3] worked intensely on the option pricing formula. At the same time, Samuelson's student, Robert Merton, was also working on it. Merton also produced fundamental results[4] that enriched the understanding of the Black–Scholes formula at the time.

---

[1]Bachelier, L (1900). Théorie de la Spéculation, *Annales de l'Ecole Normale Supérieure.*
[2]See an interesting letter describing Einstein in http://libraries.mit.edu/archives/exhibits/wiener-letter/index1-12.html.
[3]Their prize-winning Black–Scholes formula is found in Black, F and M Scholes (1973). The pricing of options and corporate liabilities. *Journal of Political Economy*, 81, 637–654.
[4]See Merton, RC (1973). Theory of rational option pricing. *Bell Journal of Economics and Management Science*, 4, Spring, 141–183; and Merton, RC (1976). Option pricing when underlying stock returns are discontinuous. *Journal of Financial Economics*, 3, 125–144.

The Chicago Board of Trade had started trading options in 1973. The Black–Scholes formula, when endorsed by the academic community, became a popular tool for traders, partly due to its analytical simplicity, and helped in the expansion of the options market.

9.1. Brownian Motion

A random walk can be used as a starting point for the study of Brownian Motion (BM). For i.i.d. X_t, $\forall t$

$$X_t = \begin{cases} +1, & \text{with probability } p = \dfrac{1}{2} \\ -1, & \text{with probability } 1 - p = \dfrac{1}{2} \end{cases}$$

where X_t occurs at time t, X_{t+1} occurs at time $t+1$, one unit time apart, and so on.

Define $S_0 = X_0 = 0$, and $S_n = \sum_{t=1}^{n} X_t$ for $n = 1, 2, \ldots$ etc. The process S_n is a symmetric random walk process that denotes the level of the random walk after n units of time have just lapsed.

The increment of the random walk between $i = a$ and $i = b > a$ is $S_b - S_a = \sum_{i=a+1}^{b} X_i$. Since $E(X_i(\omega)) = 0$, and $\text{var}(X_i(\omega)) = E(X_i^2(\omega)) = 1$, then $\text{var}(S_b - S_a) = b - a$. Variance is computed by taking the probability-weighted average of the squared terms across all paths ω at a point in time.

Now, consider a particular sample path ($\omega \in \Omega$) characterized by the realized values of $S_1, S_2, \ldots, S_n$ across time. The (sampled)[5] quadratic variation of the symmetric random walk along path ω, between $i = a$ and $i = b > a$ is a RV defined as

$$\langle S, S \rangle_{b-a} = \sum_{i=a+1}^{b} (S_i - S_{i-1})^2 = b - a.$$

For this discrete symmetric random walk, it turns out that along any sample path ω, the sampled quadratic variation is constant for a given time segment $(a, b]$: $\langle S, S \rangle_{b-a}(\omega) = b - a$. Note that the sampled quadratic variation here is just the number of time steps in the interval since each time step has a squared variation of 1 on that partition of time.

Let a scaled (making time interval between each random increment or realization of X_t smaller, and making the random steps smaller than unit

[5] "Sampled" refers to a RV taking values at discrete time partitions.

length) symmetric random walk be:

$$W_t^{1/n} = \frac{X_{1/n}}{\sqrt{n}} + \frac{X_{2/n}}{\sqrt{n}} + \cdots + \frac{X_{tn/n}}{\sqrt{n}} = \frac{1}{\sqrt{n}} \sum_{j=1}^{tn} X_{j/n}, \tag{9.1}$$

where the subscripts to W and X denote the time (in units) when those values take place, and the superscript to W denotes that intervals are taken at $1/n$ units of time.

$X_{j/n}$ is an i.i.d. RV with mean 0 and variance equal to 1, and so $\frac{X_{j/n}}{\sqrt{n}}$ is an i.i.d. RV with mean 0 and variance $1/n$. $W_1^{1/n}$ is the sum of n i.i.d. $\frac{X_{j/n}}{\sqrt{n}}$ RVs, and thus has $E(W_1^{1/n}) = 0$, and $\mathrm{var}(W_1^{1/n}) = n \times (1/n) = 1$. In addition, $\mathrm{var}(W_t^{1/n}) = t$.

For any partition of time $(0,t]$, the quadratic variation of the scaled random walk on $(0,t]$, for any $n \geq 1$, is

$$\begin{aligned} \langle W^{1/n}, W^{1/n} \rangle(t) &= \sum_{j=1}^{tn} \left(W_{j/n}^{1/n} - W_{(j-1)/n}^{1/n} \right)^2 \\ &= \sum_{j=1}^{tn} \left(\frac{1}{\sqrt{n}} X_{j/n} \right)^2 \\ &= \frac{tn}{n} = t. \end{aligned} \tag{9.2}$$

Again, this is the same value as the variance, but is a different commodity altogether. It may seem strange that the sum of RV's $\left(\frac{1}{\sqrt{n}} X_{j/n}\right)^2$ becomes a constant t. This is possible if we think of the Bernoulli $\tilde{X}_t$; now, $\tilde{X}_t^2$ is a RV, but it is also a trivial probability distribution where $P(X_t^2 = 1) = 1$.

Now, from Eq. (9.1), using the CLT

$$W_t^{1/n} = \sqrt{t} \left(\frac{\sum_{j=1}^{tn} X_{j/n}}{\sqrt{tn}} \right) \longrightarrow \sqrt{t}\, Z \sim N(0,t). \tag{9.3}$$

For $t > s$

$$W_t^{1/n} = \left(W_t^{1/n} - W_s^{1/n}\right) + W_s^{1/n},$$

and so

$$E\left(W_t^{1/n} \middle| W_s^{1/n}\right) = E\left(W_t^{1/n} - W_s^{1/n} \middle| W_s^{1/n}\right) + W_s^{1/n} = W_s^{1/n}.$$

Thus, $W_t^{1/n}$ is a martingale. Also, $\left(W_t^{1/n} - W_s^{1/n}\right)$ is independent of $\left(W_b^{1/n} - W_a^{1/n}\right)$, for $t > s > b > a$.

From Eq. (9.1), for $t > s$

$$W_t^{1/n} - W_s^{1/n} = \frac{1}{\sqrt{n}} \sum_{j=sn+1}^{tn} X_{j/n}$$

$$= \sqrt{t-s}\left(\frac{\sum_{j=sn+1}^{tn} X_{j/n}}{\sqrt{(t-s)n}}\right) \underset{n\uparrow\infty}{\longrightarrow} \sqrt{t-s}\,Z \sim N(0, t-s). \tag{9.4}$$

Definition 9.1 The process W_t, $t \geq 0$, is a $\mathcal{P}$-Brownian motion[6] iff

(1a) $W_t(\omega)$ is continuous or has continuous sample paths, and $W_0 = 0$;
(1b) For any $0 = t_0 < t_1 < t_2 < \cdots < t_n$,

$$W_{t_1} = W_{t_1} - W_{t_0}, W_{t_2} - W_{t_1}, W_{t_3} - W_{t_2}, \ldots, W_{t_n} - W_{t_{n-1}}$$

are independent, and
(1c) For any $t_a < t_b$,

$$W_{t_b} - W_{t_a} \sim N(0, t_b - t_a).$$

■

There is a strong semblance between BM W_t and the discrete scaled random walk process $W_t^{1/n}$ as indicated by observations (2a), (2b), and (2c) below.

(2a) $W_0^{1/n} = 0$ and $W_0 = 0$
(2b) From (1b), for any $0 = t_0 < t_1 < t_2 < \ldots < t_n$,

$$W_{t_1}^{1/n} = W_{t_1}^{1/n} - W_{t_0}^{1/n}, \quad W_{t_2}^{1/n} - W_{t_1}^{1/n},$$
$$W_{t_3}^{1/n} - W_{t_2}^{1/n}, \ldots, W_{t_n}^{1/n} - W_{t_{n-1}}^{1/n}$$

are independent.
(2c) Equation (9.4) shows the difference of scaled random walk is normally distributed with zero mean and variance equal to the difference in time.

[6] Processes with stationary independent increments are called Lévy processes, so it is seen below that BM is a specialization of a Lévy process to stationary Gaussian distributions. The other major class of Lévy process are the Poisson processes.

Point (2c) looks similar to property (1c), although they are really not identical. (9.4) is a limit result of the convergence in distribution, while in (1c), W_t is already a normally distributed RV. Also, the sample path of W_t in (1a) is continuous and no where differentiable — "extremely jagged at any point", whereas the process $W_t^{1/n}$ is discrete.

Therefore, to show how $W_t^{1/n}$ may converge in some sense to W_t as $n \uparrow \infty$ is beyond the scope of this book. It also involves the idea of construction of a BM to show its existence. One approach is the Lévy-Ciesielski construction of BM using Haar functions. The sense of continuity of sample path will have to invoke deeper concepts of continuity such as uniformly Hölder continuity with exponent, which we will not discuss here. However, we later give an informal and intuitive "proof" of the non-existence of slope in a BM.

Some Properties of BM

(3a) W_t is a martingale. (More generally, BM is a local martingale and also a semimartingale.)

(3b) W_t is a Markov process.
It is easy to show that BM W_t is a martingale. It is a little bit harder to show that it is a Markov process, and we leave it as an exercise.

(3c) $W_t \equiv W_t - W_0$, so $E(W_t) = E(W_t - W_0) = 0$, and $\text{var}(W_t) = \text{var}(W_t - W_0) = t$.

(3d) For $t > s \geq 0$, $\text{cov}(W_s, W_t) = E(W_s W_t) = E[W_s(W_t - W_s) + W_s^2] = E[W_s(W_t - W_s)] + E[W_s^2] = E[W_s]E[W_t - W_s] + E[W_s^2] = 0 + s = s$. Hence, $\text{cov}(W_s, W_t) = E(W_s W_t) = s \wedge t$, or $\min(s, t)$.

(3e) The quadratic variation on $(0, T]$, $\langle W, W\rangle(T) = T$ in mean square or $\mathcal{L}^2$-convergence for a BM.

This may seem to follow most easily from Eq. (9.2) if we may think of the limit of $W_t^{1/n}$ as BM W_t. However, as mentioned previously, this is a bit of a leap of faith without further technical apparatus to do the convergence. For a proof, it is more straightforward and mathematically more correct to use the definition of BM and the defined properties of BM, (3a)–(3d), to come up with the result. We show the proof of (3e) as follows.

Proof. Consider a BM W_t. Let $\triangle = T/n$, and let

$$Y_\pi = \sum_{j=0}^{n-1} \left(W_{(j+1)\triangle} - W_{j\triangle}\right)^2$$

where π refers to a partition of $(0, T]$ into time intervals $\triangle = T/n$. Each $\left(W_{(j+1)\triangle} - W_{j\triangle}\right)^2$ is an i.i.d. RV $\sim [N(0, \triangle)]^2$ (due to properties (3b) and (3c) of BM), and thus has mean $\triangle$ and variance $3\triangle^2 - \triangle^2 = 2\triangle^2$.

Then the mean of Y_π, a sum of random functions on infinitesimally small partitions of time, is

$$E(\tilde{Y}_\pi) = \sum_{i=0}^{n-1} E\left(W_{(j+1)\triangle} - W_{j\triangle}\right)^2 = n\triangle = T.$$

The variance of such a sum is

$$\operatorname{var}(\tilde{Y}_\pi) = \sum_{i=0}^{n-1} \operatorname{var}\left(W_{(j+1)\triangle} - W_{j\triangle}\right)^2 = n(2\triangle^2) = 2T\triangle.$$

Hence, for a given fixed T, as the partition gets finer or smaller

$$\lim_{\triangle \to 0} \operatorname{var}\left(\tilde{Y}_\pi\right) = 0.$$

Therefore

$$\lim_{\triangle \to 0} E\Big(Y_\pi - E(Y_\pi)\Big)^2 = \lim_{\triangle \to 0} E\Big(Y_\pi - T\Big)^2 = 0,$$

which means that

$$Y_\pi \underset{n \uparrow \infty}{\overset{\mathcal{L}^2}{\longrightarrow}} T.$$

Any sequence $Y_{\pi(n)}$ above which converges in mean square to T will have a subsequence that converges to T almost surely. Hence, selecting the time indexes n' for the subsequence to form the partitions π'

$$Y_{\pi'} \underset{n \uparrow \infty}{\overset{\text{a.s.}}{\longrightarrow}} T.$$

$\langle W, W\rangle(T) = T$ is also a convergence in probability. ■

(3f) From property (3e), $\int_0^T (dW_u)^2 = T = \int_0^T du$, hence $(dW_t)^2 = dt$. Moreover, $dW_t dt = 0$, and $(dt)^2 = 0$. Thus also, $\operatorname{var}[dW_t] = E(dW_t)^2 - [E(dW_t)]^2 = dt - 0 = dt$. In addition, $\operatorname{var}[(dW_t)^2] = \operatorname{var}[dt] = 0$. Based on definition (2c) of BM, we can informally express

$dW_t \sim N(0, dt)$. The above results implicitly uses convergence in mean square.

(3g) We can write $\int_s^t dW_u = [W(u)]_s^t = W_t - W_s$. Rearranging, therefore

$$W_t = W_s + \int_s^t dW_u.$$

Then, the conditional probability distribution of W_t given W_s, $s < t$, is normal with mean $E(W_t|W_s) = W_s + \int_s^t E(dW_u) = W_s$, and with variance $\int_s^t \text{var}\, dW_u = \int_s^t du = (t - s)$.

(3h) A BM sample path $W_t(\omega)$ or $W(t, \omega)$ is nowhere differentiable. To explain this intuitively, consider a particular continuous sample path $f_t(\omega)$ (avoiding getting into jump processes with discontinuous sample path) of a continuous-time stochastic process. If this path on $(0, T]$ has at least a finite segment, say $(s, u] \in (0, T]$ on which it is differentiable, i.e., $f'(t)$ exists $\forall t \in (s, u]$, and $f'(t)$ is continuous in $(s, u]$, then the quadratic variation of $f(t)$ on $(s, u]$ is

$$\int_s^u |df(t)|^2 = \int_s^u |f'(t)dt|^2 = \lim_{\triangle \downarrow 0} \left(\triangle \int_s^u |f'(t)|^2 \, dt \right) = 0,$$

since by the extreme value theorem, there exists a bounded from above maximum and bounded from below minimum given $f'(t)$ is continuous.[7] Hence, if a BM is differentiable on some segment $(s, u]$, its quadratic variation on $(0, T]$ is not T, but $T - (u - s) < T$. Therefore, we see that a BM is no where differentiable on $(0, T]$.

(3i) A BM over finite interval $(0, t]$ is of unbounded (total) variation, i.e.,

$$\sup_{\tau_n} \sum_{i=1}^{n} |W(t_i) - W(t_{i-1})| = \infty$$

where the supremum is taken over all partitions τ_n of [0,t] and where $n \longrightarrow \infty$.

With the above tools, we can now consider continuous time stochastic processes $\tilde{X}_t(\omega)$ constructed from BM. We give three examples of such processes below. These processes X_t with drifts and volatility coefficients are also called generalized Wiener processes. Sometimes, they are more

[7]In a rigorous proof, the additional assumption of continuous $f'(t)$ is not needed. In ordinary calculus or real analysis, it is possible for $f'(t)$ to exist for every $t \in (s, u]$ and is bounded, though not continuous, and where $\int_s^u |f'(t)|^2 dt$ has no Riemann-integral solution.

conveniently expressed in derivative form as follows and are called stochastic differential equations or SDEs. This is sometimes called a short-hand way of expressing the process $X(t,\omega)$. Note that we have not quite explained the integral form of $\int f(t,X_t)dW_t$ where the process is more appropriately defined.

(4a) $dX_t = \mu dt + \sigma dW_t$ is an arithmetic BM with drift μ and volatility (or sometimes termed "diffusion coefficient") σ of dX_t/X_t.
(4b) $dX_t = \mu X_t dt + \sigma X_t dW_t$ is a geometric BM (GBM) with drift μ and volatility (or "diffusion coefficient") σ of dX_t/X_t.
(4c) $dX_t = \kappa(\mu - X_t)dt + \sigma X_t^\gamma dW_t$ is a mean-reverting process with reverting speed of adjustment κ, long run mean μ, and volatility σX_t^γ of dX_t.

In (4c), if $\gamma = 1$, it is called the Ornstein-Uhlenbeck process. One may note that the Ornstein-Uhlenbeck process is really a GBM with a non-zero added term $\kappa\mu dt$.

More general stochastic processes allow μ and σ to be functions of time and of the underlying process itself, i.e., $\mu(t,X_t)$, $\sigma(t,X_t)$, where the functions are adapted to $\mathcal{F}_t$ (nonanticipative). Even more general cases include functions where X_t can include past histories $X_{t-1},\ldots$, etc.

Recall from ordinary calculus or real analysis that the fundamental theorem of (integral) calculus states that given $f(u)$ is a continuous function on $u \in [a,b]$, then a function $F(u)$ on $[a,b]$, if it exists, is s.t. $F'(u) \triangleq \frac{dF(u)}{du} = f(u)$, iff $F(x) - F(a) = \int_a^x f(u)du$, $a \le x < b$, where the integral is a Riemann-integral. This result is useful in the sense that once a function F is derived, any definite integral of $f(u)$ w.r.t. u can be quickly computed analytically from $F(x)-F(a)$ in the support $[a,x]$. F is called the indefinite integral, antiderivative, or primitive of f. Another related important mean value theorem for ordinary calculus with continuous $f(u)$ states that there exists a point $x_0 \in [a,b]$ such that $\frac{F(b)-F(a)}{b-a} = \frac{1}{b-a}\int_a^b f(u)du = f(x_0)$.

Thus in ordinary calculus, we see that $dF(u) = f(u)du$ for a continuous function $f(u)$, and that integral $\int_a^b f(u)du$ is well defined. By the Taylor expansion, suppose that a 2D function $F(u,v)$ is infinitely differentiable in u and in v, then

$$F(t,x) = F(t_0,x_0) + F_t(t_0,x)(t-t_0) + F_x(t,x_0)(x-x_0)$$
$$+ \frac{1}{2!}(F_{tt}(t_0,x)(t-t_0)^2$$

$$+ 2F_{tx}(t_0, x_0)(t - t_0)(x - x_0) + F_{xx}(t, x_0)(x - x_0)^2)$$
$$+ \cdots .$$

We may also use the version with (t_0, x_0) as arguments.

In the limit as $t - t_0 \to 0$ and $x - x_0 \to 0$, then we have

$$dF(t,x) = F_t dt + F_x dx + \frac{1}{2!}(F_{tt}(dt)^2 + 2F_{tx} dt\, dx + F_{xx}(dx)^2)$$
$$+ o(dt)^2 + o(dx)^2$$

In ordinary calculus, $(dt)^2 = 0$, $dtdx = 0$, and $(dx)^2 = 0$, hence

$$dF(t,x) = F_t\, dt + F_x\, dx,$$

which is also called the total derivative of F w.r.t. t and x, and whose existence and interpretation comes from the fundamental theorem of ordinary calculus where t, x are deterministic functions of time or deterministic variables.

However, suppose X_t is a stochastic process such as in the 3 examples (4a)–(4c) above, i.e., $X(t, \omega)$ at t is a RV with a probability distribution, and its SDE is generally of the form

$$dX_t = \mu(t, X_t)dt + \sigma(t, X_t)dW_t.$$

Note that in this case, we can write informally:

$$(dX_t)^2 = \mu(t, X_t)^2(dt)^2 + 2\mu(t, X_t)\sigma(t, X_t)dWdt + \sigma(t, X_t)^2(dW)^2.$$

Using BM property (3f), $(dX_t)^2 = \sigma(t, X_t)^2 dt$, the other terms on the RHS being zeros. Similarly,

$$dX_t dt = \mu(t, X_t)(dt)^2 + \sigma(t, X_t)dW_t dt = 0.$$

We use Taylor's expansion, then

$$dF(t, X_t(\omega)) = F_t dt + F_X dX_t$$
$$+ \frac{1}{2!}(F_{tt}(dt)^2 + 2F_{tX} dt\, dX_t + F_{XX}(dX_t)^2)$$
$$+ o(dt)^2 + o(dX_t)^2$$
$$= F_t\, dt + F_X\, dX_t + \frac{1}{2}F_{XX}(dX_t)^2.$$

This is sometimes informally termed the fundamental theorem of stochastic calculus. It extends the ordinary chain rule of calculus by one additional term that is key: $\frac{1}{2}F_{XX}(dX_t)^2$. The result is obtained using heuristic ideas of how the limits work. For a more formal and complete approach, see the next section.

9.2. Itô's Calculus

BM property (3f): $\int_0^t dW_u = W_t \sim N(0,t)$ has been known for a while. Back in 1908, Langevin studied equations in physics of the form (known as Langevin equation)

$$X_t = X_0 + \mu \int_0^t X_u du + \sigma \int_0^t dW_u, \quad t \in [0,T].$$

The solution was quite straightforward: $X_t = X_0 + \mu \int_0^t X_u du + \sigma \tilde{W}_t$ where $\int_0^t X_u du$ can typically be solved as an ordinary integral, and $W_t \sim N(0,t)$.

However, the difficulty arose when stochastic integrals were of the form $\int_0^t f(u, X_u) dW_u$ where $f(u, X_u)$ is not a constant. The Japanese mathematician Kiyosi Itô worked on this and provided a completely satisfactory mathematical solution called Itô's integral or Itô's process,[8] inspiring a new calculus called Itô's calculus or the beginning of important applications in stochastic calculus.

An Itô process is a stochastic process with continuous sample paths described by

$$X_t = X_0 + \int_0^t \mu(u, X_u) du + \int_0^t \sigma(u, X_u) dW_u, \quad t \in [0,T], \tag{9.5}$$

where $\mu(u, X_u)$ and $\sigma(u, X_u)$ are adapted (or non-anticipative) processes to $\mathcal{F}_u$, and W_u is a Wiener process or BM.

To make the integrals provide meaning, they must somehow be finite, so an Itô process must also have $E \int_0^t \sigma_u^2 du < \infty$ a.s. and $\int_0^t \mu_u du < \infty$ a.s. The latter integrals over time t or some deterministic function are computed using the Lesbesgue integral approach rather than Riemann-Stieltjes (RS) in order to avoid cases of nonconvergence in the RS approach.

Moreover, the integral $\int_0^t \sigma(u, X_u) dW_u$ is called Itô integral using the framework and mathematics developed by Itô, and is a.s. a martingale since the integrand is non-anticipating (a X_u-adapted-to-$\mathcal{F}_u$ in continuous

---

[8]See Itô, K (1951). On stochastic differential equations. *Memoirs American Mathematical Society*, 4, 1–51; Itô, K (1986). *Selected Papers*. DW Stroock and SRS Varadhan (eds.). New York: Springer-Verlag.

time process). The basic idea is to construct the stochastic integral where the integrand, being adapted, is evaluated at the beginning of each time interval in an infinitesimal approximation by Riemann-type sum, viz.

$$\int_0^t \sigma(u, X_u) dW_u \approx \sum_{i=0}^{t/\triangle} \sigma(i\triangle, X_{i\triangle})(W_{(i+1)\triangle} - W_{i\triangle}).$$

This aspect makes the Itô approach to solving the integrals very attractive as martingales carry a deep set of applicable results in mathematical analyses on its own rights. The quadratic variation $\langle X, X\rangle = \int_0^t \sigma_u^2 du$. If $D = \int_0^t \mu_u du$, its quadratic variation is zero, i.e., $\langle D, D\rangle = 0$.

The characterization of the integral proceeds as follows:

$$\int_0^T f(t, W_t) dW_t = \lim_{||\Pi|| \to 0} \sum_{j=0}^{n-1} f(t_j, W(t_j))(W(t_{j+1}) - W(t_j)),$$

where $\Pi = \{t_0, t_1, \ldots, t_j, \ldots, t_n\}$ is any partition of $[0, T]$, where n can $\uparrow \infty$, and $||\Pi||$ denotes the norm (or some defined distance measure) of the partition Π, and in this case is $||\Pi|| = \max_j |t_{j+1} - t_j|$. So if the norm approaches zero, it means the maximum of any partitioned intervals approaches zero (so, all intervals must approach zero). The latter is a stronger uniform convergence than if partitions are defined to be equidistant in the first place. Also, the condition of "any partition" covers all possible subintegrals, i.e., $\int_a^b f(t, W_t) dW_t$ for $0 \le a < b \le T$. Notice that in the summation, for each t_j, f is measured at t_j, not t_{j+1} or not any $t > t_j$ while it is multiplied with $(W(t_{j+1}) - W(t_j))$ that is independent of all functions at t_j. Hence, f is non-anticipating (and predictable). This characterizes Itô integration of f as taking its left end point in the interval.

Based on Eq. (9.5), let $f(t, X_t)$ be a function where partial derivatives f_t, f_X, f_{XX}, f_{tX}, f_{tt} exist. For each t, by Taylor's theorem

$$\begin{aligned} f(t_{j+1}, X_{j+1}) - f(t_j, X_j) = {} & f_t(t_j, X_j)[t_{j+1} - t_j] + f_X(t_j, X_j)[X_{j+1} - X_j] \\ & + \frac{1}{2} f_{XX}(t_j, X_j)[X_{j+1} - X_j]^2 \\ & + f_{tX}(t_j, X_j)[t_{j+1} - t_j][X_{j+1} - X_j] \\ & + \frac{1}{2} f_{tt}(t_j, X_j)[t_{j+1} - t_j]^2 + o([t_{j+1} - t_j]^2). \end{aligned}$$

Then, on a partition Π on $[0, T]$ over n intervals, ($t_0 = 0$ and $t_n = T$)

$$\begin{aligned}
f(T, X_T) - f(0, X_0) &= \sum_{j=0}^{n-1} [f(t_{j+1}, X_{j+1}) - f(t_j, X_j)] \\
&= \sum_{j=0}^{n-1} f_t(t_j, X_j)[t_{j+1} - t_j] \\
&\quad + \sum_{j=0}^{n-1} f_X(t_j, X_j)[X_{j+1} - X_j] \\
&\quad + \frac{1}{2}\sum_{j=0}^{n-1} f_{XX}(t_j, X_j)[X_{j+1} - X_j]^2 \\
&\quad + \sum_{j=0}^{n-1} f_{tX}(t_j, X_j)[t_{j+1} - t_j][X_{j+1} - X_j] \\
&\quad + \frac{1}{2}\sum_{j=0}^{n-1} f_{tt}(t_j, X_j)[t_{j+1} - t_j]^2 + o\big([t_{j+1} - t_j]^2\big).
\end{aligned}$$

Taking limits as $||\Pi|| \to 0$ on both LHS and RHS,

$$\begin{aligned}
f(T, X_T) - f(0, X_0) = &\int_0^T f_t(t, X_t)\,dt \\
&+ \int_0^T f_X(t, X_t)\,dX_t \\
&+ \frac{1}{2}\int_0^T f_{XX}(t, X_t)[dX_t]^2.
\end{aligned}$$

Considering that $\langle X, X\rangle[0, T] \equiv \int_0^T (dX_t)^2 = \int_0^T \sigma(t, X_t)^2 dt$, or $d\langle X, X\rangle = (dX_t)^2 = \sigma(t, X_t)^2 dt$, then we have the formula or new chain rule, which is what Itô gave besides the Itô integral.

Theorem 9.1 (Itô–Doeblin Formula): *For* $dX_s = \mu(s, X_s)ds + \sigma(s, X_s)dW_s$,

$$\begin{aligned}
f(t, X_t) = f(0, X_0) &+ \int_0^t f_t(s, X_s)\,ds + \int_0^t f_X(s, X_s)\,dX_s \\
&+ \frac{1}{2}\int_0^t f_{XX}(s, X_s)\sigma(s, X_s)^2\,ds.
\end{aligned}$$

∎

Its differential form is

$$df(t, X_t) = f_t(t, X_t)dt + f_X(t, X_t)dX_t + \frac{1}{2}f_{XX}(t, X_t)(dX_t)^2.$$

This is often called Itô's lemma.

Itô Integrals

First, consider solving $\int_0^1 W_t(\omega)dW_t(\omega)$ path by path (i.e., each distinct ω) as a RS integral. Clearly, this is not possible or meaningful since the length W_t has infinite variation.

Although we cannot apply the formal sense of the RS integration, we can use the idea of RS sums as follows.

Consider a partition of [0,t] into intervals at points $0 = t_0 < t_1 < t_2 < t_3 < \cdots < t_{n-1} < t_n = t$, and the RS sum

$$S_n = \sum_{i=0}^{n-1} W(t_i)\left(W(t_{i+1}) - W(t_i)\right).$$

Let $W_0 = 0$. The above becomes

$$S_n = \frac{1}{2}W(t_n)^2 - \frac{1}{2}\sum_{i=0}^{n-1}\left(W(t_{i+1}) - W(t_i)\right)^2,$$

which has the limit as $||\Pi|| \to 0$ of

$$\int_0^t W_u dW_u = \frac{1}{2}W_t^2 - \frac{1}{2}t \tag{9.6}$$

since $\int_0^t (dW_u)^2 = \int_0^t du = t$.

If we use ordinary calculus, where $Y_0 = 0$,

$$\int_0^t Y_u dY_u = \left[\frac{Y_u^2}{2} + c\right]_0^t = \frac{1}{2}Y_t^2.$$

Thus, ordinary calculus produces a different result short of the additional term $-\frac{1}{2}t$ that appears in Eq. (9.6).

Solution to GBM SDE

Suppose X_t has stochastic differential equation: $dX_t = \mu X_t dt + \sigma X_t dW_t$ where μ and σ are instantaneous mean and volatility of dX_t/X_t that are constants. Then, putting $F(X_t) = \ln X_t$, we have, applying Itô's lemma:

$$\begin{aligned} d\ln X_t &= \frac{1}{X_t} dX_t + \frac{1}{2}\left(-\frac{1}{X_t^2}\right)(dX_t)^2 \\ &= (\mu dt + \sigma dW_t) - \frac{1}{2}\sigma^2 dt \\ &= \left(\mu - \frac{1}{2}\sigma^2\right) dt + \sigma dW_t. \end{aligned} \tag{9.7}$$

If we take the definite integral on Eq. (9.7) over support [0,T] and define $W_0 = 0$,

$$\int_0^T d\ln X_t = \int_0^T \left(\mu - \frac{1}{2}\sigma^2\right) dt + \sigma \int_0^T d\tilde{W}_t.$$

Then

$$\ln X_T - \ln X_0 = \left(\mu - \frac{1}{2}\sigma^2\right) T + \sigma \tilde{W}_T,$$

and

$$X_T = X_0 \exp\left((\mu - \frac{1}{2}\sigma^2)T + \sigma\tilde{W}_T\right).$$

This is a strong solution[9] to the GBM stochastic differential equation; the solution is a stochastic process $X_t(\omega)$ which is also called a diffusion, has a probability distribution at any time point t, and in this case is also called a lognormal diffusion.

9.3. Martingale Method For Black-Scholes Formula

Define a stochastic process $\eta_t[a, b]$ as a stochastic exponential of $\int_0^t b_s dW_s$:

$$\eta_t[a, b] = \exp\left[\int_0^t \left(a_s - \frac{1}{2}b_s^2\right) ds + \int_0^t b_s \, dW_s\right]. \tag{9.8}$$

The stochastic process in the exponent of (9.8) is well-defined, provided the processes $a_s - \frac{1}{2}b_s^2 \in \mathcal{L}^1$ and $b_s \in \mathcal{L}^2$. This is achieved by assuming a_s

[9]A weak solution does not enable a mapping of BM path $W_t(\omega)$ to solution $X_t(\omega)$ though it may offer a distribution indexed by time. A strong solution would require conditions that initial X_0 has a finite second moment if it is a RV, that X_t has a continuous sample path, and the drift and diffusion coefficients satisfy the Lipschitz condition, i.e., do not change too fast, $|\mu(t, y) - \mu(t, x)| + |\sigma(t, y) - \sigma(t, x)| \leq K|y - x|$, K finite, for some small displacements in X_t from x to y.

is bounded and b_s satisfies the Novikov's condition:

$$E\left(\exp\left[\frac{1}{2}\int_0^t |b_s|^2 ds\right]\right) < \infty.$$

The stochastic exponential is a (strictly) positive stochastic process as its value at any t is strictly positive. Stochastic exponentials are sometimes called exponential martingales because they are usually martingales when certain regularity conditions are met. This type of stochastic exponential is sometimes referred to as Doléans-Dade exponential.

There are some basic properties and relationships in such a stochastic exponential.

(5a) Let

$$d\ln Y_t = \left(a_t - \frac{1}{2}b_t^2\right)dt + b_t dW_t. \tag{9.9}$$

Integrating over [0,t], $\ln Y_t = \ln Y_0 + \int_0^t (a_s - \frac{1}{2}b_s^2)ds + \int_0^t b_s dW_s$. Hence, $Y_t = Y_0\,\eta_t[a,b]$. Also, $dY_t = Y_0 d\eta_t[a,b]$, and $\frac{dY_t}{Y_t} = \frac{d\eta_t[a,b]}{\eta_t[a,b]}$. By Itô's lemma, since $d\ln Y_t = \frac{dY_t}{Y_t} - \frac{1}{2}\left(\frac{dY_t}{Y_t}\right)^2 = \frac{d\eta_t[a,b]}{\eta_t[a,b]} - \frac{1}{2}\left(\frac{d\eta_t[a,b]}{\eta_t[a,b]}\right)^2$, then

$$\frac{dY_t}{Y_t} = \frac{d\eta_t[a,b]}{\eta_t[a,b]} = (a_t dt + b_t dW_t),$$

as only then we reclaim (9.9).

(5b)

$$\begin{aligned}
\frac{\eta[a,b]}{\eta[c,d]} &= \exp\left(\int\left(a - \frac{1}{2}b^2\right)ds + \int b\,dW_s\right) \\
&\quad \times \exp\left(-\int\left(c - \frac{1}{2}d^2\right)ds - \int d\,dW_s\right) \\
&= \exp\left(\int([a-c] - \frac{1}{2}[b^2 - d^2])ds + \int [b-d]dW_s\right) \\
&= \exp\left(\int([a-c] - \frac{1}{2}[(b-d)^2 - b^2 - d^2\right. \\
&\quad \left. + 2bd + b^2 - d^2])ds + \int [b-d]dW_s\right)
\end{aligned}$$

$$= \exp\left(\int([a-c]-\frac{1}{2}[(b-d)^2 + 2d(b-d)])\,ds + \int[b-d]\,dW_s\right)$$

$$= \eta[a-c-d(b-d), b-d].$$

Earlier, the GBM process X_t, e.g., a stock price, is seen as

$$X_t = X_0\,\eta_t[\mu, \sigma].$$

For the money market or deposit account process, $M_t = e^{rt}$,

$$M_t = M_0\,\eta_t[r, 0].$$

Then, the discounted stock price can be expressed as

$$\frac{X_t}{M_t} = \frac{X_0}{M_0}\,\eta_t[\mu - r, \sigma]. \tag{9.10}$$

From Eq. (9.10) and property (5a), we obtain

$$d\frac{X_t}{M_t} = \frac{X_0}{M_0}\,d\eta_t[\mu - r, \sigma] = \frac{X_t}{M_t}[(\mu - r)dt + \sigma dW_t]. \tag{9.11}$$

One particular stochastic exponential of $-\int_0^t \sigma_s dW_s$, or $\eta_t[0, -\sigma]$ plays a significant role:

$$\eta_t[0, -\sigma] = \exp\left[-\frac{1}{2}\int_0^t \sigma_s^2\,ds - \int_0^t \sigma_s\,dW_s\right]. \tag{9.12}$$

It is an exponential martingale. For constant σ, it is easily seen as a martingale since

$$Z_t = \exp\left(-\sigma W_t - \frac{1}{2}\sigma^2 t\right)$$

$$= \exp\left(-\sigma(W_t - W_s) - \frac{1}{2}\sigma^2(t-s)\right)\exp\left(-\sigma W_s - \frac{1}{2}\sigma^2 s\right).$$

So, for $t > s$,

$$E(Z_t|Z_s) = e^0 Z_s = Z_s,$$

since Z_s is adapted to $\mathcal{F}_t$ in $(\Omega, \mathcal{F}, P)$.

Note that $E(Z_t) = E\big[\exp\big(-\sigma W_t - \frac{1}{2}\sigma^2 t\big)\big] = e^0 = 1$.

In Chap. 7, Theorem 7.2, we saw how to define a new probability measure Q for event A and $E_P(Z) = 1$ by

$$Q(A) = \int_A Z(\omega)dP(\omega).$$

Hence for all t, our exponential martingale Z_t is a Radon-Nikodým random process such that

$$dQ(\omega) = Z_t\, dP(\omega).$$

Theorem 9.2 (Girsanov theorem): *(Sometimes also called the Cameron–Martin–Girsanov theorem.) Define the process $d\tilde{W}_t = dW_t + \gamma dt$, where the RHS is an SDE under the P-measure or W_t is a Wiener process under the probability space $(\Omega, \mathcal{F}, P)$. γ satisfies the Novikov condition $E\big(\exp\big[\frac{1}{2}\int_0^t |\gamma|^2 ds\big]\big) < \infty$ (in order for stochastic integral to work here).*

Then, $\tilde{W}_t$ is a Wiener process under a different probability measure Q in probability space $(\Omega, \mathcal{F}, Q)$ where measure Q is related to P by the Radon-Nikodým derivative $\frac{dQ}{dP} = \eta_t[0, -\gamma_s]$. ∎

An important point to note is that the term γ in $d\tilde{W}_t^Q = dW_t^P + \gamma dt$ is the same γ in $\frac{dQ}{dP} = \eta_t[0, -\gamma] = \exp\big[-\frac{1}{2}\int_0^t \gamma^2 ds - \int_0^t \gamma dW_s\big]$.

What the Girsanov theorem is saying is that if we track a sample path of a standard BM $W_t(\omega)$ which has zero drift, then add a drift γt, the result is a different path belonging to BM $\tilde{W}_t(\omega)$. Let $\gamma > 0$. The probability of this augmented path is, however, changed from $P(\omega)$ to $Q(\omega)$ according to $dQ(\omega) = \exp(-\gamma W_t - \frac{1}{2}\gamma^2 t)dP(\omega)$. Thus, a smaller probability is put on paths that drifts up ($W_t > 0$) due to positive γ. On balance, the probability distribution of $\tilde{W}_t$ still has zero mean for each t due to the counter-balancing Q measures. Hence, it is remarkable that $\tilde{W}_t$ is still standard Wiener and a martingale.[10] We show intuitively the outcome of the theorem as follows. Consider $\gamma d\tilde{W}_t = \gamma dW_t + \gamma^2 dt$ or integrating over $[0, t]$, $\gamma\tilde{W}_t = \gamma W_t + \gamma^2 t$.

[10]A rigorous proof can be found in advanced textbooks or monographs such as Karatzas, I and E Shreve (1988). *Brownian Motion and Stochastic Calculus.* New York: Springer-Verlag.

Consider

$$\begin{aligned}
E_P[\exp(\gamma W_t)] &= \exp\left(E(\gamma W_t) + \frac{1}{2}\text{var}(\gamma W_t)\right) \\
&= \exp\left(0 + \frac{1}{2}\gamma^2 t\right) \\
&= \exp\left(\frac{1}{2}\gamma^2 t\right).
\end{aligned}$$

By the Girsanov theorem, this should be equal to

$$\begin{aligned}
E_Q[\exp(\gamma \tilde{W}_t)] &= E_Q[\exp(\gamma W_t + \gamma^2 t)] \\
&= \int_\Omega [\exp(\gamma W_t + \gamma^2 t)] \frac{dQ}{dP} dP \\
&= E_P\left[\exp(\gamma W_t + \gamma^2 t) \exp\left(- \gamma W_t - \frac{1}{2}\gamma^2 t\right)\right] \\
&= E_P\left[\exp\left(\frac{1}{2}\gamma^2 t\right)\right] \\
&= \exp\left(\frac{1}{2}\gamma^2 t\right).
\end{aligned}$$

The Girsanov theorem can be used to change the drift of a diffusion process and to turn it into a martingale under a different probability measure. Thus, a process $dW_t + \gamma dt$ may have a non-zero drift γdt under the P-measure. But by applying the Q-measure with density $dQ/dP = Z(\omega)$ where Q and P are equivalent probability measures and $E_P(Z) = 1$, we can turn the process into another martingale diffusion process $\tilde{W}_t$ now with zero drift under measure Q.

This idea of changing drift as well as changing probabilities can be illustrated by the following simple discrete case, where each variable X in P is shifted upward to Y but now with probability Q. Their means $E(X)$, $E(Y)$ remain the same at 0. See Table 9.1.

Table 9.1: Probability Under the Girsanov Theorem

X	$P(X)$	Y	$Q(Y)$
-1	1/3	$-1/2$	2/3
0	1/3	1/2	1/6
$+1$	1/3	3/2	1/6

In Chap. 7, we saw the First fundamental theorem of asset pricing where in no-arbitrage equilibrium, there exists an equivalent martingale measure Q to physical measure P, under which the discounted asset price process is a martingale.

In Eq. (9.10) earlier, the discounted price process $\frac{X_t}{M_t} = \frac{X_0}{M_0}\eta_t[\mu - r, \sigma]$. From Eq. (9.11),

$$d\left(\frac{X_t}{M_t}\right) = \frac{X_t}{M_t}[(\mu - r)\,dt + \sigma\,dW_t].$$

Hence, under the P-measure where X_t follows the GBM, the discounted price process has a drift $\mu - r$ which is not zero. To apply Girsanov's theorem and change the drift to zero, we use $\gamma = \frac{\mu-r}{\sigma}$ in the Radon-Nikodým derivative $\frac{dQ}{dP} = \eta_t[0, -\gamma]$ to switch to a different Wiener process $\tilde{W}_t = W_t + \frac{\mu-r}{\sigma}t$.

Now

$$\begin{aligned} d\left(\frac{X_t}{M_t}\right) &= \frac{X_t}{M_t}[(\mu - r)\,dt + \sigma\,dW_t] \\ &= \frac{X_t}{M_t}\left[(\mu - r)\,dt + \sigma\left(d\tilde{W}_t - \frac{\mu - r}{\sigma}dt\right)\right] \\ &= \frac{X_t}{M_t}[(\mu - r)\,dt + \sigma d\tilde{W}_t - (\mu - r)\,dt] \\ &= \frac{X_t}{M_t}\sigma d\tilde{W}_t. \end{aligned} \tag{9.13}$$

Thus, X_t/M_t is now a martingale under the equivalent Q-measure. Under this Q-measure or equivalently, the distribution of X_t in (9.13), all derivative prices $C(X_t)$ can be solved by taking expectation $E_0(C_t)$ w.r.t. this distribution X_t and discounting by the risk-free return e^{rt}.

Directly solving Eq. (9.13) above for the distribution of X_t,

$$\frac{X_t}{M_t} = \frac{X_0}{M_0}\exp\left(-\frac{1}{2}\sigma^2 t + \sigma\tilde{W}_t\right),$$

or

$$\ln(X_t/X_0) = \left(r - \frac{1}{2}\sigma^2\right)t + \sigma\tilde{W}_t,$$

or $\ln(X_t/X_0) \sim N((r - \frac{1}{2}\sigma^2)t, \sigma^2 t)$. Recall that the latter was also derived as the risk-neutral probability distribution on underlying stock return in

Chap. 3. Under Q, this probability distribution implies a risk-neutral SDE of X_t as follows

$$dX_t = rX_t dt + \sigma X_t d\tilde{W}_t.$$

The latter can also be directly obtained by applying the Girsanov theorem on the original GBM, changing drift μ into the risk-neutral GBM with drift r as above. (Re-emphasizing, the objective of this is to find the discounted security price X_t/M_t as a martingale process, so its expectation is a no-arbitrage price.)

The above risk-neutral probability distribution under Q, $N((r - \frac{1}{2}\sigma^2)t, \sigma^2 t)$ which applies to $\ln(X_t/X_0)$, is used to price any European derivative $C(X_t, t)$.

For the European call, price C_0, on a stock with price S_0, let $X_0 = S_0$. Time to maturity is τ:

$$\begin{aligned} C_0 &= e^{-r\tau} E[\max(S_\tau - K, 0)] \\ &= e^{-r\tau}(E[S_\tau | S_\tau \geq K]\, P[S_\tau \geq K] - K\, P[S_\tau \geq K]). \end{aligned}$$

First, $\tilde{S}_\tau \equiv S_0 \exp((r - 1/2\sigma^2)\tau + \sigma\sqrt{\tau}\,\tilde{z})$ where $\tilde{z} \sim N(0,1)$. Then,

$$\begin{aligned} E(S_\tau | S_\tau \geq K) &= E\Bigg(S_0 \exp((r - 1/2\sigma^2)\tau + \sigma\sqrt{\tau}\,\tilde{Z}) \\ &\qquad \Bigg| Z \geq \frac{\ln(K/S_0) - (r - 1/2\sigma^2)\tau}{\sigma\sqrt{\tau}} \Bigg) \\ &= \Bigg(\int_{\frac{\ln(K/S_0) - (r-1/2\sigma^2)\tau}{\sigma\sqrt{\tau}}}^{\infty} S_0 \exp((r - 1/2\sigma^2)\tau \\ &\qquad + \sigma\sqrt{\tau}\,z)\, \phi(z)\, dz \Bigg) \Bigg/ P(S_\tau \geq K) \\ &= S_0 \exp\left((r - 1/2\sigma^2)\tau + \frac{1}{2}\sigma^2\tau \right) \Bigg[\int_{\frac{\ln(K/S_0) - (r-1/2\sigma^2)\tau}{\sigma\sqrt{\tau}}}^{\infty} \\ &\qquad \exp\left(-\frac{1}{2}\sigma^2\tau + \sigma\sqrt{\tau}z - \frac{1}{2}z^2 \right) \frac{1}{\sqrt{2\pi}} dz \Bigg] \Bigg/ P(S_\tau \geq K) \end{aligned}$$

$$= S_0 \exp\left((r - 1/2\sigma^2)\tau + \frac{1}{2}\sigma^2\tau\right)\left[\int_{\frac{\ln(K/S_0)-(r-1/2\sigma^2)\tau}{\sigma\sqrt{\tau}}}^{\infty}\right.$$

$$\left.\exp\left(-\frac{1}{2}(\sigma\sqrt{\tau} - z)^2\right)\frac{1}{\sqrt{2\pi}}dz\right] \Big/ P(S_\tau \geq K)$$

(doing a change of variable $y = \sigma\sqrt{\tau} - z$)

$$= S_0 \exp(r\tau)\left[\int_{-\infty}^{\frac{\ln(S_0/K)+(r+1/2\sigma^2)\tau}{\sigma\sqrt{\tau}}} \exp\left(-\frac{1}{2}y^2\right)\right.$$

$$\left.\times \frac{1}{\sqrt{2\pi}}dy\right] \Big/ P(S_\tau \geq K).$$

Then

$$E[S_\tau | S_\tau \geq K]P[S_\tau \geq K] = S_0 e^{r\tau} N(d_1),$$

where $d_1 = \frac{\ln(S_0/K)+(r+1/2\sigma^2)\tau}{\sigma\sqrt{\tau}}$.

Next,

$$P(S_\tau \geq K) = P(\ln S_\tau - \ln S_0 \geq \ln K - \ln S_0) = P\left(\ln\frac{S_\tau}{S_0} \geq \ln\frac{K}{S_0}\right).$$

Since $\ln\frac{S_\tau}{S_0} \sim N((r - 1/2\sigma^2)\tau, \sigma^2\tau)$, then

$$\begin{aligned}
P\left(\ln\frac{S_\tau}{S_0} \geq \ln\frac{K}{S_0}\right) &= P\left(\frac{\ln(S_\tau/S_0) - (r - 1/2\sigma^2)\tau}{\sigma\sqrt{\tau}}\right. \\
&\qquad \left.\geq \frac{\ln(K/S_0) - (r - 1/2\sigma^2)\tau}{\sigma\sqrt{\tau}}\right) \\
&= \int_{\frac{\ln(K/S_0)-(r-1/2\sigma^2)\tau}{\sigma\sqrt{\tau}}}^{\infty} \phi(z)dz \\
&= \int_{-\infty}^{\frac{\ln(S_0/K)+(r-1/2\sigma^2)\tau}{\sigma\sqrt{\tau}}} \phi(z)dz \\
&= N(d_2)
\end{aligned}$$

where $d_2 = \frac{\ln(S_0/K)+(r-1/2\sigma^2)\tau}{\sigma\sqrt{\tau}}$.

Therefore, $C_0 = e^{-r\tau}(S_0\, e^{r\tau}\, N(d_1) - K\, N(d_2))$, or $C_0 = S_0\, N(d_1) - Ke^{-r\tau}\, N(d_2)$.

9.4. Black–Scholes SDE

Assume stock price follows GBM

$$\frac{dS}{S} = \mu dt + \sigma dW$$

where we left out subscripts of time t to S_t and W_t for convenience. In the BS GBM, drift or instantaneous (over dt) expected return on the stock μ, and diffusion coefficient or instantaneous (over dt) stock return volatility σ, are both constants.

Suppose there exists a derivative asset (e.g., a call or put) whose value at time t, $f(S,t)$, is dependent on S_t and t. Applying Itô's lemma,

$$\begin{aligned} df &= f_t dt + f_S dS + \frac{1}{2} f_{SS}(dS)^2 \\ &= \left[f_t + f_S S\mu + \frac{1}{2} f_{SS} S^2\sigma^2 \right] dt + f_S S\sigma dW. \end{aligned} \tag{9.14}$$

At time point t, consider a hedge portfolio consisting of 1 unit of the derivative asset and $-f_S$ units of the underlying stock. The hedge portfolio weights will change continuously at each t.

The hedge portfolio cost is $P = f - f_S S$. By Itô's lemma, the SDE for the portfolio value is, with $-f_S$ units of stocks as constant over the instant dS takes place. (If $-f_S$ were not treated as a constant hedge at t, then the total derivative below would be risky, not riskless.) Then,

$$\begin{aligned} dP &= df - f_S dS \\ &= \left[f_t + f_S S\mu + \frac{1}{2} f_{SS} S^2\sigma^2 \right] dt + f_S S\sigma dW - f_S S\mu dt - f_S S\sigma dW \\ &= \left[f_t + \frac{1}{2} f_{SS} S^2\sigma^2 \right] dt. \end{aligned}$$

The hedge portfolio instantaneous gain thus provides an instantaneous risk-free return over dt of $\left[f_t + \frac{1}{2} f_{SS} S^2\sigma^2\right]$. If there exists a risk-free asset in the economy with rate of return r, then to prevent arbitrage,[11] the portfolio

[11] This is deterministic arbitrage or sure gain w.p.1, and should not be confused with the entity "statistical arbitrage" practiced in the hedge-fund industry. Statistical arbitrage is risky arbitrage by buying underpriced market instruments and selling overpriced ones. The degree of underpricing or overpricing is determined by proprietary models, which are of course subject to model errors and are based on empirical and statistical assessments. The belief is that in the long run, there should be convergence to positive profits collected over time. One caveat is that the practice of statistical arbitrage through automated

value can be invested in the risk-free asset to bring about an identical increase in value risklessly, i.e.,

$$\left[f_t + \frac{1}{2} f_{SS} S^2 \sigma^2\right] dt = [f - f_S S] r dt,$$

so

$$\frac{1}{2}\sigma^2 S^2 f_{SS} + rS f_S - rf + f_t = 0. \tag{9.15}$$

Equation (9.15) is the Black–Scholes fundamental PDE for no-arbitrage derivative pricing. The equation is a second-order PDE of the parabolic type which is often used in solving heat transfer problems in physics and engineering. Notice that the actual physical or empirical drift or expected return μ does not appear in the PDE. Risk preferences of individuals also do not matter. When boundary condition is provided, the equation may be solvable to obtain the price function $f(S,t)$ for any given S_t and t.

For example, in a European call option with no-arbitrage equilibrium price $f(S,t)$, the boundary condition is: $f(S,T) = \max(S_T - K, 0) = (S_T - K)^+$ where T is the option maturity and K is the strike price of the option. The initial conditions are $f(0,t) = 0$, and $\lim_{S\to\infty} f(S,t) = \infty$. For a European put, the boundary condition is $f(S,T) = \max(K - S_T, 0) = (K - S_T)^+$. The initial conditions are $f(0,t) = K$, and $\lim_{S\to\infty} f(S,t) = 0$.

The fundamental BS pricing PDE in Eq. (9.15) can also be obtained using an equilibrium argument as follows.

First, we show that any two securities whose instantaneous returns are perfectly correlated must have the same Sharpe ratio iff there is no arbitrage opportunity.

Lemma 9.1: *Suppose two assets' returns are perfectly correlated:*

$$dX_1 = X_1(\mu_1 dt + \sigma_1 dW),$$

$$dX_2 = X_2(\mu_2 dt + \sigma_2 dW),$$

so their instantaneous correlation coefficient is

$$\frac{\sigma_1 \sigma_2 dt}{(\sigma_1\sqrt{dt})(\sigma_2\sqrt{dt})} = 1,$$

---

trading, since it involves huge portfolios in order to aggregate over tiny pockets of positive alphas, must be able to withstand sudden aberrations and thus the danger of illiquid markets and huge margin calls. The importance of this is illustrated by the bailout and subsequent closure of LTCM in 1998, and the unsustainable oil bets of Metallgesellschaft AG in 1993 when it lost over $1.4 billion.

then, the assets' Sharpe ratios are the same constant $\frac{\mu_j - r}{\sigma_j}$ *when a risk-free rate* r *exists.*

Proof. Buy $1/(\sigma_1 X_1)$ shares of asset 1 and short $1/(\sigma_2 X_2)$ shares of asset 2. Net cashflow inflow $X_2/(\sigma_2 X_2) - X_1/(\sigma_1 X_1) = 1/\sigma_2 - 1/\sigma_1$ is then invested in a risk-free bond with rate r.

Thus, the zero cost portfolio has payoff over dt of

$$\frac{dX_1}{\sigma_1 X_1} - \frac{dX_2}{\sigma_2 X_2} + \left(\frac{1}{\sigma_2} - \frac{1}{\sigma_1}\right) r dt = \left(\frac{\mu_1 - r}{\sigma_1} - \frac{\mu_2 - r}{\sigma_2}\right) dt.$$

Since it is zero cost, to prevent any arbitrage opportunity, the coefficient of dt on the RHS must equal to zero, which means

$$\frac{\mu_1 - r}{\sigma_1} = \frac{\mu_2 - r}{\sigma_2}.$$

The two assets may have different risk premia $\mu_j - r$, but they have the same Sharpe ratio. ■

Now, we can derive the fundamental BS PDE (9.15) in another way based on Lemma 9.1. For derivative security $f(S, t)$, its SDE is Eq. (9.14). Equation (9.14) shows its return has perfect correlation with the underlying stock return since both are driven by a single source of innovation dW. Then, the Sharpe ratio of its return df/f is equal to that of the stock to avoid arbitrage opportunity.

For df/f, its expected instantaneous return per dt is

$$\frac{E(df/f)}{dt} = \frac{(f_t + f_S S\mu + \frac{1}{2} f_{SS} S^2 \sigma^2)}{f}.$$

Its instantaneous diffusion coefficient of df/f is $\frac{f_S S\sigma}{f}$. Hence, the Sharpe ratio of the derivative return is:

$$\frac{(f_t + f_S S\mu + \frac{1}{2} f_{SS} S^2 \sigma^2) - rf}{f_S S\sigma} = \frac{\mu - r}{\sigma}$$

or

$$f_t + rSf_S + \frac{1}{2} f_{SS} S^2 \sigma^2 - rf = 0,$$

which is the fundamental equation in (9.15).

Note that the above is obtained without explicitly using the continuous hedging argument employed earlier. It was obtained by employing the condition of no-arbitrage at any moment of time t. It is thus purely an equilibrium argument, and Black–Scholes equation in a sense is a meaningful equilibrium fair pricing. The equivalence of the Sharpe ratio across any two trading securities in continuous time is obtainable without existence of options. However, the fundamental BS PDE (9.15) with the existence of options, even if it can be derived by just using the notion of equal Sharpe ratios, really cannot avoid the fact that with continuous-time trading under the no-arbitrage argument to derive equal Sharpe ratio, it amounts to the existence of continuous-time hedging as well.

The true sense in which the BS option prices can still be obtained without having continuous-time hedging, is when the securities do not trade continuously. In this case, there has to be additional assumptions such as specific preference function and log-normally distributed prices as shown in Rubinstein (1976)[12] in order to derive the exact BS formula.

For the result on the equality of instantaneous Sharpe ratios across securities that are perfectly correlated, the proof can be obtained (without using an arbitrage argument) by employing Merton's intertemporal CAPM[13] result where over *dt*,

$$\mu_i - r = \rho_{iM} \frac{\sigma_i}{\sigma_M} (\mu_M - r).$$

All quantities are instantaneous and different security returns have different volatilities. But if $\rho_{iM} = \rho_{jM}$ for securities i and j, due to perfect correlation, then

$$\begin{aligned} \frac{\mu_i - r}{\mu_j - r} &= \frac{\rho_{iM} \frac{\sigma_i}{\sigma_M}}{\rho_{jM} \frac{\sigma_j}{\sigma_M}} \\ &= \frac{\sigma_i}{\sigma_j}. \end{aligned}$$

9.5. Fourier Transform Method

In the last section, we saw how a derivative price $f(S, t)$ is characterized by the fundamental PDE in Eq. (9.15). We can obtain the Black–Scholes formula of a derivative by employing the boundary conditions. In this section, we show how to solve the PDE using the Fourier transform method. This is not the only method, but is a general method that can be used to solve other PDEs.

---

[12]Rubinstein, M (1976). The valuation of uncertain income streams and the pricing of options. *Bell Journal of Economics*, 7(2), 407–425.

[13]Merton, RC (1973). An intertemporal capital asset pricing model. *Econometrica*, 41, 867–887.

The modus operandi of the method is to transform the orginal PDE into a simpler PDE, solve its Fourier transform, and then use the inverse transform to convert the solution back to the solution of the original PDE.

We begin by providing the notion of the Fourier transform and an important convolution theorem which we require to use.

Definition 9.2 The function

$$\hat{f}(\lambda) = \int_{-\infty}^{\infty} f(x) e^{i\lambda x}\, dx$$

where i is the complex number ($i^2 = -1$), is called the Fourier transform of function $f(\cdot)$. ■

Note that the definition of Fourier transform in engineering convention is a bit different, though the corresponding inverse transform nevertheless produces the same result.

Given a piecewise continuous (or else continuous) function $f(x)$, and existence of its Fourier transform $\hat{f}(\lambda)$, the inverse transform of the latter reproduces

$$f(x) = \frac{1}{2\pi} \lim_{a \uparrow \infty} \int_{-a}^{a} e^{-i\lambda x} \hat{f}(\lambda) d\lambda.$$

Now, we can prove a theorem involving the convolution of two functions f and g. Suppose function h is the convolution of f and g, so that

$$h(x) \equiv (f * g)(x) \triangleq \int_{-\infty}^{\infty} f(x-u) g(u) du.$$

Assume the functions f and g are bounded $\mathcal{L}^1$ so that integrals taken over them are absolutely convergent, and that we can apply Fubini's theorem to interchange the positions of the integrals.

Theorem 9.3 *If f and g are bounded $\mathcal{L}^1$, then the Fourier transform of their convolution is the product of their Fourier transforms: $\widehat{f * g}(\lambda) = \hat{f}(\lambda)\hat{g}(\lambda)$.* ■

Proof.

$$\begin{aligned}
\hat{h}(\lambda) &= \int_{-\infty}^{\infty} e^{i\lambda x}\left[\int_{-\infty}^{\infty} f(x-u)g(u)du\right]dx \\
&= \int_{-\infty}^{\infty} g(u)\left[\int_{-\infty}^{\infty} e^{i\lambda x} f(x-u)\,dx\right]du \\
&= \int_{-\infty}^{\infty} g(u)\left[\int_{-\infty}^{\infty} e^{i\lambda(u+w)} f(w)\,dw\right]du \\
&\qquad \text{(by letting } x = u+w\text{)} \\
&= \int_{-\infty}^{\infty} g(u)e^{i\lambda u}\hat{f}(\lambda)du \\
&= \hat{f}(\lambda)\int_{-\infty}^{\infty} g(u)e^{i\lambda u}du \\
&= \hat{f}(\lambda)\hat{g}(\lambda).
\end{aligned}$$

■

When the Fourier transform is applied to a density function, we obtain the special case of characteristic function. The characteristic function of a normal density function $\phi(z)$ of z with mean zero and variance σ^2 can be easily shown to be $\hat{\phi}(\lambda) = \exp(-\frac{1}{2}\lambda^2\sigma^2)$.

In solving PDE in Eq. (9.15) subject to the following terminal condition (time T) (6a) and boundary conditions (6b) and (6c),

(6a) $f(S,T) = \max(S-K,0)$
(6b) $f(0,t) = 0$ for all $t \leq T$, and
(6c) $\lim_{S\to\infty} f(S,t) = S$,

we are trying to solve for the price of a European call. The PDE is a parabolic type PDE similar to heat transfer problems in physics. There the initial condition (at time $t = 0$) $f(x,0)$ where the distribution of temperature or heat along a unit direction with distance x from one origin is a critical input to determining the solution of $f(x,t)$ at some time later. Such heat equations typically have at least two other boundary conditions (at specific points of value x). By changing the time index of our option problem from t to time-to-maturity $T-t$, the option terminal condition $f(S,T)$ can then be turned into an initial condition $f(S,\tau' = 0)$ where $\tau' = T - t$.

First, consider the following transforms or change of variables in order to simplify Eq. (9.15).

Let $S = Ke^z$, $f = K\,g(z,\tau)$, and $\tau = \frac{1}{2}\sigma^2(T-t)$. Equation (9.15) is then transformed into the the following PDE with the new set of variables:

$$-g_\tau + g_{zz} + (\theta - 1)g_z - \theta g = 0, \tag{9.16}$$

where $\theta = \frac{2r}{\sigma^2}$, subject to initial condition (using τ means moving backward from T instead of moving forward from t to T; here, τ increases from value 0 to value $\frac{1}{2}\sigma^2\tau'$, and is *de facto* an initial condition) $g(z,0) = \max(e^z - 1, 0)$, and boundary conditions

$$\lim_{z\to-\infty} g(z,\tau) = 0, \quad \text{and} \quad \lim_{z\to+\infty} g(z,\tau) = e^z \to \infty.$$

Now, let $g(z,\tau) = \exp\left(-\frac{1}{4}\left[2(\theta-1)z + (\theta+1)^2\tau\right]\right)u(z,\tau)$, then the parabolic PDE becomes

$$u_\tau = u_{zz} \tag{9.17}$$

with initial condition $u(z,0) = \max(e^{(\theta+1)z/2} - e^{(\theta-1)z/2}, 0)$, for $\tau \geq 0$ and $z \in (-\infty, +\infty)$.

To solve the initial value problem in diffusion equation (9.17), take the Fourier transform of both sides of the equation, and by interchanging the integration and differentiation operations, the LHS is

$$\int_{-\infty}^{\infty} e^{i\lambda z} u_\tau(z,\tau)dz = \frac{\partial}{\partial\tau}\hat{u}(\lambda,\tau).$$

The RHS is, after integrating by parts,

$$\begin{aligned}\int_{-\infty}^{\infty} e^{i\lambda z} u_{zz} dz &= \int_{-\infty}^{\infty} e^{i\lambda z} du_z \\ &= -\lambda^2\hat{u}(\lambda,\tau).\end{aligned}$$

Therefore, we now have an ODE to solve:

$$\hat{u}_\tau(\lambda,\tau) = -\lambda^2\hat{u}(\lambda,\tau).$$

The solution to the ODE is

$$\hat{u}(\lambda,\tau) = h(\lambda)e^{-\lambda^2\tau}. \tag{9.18}$$

By putting $\tau = 0$, we see that $h(\lambda) = \hat{u}(\lambda, 0)$. Without loss of generality, let $\hat{v}(\lambda) = \hat{u}(\lambda, 0)$. Then, Eq. (9.18) RHS is the product of two Fourier transforms, viz.

$$\hat{u}(\lambda, \tau) = \hat{v}(\lambda)\hat{\phi}(\lambda),$$

where the transform $\phi(\lambda)$ is of the normal density function of a RV $N(0, 2\tau)$. Invoking Theorem 9.3, the Fourier transform of the convolution

$$\int_{-\infty}^{\infty} \phi(z-s)v(s)\,ds = \frac{1}{2\sqrt{\pi\tau}} \int_{-\infty}^{\infty} e^{-(z-s)^2/4\tau} u(s,0)\,ds$$

is the Fourier transform of function $u(z, \tau)$ on the LHS of Eq. (9.18).

Therefore,

$$u(z,\tau) = \frac{1}{2\sqrt{\pi\tau}} \int_{-\infty}^{\infty} u(s,0)\, e^{-(z-s)^2/4\tau}\, ds, \tag{9.19}$$

where $u(s,0) = \max(e^{(\theta+1)s/2} - e^{(\theta-1)s/2}, 0)$, for $\tau \geq 0$ and $s \in (-\infty, +\infty)$.

If we substitute $s = x\sqrt{2\tau} + z$, for constant z, into (9.19), we obtain

$$\begin{aligned} u(z,\tau) &= \frac{1}{\sqrt{2\pi}} \int_{-z/\sqrt{2\tau}}^{\infty} e^{\frac{1}{2}(\theta+1)(z+x\sqrt{2\tau})}\, e^{-\frac{1}{2}x^2}\, dx \\ &\quad - \frac{1}{\sqrt{2\pi}} \int_{-z/\sqrt{2\tau}}^{\infty} e^{\frac{1}{2}(\theta-1)(z+x\sqrt{2\tau})}\, e^{-\frac{1}{2}x^2}\, dx. \end{aligned}$$

By reversing all the substituted variables to their original variables in S, K, T, t, r, and σ^2, we will obtain the Black–Scholes call price as shown in Chap. 3, Eq. (3.9) and also in Section 9.3.

It turns out that the solution method we just completed is closely related to the Feynman–Kac theorem where we show a simple version below. It is also closely related to the martingale method explained in Section 9.3. In Chap. 3, we also saw how the numerical methods of a lattice tree modeling the stochastic process itself and of a finite difference modeling a PDE can both converge to a same solution in Feynman–Kac here. The Feynman–Kac theorem basically connects the parabolic PDE to an expectation operator over a random function of a related SDE. It provides a stochastic representation of the solution of a parabolic PDE in terms of the unique expectation.

Theorem 9.4 (Feynman–Kac Theorem): *Assume the drift and diffusion coefficients are Borel-measurable, adapted to $\mathcal{F}_t$, and suitably bounded* *to allow existence of solution to the SDE:*

$$ds_t = \mu(S_t, t)dt + \sigma(S_t, t)dW_t.$$

Suppose there exists a parabolic PDE

$$\frac{1}{2}\sigma^2(S_t, t)f_{SS} + \mu(S_t, t)f_S + f_t = b(S_t, t)f$$

defined for all t, S_t *over* $[0, T]$, *and subject to terminal condition* $f(S_T, T) = V(S_T)$.

Then, f *admits a stochastic representation or solution in the form*

$$f(S_t, t) = E_t\left[e^{-\int_t^T b(S_u)du} V(S_T)\right]. \qquad \blacksquare$$

It is important to note that all three connected entities, the SDE of the underlying, the PDE of the underlying and the derivative of the underlying, and the price of the derivative as an expectation operator must be based on the same probability measure. They are originally under the empirical or physical P-measure. However, they can all be transformed to operate under the risk-neutral or EMM or Q-measure using the Girsanov theorem 9.2. Then, the expectation is taken w.r.t. a normalized derivative price that is a martingale (which gives rise to the idea of "as if" risk neutrality since a risk-neutral preference values current price as expected future price). Similarly, the PDE and SDE are under the Q-measure and so the drift term or coefficient of dt in the SDE, and coefficient term of f_S in the PDE must be adjusted accordingly under the Girsanov transform. Likewise, the Wiener process dW^P under P-measure originally becomes dW^Q under the Q-measure.

As an immediate application of the Feynman–Kac theorem in our solution of the parabolic Black–Scholes PDE, note that putting $\mu(S_t, t) = rS_t$, $\sigma(S_t, t) = \sigma S_t$, and $b(S_t, t) = r$ yields the fundamental Black–Scholes PDE, both under the Q-measure. To price a European call option, let the terminal condition be $f(S_T, T) \equiv V(S_T) = (S_T - K)^+$. Applying the theorem, then

$$f(S_t, t) = E^Q\left[e^{-\int_t^T r du}(S_T - K)^+ | S_t\right].$$

We still need, of course, to solve for the probability distribution of S_T at T in order to obtain an explicit analytical solution of $f(S_t, t)$, $\forall t \in [t, T]$. If we let $f(S_t, t)$ be the probability function $P(S_t, t|S_0, 0)$, then the Feynman–Kac

PDE, with $b(S_t, t) = 0$, is also the Kolmogorov backward equation in the transition probability of S_t.

The solution to the GBM, or a SDE under Q-measure, resulting in a probability distribution (from which expectations can be taken), may be obtained by either solving the Kolmogorov backward transition equation or by the Kolmogorov forward equation on transition probability shown below

$$\frac{\partial p(t,x,y)}{\partial t} = \frac{1}{2}\frac{\partial^2[\sigma^2(y)p(t,x,y)]}{\partial y^2} - \frac{\partial[\mu(y)p(t,x,y)]}{\partial y}$$

where $p(t,x,y)$ is the transition probability density function of state x at t, and state y at $s > t$ for the underlying process of S_t. The Kolmogorov forward equation is also called the Fokker–Planck equation.[14]

For the GBM, the solution is quite simple by using Itô's lemma directly, and S_T is lognormally distributed as $S_t e^Y$ where $Y \sim N((r - 1/2\sigma^2)\tau, \sigma^2\tau)$.

9.6. Problem Set 9

1. Show that BM W_t is (a) a martingale, and (b) a Markov process.
2. Find the covariance matrix for multi-dimensional vector BM

$$(W_{t_1}, W_{t_2}, \ldots, W_{t_n})^T, \quad \text{for } t_1 < t_2 < \cdots < t_n$$

3. Evaluate $\int_0^T |dW_t|\, dt$ and $\int_0^T (dt)^2$.
4. Find the distribution of $W_b - W_a$ where $a < b$.
5. Consider the discrete symmetric random walk process $W_t^{1/n}$ in Eq. (9.1) which converges toward the BM process $W_t \sim N(0,t)$ as $n \to \infty$. What is the variation of $W_t^{1/n}$ as $n \to \infty$?
6. If $dX = aX\,dt + bX\,dW$ and $dY = eY\,dt + fY\,dW$, find using Itô's lemma, dXY.
7. If $dY = Y(\mu dt + \sigma dW)$, find the Itô process in terms of SDE for Y^{-1}.
8. Show $\int_{-\infty}^{\infty} e^{i\lambda z} u_{zz} dz = -\lambda^2 \hat{u}(\lambda, \tau)$ in detail.

[14] James, P (2003). *Option Theory*. West Sussex: John Wiley contains a detailed discussion on the use of the Kolmogorov equations in solving option pricing. For more mathematical details of the Kolmogorov equations, one may refer to Karlin, S and H Taylor (1975). *A First Course in Stochastic Processes*, and also (1981). *A Second Course in Stochastic Processes*. California: Academic Press.

Chapter 10

HEDGING AND MORE OPTION PRICING

In this chapter, we consider the scenario in which a sell-side banker has sold some derivatives (e.g., calls or puts), and then attempts to optimally hedge his exposure to potential losses should the underlying asset price move adversely. The flipside of the same coin in terms of technique, is the situation of a buy-side investor or an export firm with an exposure to the underlying asset such as a stock portfolio or a currency receivable, whereby the investor or firm is trying to hedge by using a dynamic option strategy.

For investors or firms that have made money and have decided to lock in the profit or payments or costs, a one-time-only forward or futures contract at an agreeable price would suffice. This is a static hedge. The investors or firms can also use a one-time-only derivative purchase or static hedge such as buying options to lock in a minimum return or else a maximum cost. Options are like forwards or futures that provide a hedge to ensure a minimum return or a maximum loss, whichever the case, but are unlike forwards and futures in that they allow participation in some profits, should the market move in a particular direction. The cost of such a profit participation is the option premium or price that has to be paid upfront. In principle, there is no upfront payment for forwards or futures, excepting margin requirements.

Dynamically hedging an exposed option position is also called "replication", "replicating the option", or "mimicking the option". A dynamic hedge involves adjusting the hedge positions frequently, if not continuously. If replication is exact, then *de facto* it is a way of creating an option that can be sold. In this sense, the option is said to be a "redundant" asset since it can be replaced by a portfolio of hedges. However, in actual market situations, the replication is like a new technology, and the resulting option is a new product that the sell-side with the technology can market at a premium to realize handsome profits. Therefore, in some sense, hedging is bound up in profit-making as well.

Besides considering the dynamic hedging of option exposures under diffusion, we will also consider exposures when the underlying asset price exhibits stochastic volatility and jumps. The chapter concludes with a section on American-style options.

10.1. Discrete Time Hedging

In Chap. 3, we discussed the binomial pricing model of Cox *et al.* We shall now examine the hedging aspects in more detail.

Figure 10.1 shows the binomial tree of stock price evolution in time having two states in each period or interval. The stock prices are ex-dividend prices, i.e., after dividends are issued. Dividend yield is continuously compounded over the interval based on the last stock price and accumulated at the end of each interval.

The time interval at each decision point is h number of years, where h is a small fraction. At $t = 0$, an investor buys $\triangle_0$ shares of stock, costing \$ $\triangle_0 S_0$, and also buys \$$B_0$ of risk-free bonds (lends \$ B_0) that pays continuously compounded interest rate r p.a. If $B_0 < 0$, then it is borrowing to finance the share purchase at a cost of \$ r p.a. The stock issues a continuous dividend yield of δ p.a.

At $t = 0$, total portfolio cost is \$ $\triangle_0 S_0 + B_0$. At $t = h$, portfolio value can take one of two possible outcomes. In the up-state U, the portfolio value becomes

$$U: \quad \triangle_0(uS_0 e^{\delta h}) + B_0 e^{rh}.$$

In the down-state D, the portfolio value becomes

$$D: \quad \triangle_0(dS_0 e^{\delta h}) + B_0 e^{rh}.$$

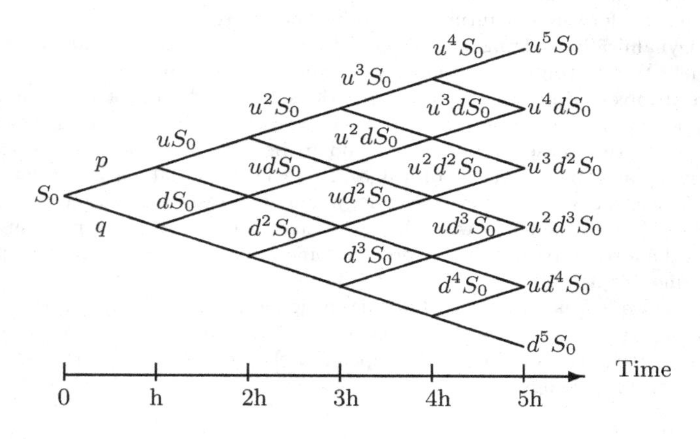

Fig. 10.1: Ex-Dividend Stock Price Evolution

If the portfolio is chosen to replicate an option outcome at $t = h$ of prices C_u at U and C_d at D, then we equate

$$U: \quad \triangle_0(uS_0e^{\delta h}) + B_0e^{rh} = C_u, \tag{10.1}$$

$$D: \quad \triangle_0(dS_0e^{\delta h}) + B_0e^{rh} = C_d. \tag{10.2}$$

Given $u, d, r, \delta, S_0, C_u, C_d$, we solve Eqs. (10.1) and (10.2) to obtain

$$\triangle_0 = e^{-\delta h}\frac{C_u - C_d}{S_0(u-d)}, \tag{10.3}$$

$$B_0 = e^{-rh}\frac{uC_d - dC_u}{u-d}. \tag{10.4}$$

The solutions in (10.3) and (10.4) allow the portfolio to replicate the option outcomes at $t = h$. To prevent arbitrage, therefore the total portfolio cost must equal the price of the option at $t = 0$, i.e.,

$$C_0 = \triangle_0 S_0 + B_0 = e^{-rh}\left(\frac{e^{(r-\delta)h} - d}{u-d}C_u + \frac{u - e^{(r-\delta)h}}{u-d}C_d\right). \tag{10.5}$$

To prevent arbitrage, we must also have

$$u > e^{(r-\delta)h} > d.$$

Equation (10.5) shows that we can put pseudoprobability measure $p = \frac{e^{(r-\delta)h}-d}{u-d} \in (0,1)$ and $1 - p = \frac{u-e^{(r-\delta)h}}{u-d} \in (0,1)$. Let us call these risk-neutral or Q-probability measures. Then

$$C_0 = e^{-rh}\, E_0^Q(\tilde{C}_h),$$

where $C_h = C_u$ with probability p, and $C_h = C_d$ with probability $1-p$. Or

$$\frac{C_0}{e^{r\times 0}} = E_0^Q\left(\frac{\tilde{C}_h}{e^{rh}}\right),$$

which indicates that $\tilde{C}_h/e^{rh}$ is a martingale. According to the second fundamental theorem of asset pricing, under this complete market (two states and two securities) where there is no arbitrage, a unique martingale measure exists — in this case, $P_Q(U) = p$ and $P_Q(D) = 1 - p$. These probabilities for the U and D states are unique because of market completeness.

Suppose the up-state at $t = h$ was U, and we move to the second interval at $t = 2h$. Again invoking the second fundamental theorem here, with no arbitrage there exists Q-measure s.t.

$$C_u = e^{-rh} E_h^Q(\tilde{C}_{2h}), \tag{10.6}$$

where $C_{2h} = C_{uu}$ or C_{ud}, and

$$P_Q(UU) = \frac{e^{(r-\delta)h} - d}{u - d} \in (0,1),$$

$$P_Q(UD) = 1 - P_Q(UU) = \frac{u - e^{(r-\delta)h}}{u - d} \in (0,1).$$

In this binomial tree where the up and down factors, u and d, on ex-dividend stock price are constant, the Q-measure probabilities are also constant at each node.

From Eq. (10.6), if C_{uu} and C_{ud} are the two attainable options (UU and UD) prices, then

$$C_u = e^{-rh}\left(\frac{e^{(r-\delta)h} - d}{u - d} C_{uu} + \frac{u - e^{(r-\delta)h}}{u - d} C_{ud}\right).$$

By a similar argument, if C_{du} and C_{dd} are the other two attainable option (DU and DD) prices, then

$$C_d = e^{-rh}\left(\frac{e^{(r-\delta)h} - d}{u - d} C_{du} + \frac{u - e^{(r-\delta)h}}{u - d} C_{dd}\right).$$

Hence

$$\begin{aligned}
E^Q & \left(E^Q \left(\frac{\tilde{C}_{2h}}{e^{2rh}} \middle| \mathcal{F}_h \right) \middle| \mathcal{F}_0 \right) \\
&= E^Q \left(\frac{\tilde{C}_h}{e^{rh}} \middle| \mathcal{F}_0 \right) \\
&= C_0.
\end{aligned}$$

We can always find the present price of an option that matures nh intervals from now by taking suitable expectations based on the Q-EMM. This is basically the derivation that goes on in Chap. 3 on the binomial CRR option pricing.

However, Lemma 7.2 in Chap. 7 gives us another tool to use in this context. With no-arbitrage, we can find a unique self-financing strategy that can replicate the attainable contingent derivative payoffs in this case in every of the complete states.

At $t = 0$, portfolio $(\triangle_0, B_0)$ comprising $\triangle_0$ number of shares and $\$B_0$ amount of bonds costs $\triangle_0 S_0 + B_0$. At $t = h$, if state U occurs, the portfolio value would adjust to

$$\triangle_0(uS_0e^{\delta h}) + B_0e^{rh} = C_u.$$

At $t = h$, if state D occurs, the portfolio value would adjust to

$$\triangle_0(dS_0e^{\delta h}) + B_0e^{rh} = C_d.$$

Hence, $\triangle_0 = \frac{C_u - C_d}{S_0e^{\delta h}[u-d]}$, and $B_0 = \frac{uC_d - dC_u}{e^{rh}[u-d]}$. The first equation gives the lattice tree discrete "delta" which we see is an approximation to the Black–Scholes delta in the next section.

If at $t = h$, it is state U, after the portfolio assumes the new value under the new price uS_0, an individual holding this portfolio can then rebalance it in such a way that it is a self-financing portfolio, in the sense that there is no cash withdrawal (received dividends are ploughed back to buy more shares or else reduce borrowing) and no fresh capital input. Under state U at $t = h$, the portfolio would be rebalanced to a new portfolio $(\triangle_1^U, B_1^U)$ such that

$$\triangle_1^U(uS_0) + B_1^U = C_u. \tag{10.7}$$

On the other hand, if it is state D, this portfolio would be rebalanced in a self-financing way to a new portfolio $(\triangle_1^D, B_1^D)$ such that

$$\triangle_1^D(dS_0) + B_1^D = C_d.$$

Moreover, if it is state U, the portfolio $(\triangle_1^U, B_1^U)$ is selected in such a way that

$$\triangle_1^U(u^2S_0e^{\delta h}) + B_1^U e^{rh} = C_{uu}, \tag{10.8}$$

and

$$\triangle_1^U(udS_0e^{\delta h}) + B_1^U e^{rh} = C_{ud}. \tag{10.9}$$

On the other hand, if it is state D, the portfolio $(\triangle_1^D, B_1^D)$ is selected in such a way that

$$\triangle_1^D(duS_0e^{\delta h}) + B_1^U e^{rh} = C_{du},$$

and

$$\triangle_1^D(d^2 S_0 e^{\delta h}) + B_1^U e^{rh} = C_{dd}.$$

Looking at the three Eqs. (10.7), (10.8), and (10.9), once we solve for $\triangle_1^U$ and B_1^U in Eqs. (10.8) and (10.9), putting the solutions into (10.7) gives $C_u = e^{-rh}(pC_{uu} + (1-p)C_{ud})$, showing how the option prices are attained each state under self-financing rebalancing.

For a European call for example, the self-financing strategy should end at maturity T with value $C_T = \max(S_T - K, 0)$ where K is the strike price and C_T, S_T are respectively the call and stock values according to the state at T.

10.2. Greeks

Greeks refer to the Greek alphabets used to denote the various partial derivatives of analytical option prices to underlying parameter shifts including changes in the underlying variable. Greeks are (variable) hedging ratios.

Consider the European Black–Scholes call price C_t at time t with exercise or strike price K, maturity at time $T > t$, and underlying asset price following GBM $dS = S(\mu dt + \sigma dW)$. The continuously compounded risk-free rate over $[t, T]$ is r:

$$C(S,t) = SN(d_1) - Ke^{-r\tau}N(d_2), \tag{10.10}$$

where $N(\cdot)$ is the CDF of a standard normal RV and $\phi(\cdot)$ is its PDF,

$$\begin{aligned} \tau &= T - t \\ d_1 &= \frac{\ln\left(\frac{S}{K}\right) + \left(r + \frac{1}{2}\sigma^2\right)\tau}{\sigma\sqrt{\tau}} \\ d_2 &= d_1 - \sigma\sqrt{\tau}. \end{aligned}$$

Using put-call parity, the European put price is

$$P(S,t) = Ke^{-r\tau}N(-d_2) - SN(-d_1). \tag{10.11}$$

We have employed the Black–Scholes option formula so far, and this is in the context of European stock options. There are other types of European-style options involving commodities, currencies, and futures, which are also based on the Black–Scholes formulation.

Consider a forward contract with current price F for future delivery of commodities, including stock, at time-to-maturity τ. By the cost-of-carry no-arbitrage model, $F = Se^{r\tau}$, or $S = Fe^{-r\tau}$. Substituting this into (10.10), we have the price of a call option on a forward contract (if we ignore the marked-to-market stochastic interest gains or losses of an otherwise similar futures contract, then this is also the price of a futures call option):

$$C(F,t) = e^{-r\tau}(FN(d_1) - KN(d_2)) \tag{10.12}$$

where

$$d_1 = \frac{\ln\left(\frac{F}{K}\right) + \frac{1}{2}\sigma^2\tau}{\sigma\sqrt{\tau}}$$
$$d_2 = d_1 - \sigma\sqrt{\tau}.$$

We provide a slightly more general version of the Black–Scholes formula of calls and puts on underlying GBM asset price Z with cost of carry η. Then,

$$C(Z,t) = e^{(\eta-r)\tau} Z\, N(d_1) - e^{-r\tau} K\, N(d_2), \tag{10.13}$$

where

$$d_1 = \frac{\ln\left(\frac{Z}{K}\right) + \left(\eta + \frac{1}{2}\sigma^2\right)\tau}{\sigma\sqrt{\tau}}$$
$$d_2 = d_1 - \sigma\sqrt{\tau}.$$

The put formula is

$$P(Z,t) = e^{-r\tau} KN(-d_2) - e^{(\eta-r)\tau} ZN(-d_1). \tag{10.14}$$

Using Eqs. (10.13) and (10.14), if it is a stock option without dividend, $\eta = r$ and $Z = S$. If it is a stock option with continuous dividend yield δ, then $\eta = r - \delta$ and $Z = S$. If it is a futures option, $\eta = 0$ and $Z = F$ since there is no carry cost (as the price for futures is not paid upfront but is settled only at maturity). If it is a currency option priced in US$ for possible gain in currency Y where the domestic US interest rate is r and the foreign currency Y interest rate is r_Y, then $\eta = r - r_Y$ (this amounts to a replication position of borrowing US$ and lending in currency Y, thus receiving payout in Y interest while paying $ interest) and Z is the spot exchange rate in $ per Y.

In the following, we develop the Greeks for the general version of Black–Scholes in (10.13) and (10.14).

From (10.13) and (10.14),

$$\frac{\partial d_1}{\partial Z} = \frac{\partial d_2}{\partial Z} = \frac{1}{Z\sigma\sqrt{\tau}}$$

$$\begin{aligned}\phi(d_2) &= \frac{1}{\sqrt{2\pi}} \exp\left(-\frac{1}{2}(d_1^2 - 2d_1\sigma\sqrt{\tau} + \sigma^2\tau)\right) \\ &= \phi(d_1)\exp\left(d_1\sigma\sqrt{\tau} - \frac{1}{2}\sigma^2\tau\right) \\ &= \phi(d_1)\frac{Z}{K}e^{\eta\tau}. \end{aligned} \tag{10.15}$$

Delta; δ

$$\begin{aligned}\frac{\partial C}{\partial Z} &= e^{(\eta-r)\tau}N(d_1) + e^{(\eta-r)\tau}Z\frac{\phi(d_1)}{Z\sigma\sqrt{\tau}} - Ke^{-r\tau}\frac{\phi(d_2)}{Z\sigma\sqrt{\tau}} \\ &= e^{(\eta-r)\tau}N(d_1) + e^{(\eta-r)\tau}\frac{\phi(d_1)}{\sigma\sqrt{\tau}} - e^{(\eta-r)\tau}\frac{\phi(d_1)}{\sigma\sqrt{\tau}} \\ &\quad \text{using Eq. (10.15)} \\ &= e^{(\eta-r)\tau}N(d_1) > 0. \end{aligned}$$

$$\frac{\partial P}{\partial Z} = -e^{(\eta-r)\tau}N(-d_1) < 0.$$

Rho; ρ

$$\begin{aligned}\frac{\partial C}{\partial r} &= e^{(\eta-r)\tau}Z\phi(d_1)\frac{\partial d_1}{\partial r} - Ke^{-r\tau}\phi(d_2)\frac{\partial d_2}{\partial r} - KN(d_2)[-\tau e^{-r\tau}] \\ &= e^{(\eta-r)\tau}Z\phi(d_1)\left[\frac{\partial d_1}{\partial r} - \frac{\partial d_2}{\partial r}\right] + \tau Ke^{-r\tau}N(d_2) \\ &= \tau Ke^{-r\tau}N(d_2) > 0. \end{aligned}$$

$$\frac{\partial P}{\partial r} = -\tau Ke^{-r\tau}N(-d_2) < 0.$$

Vega; ν

$$\begin{aligned}\frac{\partial C}{\partial \sigma} &= e^{(\eta-r)\tau}Z\phi(d_1)\frac{\partial d_1}{\partial \sigma} - Ke^{-r\tau}\phi(d_2)\frac{\partial d_2}{\partial \sigma} \\ &= e^{(\eta-r)\tau}Z\phi(d_1)\left[\frac{\partial d_1}{\partial \sigma}\right] - Ke^{-r\tau}\phi(d_2)\left[\frac{\partial d_1}{\partial \sigma} - \sqrt{\tau}\right] \\ &= e^{(\eta-r)\tau}Z\phi(d_1)\sqrt{\tau} > 0. \end{aligned}$$

Note that vegas for the BS European call and put are identical.

Gamma; Γ or γ

$$\begin{aligned}\frac{\partial^2 C}{\partial Z^2} &= \frac{\partial e^{(\eta-r)\tau} N(d_1)}{\partial Z} \\ &= e^{(\eta-r)\tau}\phi(d_1)\frac{\partial d_1}{\partial Z} \\ &= e^{(\eta-r)\tau}\frac{\phi(d_1)}{Z\sigma\sqrt{\tau}} > 0.\end{aligned}$$

Note that gammas for the BS European call and put are identical. Next, the time decay, or theta, is shown as follows.

Theta; θ

$$\begin{aligned}\frac{\partial C}{\partial t} &= -\frac{\partial C}{\partial \tau} \\ &= -\Bigg((\eta - r)e^{(\eta-r)\tau} Z\, N(d_1) + e^{(\eta-r)\tau} Z\phi(d_1)\frac{\partial d_1}{\partial \tau} \\ &\quad - KN(d_2)[-re^{-r\tau}] - e^{-r\tau}K\phi(d_2)\frac{\partial d_2}{\partial \tau}\Bigg) \\ &= -(\eta - r)e^{(\eta-r)\tau} Z\, N(d_1) - re^{-r\tau}K\, N(d_2) - e^{(\eta-r)\tau} Z\phi(d_1)\frac{\sigma}{2\sqrt{\tau}}.\end{aligned}$$

$$\begin{aligned}\frac{\partial P}{\partial t} &= -\frac{\partial P}{\partial \tau} \\ &= (\eta - r)e^{(\eta-r)\tau} Z\, N(-d_1) + re^{-r\tau}K\, N(-d_2) - e^{(\eta-r)\tau} Z\phi(d_1)\frac{\sigma}{2\sqrt{\tau}}.\end{aligned}$$

Intuitively, most options have negative θ or time decay, reflecting the fact that their value decreases as the expiry date nears. Time decay is especially fast for at-the-money options. However, unlike other major Greeks, there are exceptions when time decay or θ can become positive. For example, deep-in-the-money European options that cannot be exercised till maturity behave like holding ZC bonds, where the value increases as expiry date gets closer. Thus the signs of θ may be negative or positive.

Delta Hedging

Delta hedging refers to hedging an open option position by taking an opposite position in the underlying with a quantity equal to delta, thus creating a delta-neutral portfolio. Hedging a short (long) position in the underlying can be done by taking a long (short) position in the option by a quantity equal to 1/delta.

Suppose the portfolio is a long (short) position in a stock call option. Delta hedging this long (short) option position consists of short-selling (buying) δ number of stocks. Thus, over a small time interval, the change in value of the delta-neutral portfolio of value $\Pi = C - \delta S$ is close to zero:

$$|\triangle C - \delta \triangle S| = \left|\triangle C - \frac{\partial C}{\partial S}\triangle S\right| \approx 0.$$

Dynamic delta hedging refers to a trading strategy that continuously maintains a delta-neutral portfolio through the life of the option portfolio. In this case, the delta of the option portfolio is changing continuously and requires the number of stocks to be continuously adjusted. By continuously maintaining a delta-neutral hedge, the idea is to keep the portfolio value $\Pi = C - \delta S$ approximately constant till the option reaches maturity. The motivation for doing this may be that at time t after the option has begun, the owner has gained value in the option equal to $C_t - C_0 > 0$. To lock in this value, without having to liquidate the options for whatever reasons (including reason of options market being illiquid or having the options as a collateral pledge, and so on), the owner uses delta hedging to try to lock-in $C_t - C_0$ till maturity of option at T.

Consider the call option price $C(S, t)$ at time t. Taking the total derivative and using Itô's lemma gives

$$dC_t = C_t dt + C_S dS_t + \frac{1}{2}C_{SS}(dS_t)^2.$$

Hence, the hedged portfolio infinitesimal value change (given fixed δ over dt) is

$$\begin{aligned} d\Pi_t &= dC_t - \delta dS_t \\ &= C_t dt + C_S dS_t + \frac{1}{2}C_{SS}(dS_t)^2 - \delta dS_t \\ &= C_t dt + \frac{1}{2}C_{SS}(dS_t)^2. \end{aligned} \tag{10.16}$$

There are two aspects of Eq. (10.16). Over small discrete trading interval $\triangle$,

$$\triangle \Pi_t \approx \theta \triangle + \frac{1}{2}\Gamma(\triangle S_t)^2.$$

Taking expectations on both sides

$$E(\triangle\Pi_t) \approx \theta\triangle + \frac{1}{2}\Gamma E(\triangle S_t)^2$$
$$= \theta\triangle + \frac{1}{2}\Gamma\sigma^2 S_t^2\triangle, \qquad (10.17)$$

so a delta-hedged portfolio (e.g., long option short stocks) will still experience some drift. This is shown in Fig. 10.2.

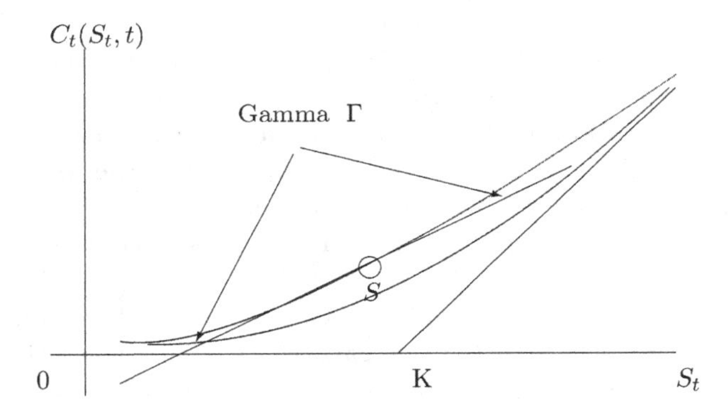

Fig. 10.2: Illustration of Delta Hedging

Figure 10.2 shows the slope of the tangent to the price point S on the upper curve as δ or C_S. At S, if the stock price either increases or decreases, there will be a gap between the new price C_t' and the tangent line representing $C_t + C_S\triangle S_t$. The gap is $\triangle C_t - \delta\triangle S_t$, which is driven by Γ or C_{SS} of the call as seen in Eq. (10.16). The gap is sometimes called the delta-hedging error. Note that whichever way the price S_t may deviate, Γ is positive and is profitable to the trader who is long call and short stock. As time passes, the price curve shifts to the lower one with negative time decay as the call is not deep in-the-money. The difference between the two price curves represents time decay or $\theta < 0$, *ceteris paribus.*

Positive Γ is similar in concept to the convexity in the price-yield relationship of a bond curve. Theoretically, given a stock price $S_t \sim$ GBM, then

$$d\Pi_t = dC_t - \delta dS_t$$
$$= C_t dt + \frac{1}{2}C_{SS}(dS_t)^2$$

$$
\begin{aligned}
&= \theta dt + \frac{1}{2}\Gamma\sigma^2 S_t^2 dt \\
&= -\left[S_t\phi(d_1)\left[\frac{\sigma}{2\sqrt{\tau}}\right] + rKe^{-r\tau}N(d_2)\right]dt + \left[\frac{1}{2}\frac{\phi(d_1)}{S_t\sigma\sqrt{\tau}}\sigma^2 S_t^2\right]dt \\
&= -rKe^{-r\tau}N(d_2)dt. \qquad (10.18)
\end{aligned}
$$

Hence, a long option delta-neutral portfolio has positive gamma, negative theta, and net negative drift as seen in (10.18). On the contrary, sell-side bankers who are short options and hold a delta-neutral portfolio face negative gamma, positive theta, and a net positive drift.

When the interval of delta-hedging $\triangle$ is not nearly zero, and/or if GBM is not exactly the underlying stock price process, then we can only use the approximation of (10.17) where on average $E(\triangle\Pi_t) \approx 0$, or $\theta = -\frac{1}{2}\Gamma\sigma^2 S_t^2 < 0$. Thus, the time decay is approximately the negative of the effect by gamma. However, the approximation of $\triangle\Pi_t \approx \theta\triangle + \frac{1}{2}\Gamma(\triangle S_t)^2$ for discrete $\triangle$ implies that the square of change of price over the interval is $(\triangle S_t)^2 \approx \sigma^2 S_t^2 \triangle Z_t^2$ where $Z_t \sim N(0,1)$. Thus, $\triangle\Pi_t$ is distributed as χ^2. Hence, there is a positive probability that $\triangle\Pi_t > 0$. Therefore, investors may prefer a portfolio with positive gamma, especially when the time-to-maturity is long so that the discrete interval time decay is small.

It is seen that Γ is small when the stock is deep in-the-money or deep out-of-the-money, i.e., $\phi(d_1)$ is very small. Thus, a banker holding a delta-neutral portfolio and short in options would prefer if the options did not end close to being at-the-money near maturity, so that the negative gamma would be small in magnitude. On the other hand, an investor holding a delta-neutral portfolio and long in options would prefer if the options ended close to being at-the-money near maturity, so that the positive gamma would be large in magnitude and bring about a positive portfolio return.

As shown in Eq. (10.18), delta-hedging on a continuous-time basis still leaves a negative drift $-rKe^{-r\tau}N(d_2)dt$. Where does this come from?

If we examine the Black–Scholes European call option pricing formula

$$C(S,t) = SN(d_1) - Ke^{-r\tau}N(d_2),$$

we find that there are two components making up or replicating a call:

(1a) long position of $N(d_1)$ or C_S or δ number of underlying shares; and

(1b) short position or borrowing of dollar amount of risk-free bond $Ke^{-r\tau}$ $N(d_2)$.

Delta hedging constitutes only the first component above. That is why there is still a remaining drift term in Eq. (10.18).

If we want to replicate a call option exactly in a continuous-time dynamic way, we should buy δ or $N(d_1)$ number of underlying shares and short $Ke^{-r\tau}N(d_2)$ amount of risk-free bonds. If we wish to hedge a long call option, we should short δ or $N(d_1)$ number of underlying shares and buy (lend) $Ke^{-r\tau}N(d_2)$ dollar amount of risk-free bonds. In the latter, the infinitesimal change in the value of the portfolio (now comprising an additional amount of risk-free bond) is

$$d\Pi_t^* = dC_t - \delta dS_t + r[Ke^{-r\tau}N(d_2)]dt = 0.$$

Now, we obtain a dynamically perfect hedge, provided that at every instance t, we rebalance our portfolio with a new $\delta = N(d_1)$ and a new amount of risk-free bond equal to $Ke^{-r\tau}N(d_2)$. It is seen in the replicating portfolio $[N(d_1)S_t, -Ke^{-r\tau}N(d_2)]$ for a call option $C(S_t, t)$ that as maturity approaches, if the call is out-of-the-money (OTM) or S_t is below K, then the replicating portfolio behaves more like a bond since $N(d_1)$ is close to zero, and in fact the bond portion is also close to zero. If the call is in-the-money (ITM) when maturity approaches, i.e., $S_t > K$, then $N(d_1)$ approaches 1 while $N(d_2)$ also approaches 1, so that the call value tends towards $S_T - K$.

In some practice, delta-hedging or any other Greek-hedge is supposed to be a self-financing hedge in order to avoid the theta effect. What this means is that in the delta-hedging case, if the exposure is one call (with delta δ), the hedge is using a stock plus a risk-free bond. Delta-hedging means eliminating delta-risk, so one equation is

$$\delta + X(1) = 0,$$

where X number of shares are sold in this case with each share having a delta of one. Clearly, $X = -\delta$. The self-financing additional constraint, however, implies that if the prices of the call and the stock are C and S now, then another equation is required:

$$C + XS + B = 0,$$

which means that if $C + XS < 0$ (where $X = -\delta$), then $B > 0$. The latter implies that the self-financing hedge (i.e., using the total value, C, of the exposure to hedge) requires in addition to the delta-neutral hedge discussed earlier in the section, a long position or investment of remnant value in a risk-free bond. This explains how this would counter the negative time decay problem.

Delta-Gamma-Hedging

With regard to the perfect replication as seen earlier of the derivatives under Black–Scholes GBM, it is noted that under GBM or log-normal diffusion, there is only one source of risk or innovation in the single Brownion motion or Wiener process dW_t. When there is only one source of risk, we require only one traded security with underlying stochastic process adapted to W_t in order to perform perfect dynamic replication; the other security being the risk-free bond. This perfect dynamic hedge is possible as the Itô's lemma implies a balancing theta and gamma risk, excepting a constant, as the interval $\triangle$ approaches zero. For discrete $\triangle$ under one source of risk or state variable, the hedging error can be made smaller with two securities instead of just one. This would be called delta-gamma hedging, and the resulting delta-gamma hedging error would be smaller and closer to zero than the delta-hedging error. We illustrate this in the following.

First, note that for any ith call option with τ_i time-to-maturity, the delta increases in the underlying asset price S_t at a decreasing rate and especially so when the call option gets far into-the-money. The call's delta approaches 1 asymptotically. The call's gamma is always positive but decreases as the underlying asset price increases.

Suppose a call 1 at time t has $\delta_1 = 0.28$ and $\gamma_1 = 0.54$. A call 2 at time t on the same underlying stock of price $S_t = \$10$ has $\delta_2 = 0.85$ and $\gamma_2 = 0.37$. An investor with a long position on one call 1 wishes to hedge it so that it becomes delta-neutral as well as gamma-neutral, i.e., with zero delta and gamma risk over period $\triangle$. We can find the number of shares of the stock and the number of call 2 that will enable such a hedge.

Let X be the number of shares and Y be the number of call 2. Then, portfolio of (1 call 1, X share, Y call 2) has a portfolio delta of

(2a) $0.28 + X(1) + Y(0.85) = 0;$

and a portfolio gamma of

(2b) $0.54 + X(0) + Y(0.37) = 0.$

Note we use the fact that a stock has a delta of 1 and gamma of 0. Solving Eqs. (2a) and (2b), we find that $X = 0.96$ and $Y = -1.46$. Hence, for every two call 1, we should delta-gamma-neutral hedge by buying 2 stocks and going short 3 call 2.

The above discrete period hedging basically relies on the discrete approximation to $dC_t = C_t dt + C_S dS_t + \frac{1}{2}C_{SS}(dS_t)^2$ which we replace with

$$\triangle C_i = \theta_i \triangle + \delta_i \triangle S + \gamma_i \triangle S',$$

for different call i, and where $\triangle S' \approx \frac{1}{2}(dS_t)^2$. Then, a hedge portfolio of (1 call 1, X share, Y call 2) produces portfolio change $\triangle C_1 + X\triangle S + Y\triangle C_2$ that should have zero coefficients on $\triangle S$ and $\triangle S'$. The latter conditions are exactly the two Eqs. (2a) and (2b).

The put-call parity also suggests a delta-gamma neutral hedge of a long call using short stock and short puts.

If we use a self-financing delta-gamma hedging, then suppose the prices of the first and second calls are respectively C_1 and C_2, and investment in a risk-free bond is $\$B$, then a third equation is required, viz.

(2c) $C_1 + 10X + YC_2 + B = 0$.

In a market when the volatility is high, a hedger may also employ a third security for delta-gamma-vega hedging, eliminating vega risk during the short interval by the portfolio of 4 assets, or 5 assets if self-financing is included.

10.3. Other Risks

In general, even when continuous dynamic hedging is feasible, when there are n sources of independent random risks over time, we require n risky securities that are not perfectly correlated in their returns processes in order to achieve perfect dynamic replication. When perfect dynamic replication is possible, the market is said to be dynamically complete. Thus, it is seen that spanning an infinite number of states at the end of period T can be achieved with only a few securities even in continuous time environment provided it is under a finite number of random risk factors or W_t's.

We saw that in the Black–Scholes world with GBM, when there is only one factor noise in W_t, then we only require a risky stock, price S_t, (and a risk-free bond) and continuous trading or continuous dynamic rebalancing to attain every state of the world over time, i.e., attain any value a derivative, price $f(S_t)$, will take over time t. Suppose the underlying asset prices are driven by two noises e.g., $dS_t = \mu dt + \sigma dW_t + \zeta dZ_t$, where W_t and Z_t are two correlated Wiener processes, with $\text{cov}(dW_t, dZ_t) = \rho dt$, and $\rho \in (0, 1)$. Then, in general, we can dynamically replicate a derivative price $f(S_t)$ at each time t using two risky assets (and a risk-free bond) that are not perfectly correlated.

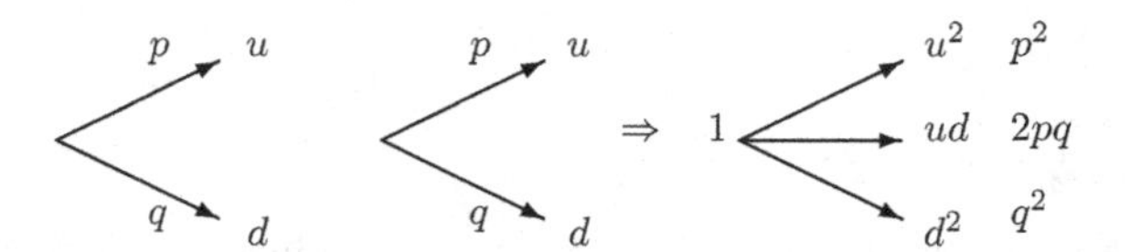

Fig. 10.3: Lattice with Two Noises

In a discrete time world such as the binomial tree (or in general a lattice tree where the number of branches from each node can be more than 2), a noise or innovation is represented by a branching into two next-period states with probabilities p and q. There we require only one risky stock (and a risk-free bond) to replicate and attain a derivative price at each discrete period time t. We saw this is algebraically solving two replication equations in the two unknowns of the risky stock and the risk-free bond quantities.

In the same discrete setting, suppose now there are two independent multiplicative noises represented by two sets of dual branching shown in Fig. 10.3. The right side of Fig. 10.3 shows how the two noises combine together such that \$1 in asset 1's value has a probability distribution in 3 states, viz. \$ u^2, \$ ud, and \$ d^2, with corresponding probabilities p^2, $2pq$, and q^2 respectively in the next period. A second asset in this two-noise economy could have a payoff probability distribution of \$ u'^2, \$ $u'd'$, and \$ d'^2, with corresponding probabilities p^2, $2pq$, and q^2 respectively in the next period. (Note $u \neq u'$, $d \neq d'$.)

To replicate a derivative of such an asset in this two-noise or trinomial branching economy will require not one, but two, risky assets whose prices are not perfectly correlated (and a risk-free bond). If the derivative takes values C_u, C_m, and C_d in the 3 nodes respectively, then perfect replication using δ_1, δ_2 of risky assets 1 and 2 respectively, and \$B amount of risk-free bond (with risk-free payoff r) produces 3 equations and 3 unknowns δ_1, δ_2, B to be solved (given u, d, u', d') in

$$\delta_1 u^2 + \delta_2 u'^2 + Br = C_u$$

$$\delta_1 ud + \delta_2 u'd' + Br = C_m$$

$$\delta_1 d^2 + \delta_2 d'^2 + Br = C_d.$$

Thus, we see the analogy with the continuous-time situation discussed earlier.

GBM and other stochastic processes (Itô processes) based on diffusion are processes with continuous sample paths. There are stochastic price processes with discontinuous sample paths that may not be perfectly hedged even under continuous time dynamic hedging (or for the same reason be perfectly replicable in reproducing some derivative price). One common situation is when the underlying price process is a mixed jump-diffusion process.

Another situation akin to having more noise is when the underlying price process contains stochastic volatility driven by another non-traded state variable that is not perfectly correlated with W_t. The volatility could be induced by illiquidity risk and is not necessarily perfectly correlated with the traded prices of VIX (an index measuring distribution-free implied volatility that is traded at the Chicago Board Options Exchange since 2004). When a state or noise is not traded, then it is not possible to use additional securities such as in the case of Fig. 10.3 to replicate the noises perfectly. We discuss in Chap. 5 that in complete market setting, the derivative prices are functions of underlying traded security prices spanning the states, and the latter are sufficient statistics. When derivative prices are dependent on other noise that is not traded and is thus unattainable in all states, then the derivative prices occur in an incomplete market. By the first fundamental law of asset pricing in Chap. 7, if no-arbitrage equilibrium is assumed, then there exists a non-unique EMM Q-measure. Derivative prices in incomplete market will typically depend on investor preference that will determine which EMM measure is applicable. The preference is sometimes hidden in some term called the price of risk or risk premium. The latter is not a traded security price, and can be shown to depend on preference parameters such as risk-aversion coefficients in a general equilibrium setup.

The two cases of jump-diffusion (JD) and of stochastic volatility (SV) are common in financial modeling. The inability to perfectly hedge or replicate these risks also implies (the flipside of the coin, really) that derivatives of underlying assets based on these driving or generating processes are not redundant (i.e., cannot be perfectly replicated by other securities or assets). This leads to interesting models of such derivative prices in their own right, and we shall discuss briefly the models involving JD and SV before touching the subject of American-style options.

A key remark is in order before we leave the subject of hedging risk of derivative exposure. The delta- and delta-gamma hedging described in the previous section are possible by using the Greeks calculated from a model. In those cases, it was the Black–Scholes European option model. If the model is indeed correct or which describes well the empirical option prices, then the model Greeks would perform well. However, those Greeks are model specific and are therefore subject to model risks. If the model is not accurate and does not explain empirical option prices well, at least during the sample period when the hedging takes place, then we are in trouble. The hedging errors will be large, not due to minor slippages or second-order Taylor approximation errors, but due to a wrong model producing Greeks that do not work.

One approach in avoiding model risk or error is to assume a regression model (that may be nonlinear) and estimate using time series data the sample coefficients of observed derivative price w.r.t. the observed underlying asset price or factors. However, this is *de facto* using an empirical regression model and is still subject to the same model risk problem, though using a linear regression model could be very easy to implement, especially in cases where the market movement of the underlying asset price is small over short intervals. There is another approach to hedge exposures in derivatives with final payoff condition e.g., $\max(S_T - K, 0)$, by rebalancing a self-financing portfolio of stocks and bonds starting at $t = 0$ in such a way that the final portfolio value at horizon maturity T, P_T, is as close to the derivative exposure $\max(S_T - K, 0)$ as possible, such as in a call. This approach makes sense only in discrete-time hedging, and therefore under incomplete markets, which is the pragmatic approach, since continuous hedging is not only physically impossible but infinitely expensive with transaction costs. (If it is continuous-time rebalancing, then we should obtain the model anyway.)

The portfolio rebalancing is optimized so as to minimize the final objective function which is the distance between P_T and $\max(S_T - K, 0)$, e.g., $\min[P_T - \max(S_T - K, 0)]^2$. Issuers of new derivatives often develop such hedging strategies and then appraise the cost of the risk $\min[P_T - \max(S_T - K, 0)]^2$ they have to bear. If the initial self-financing outlay to replicating the exposure as close as possible is \$ P_t, then the issuers would sell the derivatives at a premium price of \$ $P_t + \pi_t$ where π_t reflects the remnant risk the issuer has to bear and also including the premium for first mover advantage or innovation rent (provided the derivative is demanded). In the above process, the solution for P_t usually will require knowledge of the stochastic process of the underlying stock price (or some robust specification of it, which will then add more premium to the price due to the uncertainty risks of the parameters of the process) so that optimization can be carried out, often in a dynamic programming fashion.[1]

---

[1]One example of such a study is Bertsimas, D, L Kogan and A Lo (2001). Hedging derivative securities and incomplete markets: An ϵ-arbitrage approach, *Operations Research*, 49(3), 372–397.

We have seen the Poisson and exponential distributions earlier in Chap. 1. Starting from a basic informal description that the probability of a jump (discontinuity) in a price process S_t in the next interval $\triangle$ is essentially $\lambda\triangle$, we saw how Poisson events (jumps, in this case) occur at times τ_1, τ_2, τ_3, etc. where the interarrival times $\tau_1 - 0$, $\tau_2 - \tau_1$, $\tau_3 - \tau_2$, etc. are distributed as an exponential RV with mean $1/\lambda$. The jumps are arriving at an average rate, or intensity, of λ per unit time. The Poisson process is a right-continuous process so that when a jump occurs at t, it is adapted to $\mathcal{F}_t$, and the market knows at t that a jump has occurred or not.

Suppose λ is the intensity rate for a Poisson process. If the number of jumps by time t_i or during period $[0, t_i]$ is a counting process $N(t_i)$, and if $N(t_1) - N(t_0)$, $N(t_2) - N(t_1), \ldots$, etc. are stationary and independent, then

$$P[N(t_k) - N(t_{k-1}) = j] = \frac{(\lambda(t_k - t_{k-1}))^j}{j!} e^{-\lambda(t_k - t_{k-1})},$$

where we essentially now treat $\lambda(t_k - t_{k-1})$ as the intensity parameter but over the period $t_k - t_{k-1}$ as a time unit instead.

The mean and variance of the above independent Poisson increments, $N(t_k) - N(t_{k-1})$, are $\lambda(t_k - t_{k-1})$.

To obtain a martingale from a Poisson process, we subtract its mean. If $N(t)$ is a Poisson process, with initial condition $N(0) = 0$, its mean is λt. Define

$$M(t) = N(t) - \lambda t$$

as a compensated Poisson process, where $E_0[M(t)] = E_0[N(t)] - \lambda t = 0$.

Theorem 10.1 *M(t) is a martingale.* ■

Proof. For $t < s$

$$\begin{aligned} E(M(s)|\mathcal{F}_t) &= E(M(s) - M(t)|\mathcal{F}_t) + E(M(t)|\mathcal{F}_t) \\ &= E(N(s) - N(t)) - \lambda(s - t) + M(t)) \\ &= M(t). \end{aligned}$$

■

A compound Poisson process is more than a counting process, and is a sum of i.i.d. Poisson-jump sizes that adds up in time over the number of jumps. Suppose y_i are i.i.d. jump sizes with mean $E(y_i) = \theta$ and variance

$\text{var}(y_i) = \zeta$. The compound Poisson process is

$$\tilde{Y}(t) = \sum_{i=1}^{\tilde{N}(t)} \tilde{y}_i,$$

where $\tilde{N}(t)$ is the random number of jumps that occur over time t.

$$\begin{aligned}
E(Y(t)) &= \sum_{k=0}^{\infty} E\left[\sum_{i=1}^{k} y_i \middle| N(t) = k\right] P\{N(t) = k\} \\
&= \sum_{k=0}^{\infty} k\theta \left(\frac{(\lambda t)^k}{k!} e^{-\lambda t}\right) \\
&= \theta \lambda t e^{-\lambda t} \sum_{k=1}^{\infty} \frac{(\lambda t)^{k-1}}{(k-1)!} \\
&= \theta \lambda t. \\
\text{var}(Y(t)) &= \sum_{k=0}^{\infty} \text{var}\left[\sum_{i=1}^{k} y_i \middle| N(t) = k\right] P\{N(t) = k\} \\
&= \sum_{k=0}^{\infty} k\zeta \left(\frac{(\lambda t)^k}{k!} e^{-\lambda t}\right) \\
&= \zeta \lambda t e^{-\lambda t} \sum_{k=1}^{\infty} \frac{(\lambda t)^{k-1}}{(k-1)!} \\
&= \zeta \lambda t.
\end{aligned}$$

(Mixed) Jump Diffusion Process

Suppose the instantaneous stock return is driven by a (mixed) jump-diffusion process (or more appropriately termed an additive jump-diffusion process), then

$$\frac{dS_t}{S_t} = (\mu - \lambda[\theta - 1])dt + \sigma dW_t + (\tilde{y}_t - 1)dN_t \tag{10.19}$$

where instantaneous incremental return rate in the event of a jump is $y_t - 1$ with expected jump size factor $E(\tilde{y}_t) = \theta$; σ is the instantaneous constant volatility conditional on no jumps, i.e., it does not incorporate the volatility effect due to the jump component, and the Wiener process W_t and Poisson process N_t are independent. The mean rate of jump is λ per unit time.

The continuous-time Poisson process is such that

$$dN_t = \begin{cases} 1 & \text{with probability } \lambda dt \\ 0 & \text{with probability } 1 - \lambda dt \end{cases}.$$

A quick comment on the specification of (10.19) with the additional term involving λ in the drift is in order. If we take an unconditional expectation on (10.19), we obtain

$$E\left(\frac{dS_t}{S_t}\right) = (\mu - \lambda[\theta - 1])dt + [\theta - 1]\lambda dt = \mu dt$$

that indicates compensation for the jump drift (removing it) and would allow risk-neutral measures to be used later.

Let a call option on the underlying stock price S_t have price $f(S, t) \in C^2$, i.e., twice-continuously differentiable, with SDE:

$$\frac{df_t}{f_t} = (\mu_f - \lambda[\theta_f - 1])dt + \sigma_f dW_t + (\tilde{y}_{f,t} - 1)dN_t, \qquad (10.20)$$

where $y_{f,t}$ is the incremental jump size factor of the option price with mean θ_f; μ_f and σ_f are the instantaneous mean and volatility of the option rate of return, and the jumps occur exactly when the jumps occur on the underlying stock price.

Applying Itô's lemma, and dropping the time index t subscripts, replacing with subscripts as indications of partial derivative variables:

$$df = \left[f_t dt + f_S dS + \frac{1}{2} f_{SS}(dS)^2\right] + [(f(yS) - f(S))dN] \qquad (10.21)$$

where the first term on the RHS is the usual diffusion calculus, and the second term on the RHS shows the Poisson jump event that is incremental to the diffusion as and when the event happens.

If we take the expectation at t, of df/f in (10.20), and similarly of df/f in (10.21), we should be able to equate their resulting means, $E(df/f)$, since they are identical entities. Thus

$$\mu_f = \frac{\frac{1}{2} f_{SS}\sigma^2 S^2 + f_S(\mu - \lambda[\theta - 1])S + f_t + \lambda E[f(yS) - f(S)]}{f}. \qquad (10.22)$$

Similarly equating the diffusion coefficients

$$\sigma_f = \frac{f_S \sigma S}{f}. \qquad (10.23)$$

Suppose we form a portfolio of the stock, the call option, and a risk-free bond in proportions of some part of wealth, w_1, w_2, and w_3 so that $\sum_{i=1}^{3} w_i = 1$. (Other parts of wealth can be invested in the market portfolio, for example.) Let the risk-free rate of the risk-free bond be constant r over period now till maturity, i.e., from t to T, and we define $\tau = T - t$.

This portfolio's price is P and its instantaneous return is dP/P. Without loss of generality, we can define

$$\frac{dP}{P} = (\mu_p - \lambda[\theta_p - 1])dt + \sigma_p dW + (\tilde{y}_p - 1)dN_t. \tag{10.24}$$

From Eqs. (10.19) and (10.20),

$$w_1(dS/S) + w_2(df/f) + (1 - w_1 - w_2)rdt = dP/P.$$

Taking expectation at t,

$$w_1 E(dS/S) + w_2 E(df/f) + (1 - w_1 - w_2)r = E(dP/P),$$

or

$$w_1\mu + w_2\mu_f + (1 - w_1 - w_2)r = \mu_p.$$

Hence

$$\mu_p = w_1(\mu - r) + w_2(\mu_f - r) + r.$$

Similarly, equating the diffusion components,

$$\sigma_p = w_1\sigma + w_2\sigma_f.$$

Next comes the big picture in solving the problem. We said earlier that only a part of the market wealth is invested in this small portfolio — the rest are in the huge market portfolio. When the total combination of portfolios are considered, the jump risk in the small portfolio P is assumed to be diversifiable away to zero. This means there are so many stocks and options with jump components that they cancel all the jumps out, as some jumps are positive and some are negative, and each with different sizes. Some other MJD models try to assume that there exists another security with identical jump component $(y - 1)dN$ as the stock we are considering. But this is merely a convenient way of passing the problem onto another observed price in the market and making any derivative now a function of that price as well.

Of course, jumps, especially compound Poisson processes, may require an overwhelmingly large market to perform full diversification away of the risk, so the current model is perhaps best viewed as an approximation. With the diversification assumption, then as in CAPM, the beta of the portfolio P is now zero, i.e., $dP = Prdt$, where $E(dP/P) = \mu_p dt = rdt$ and $\sigma_p = 0$ in (10.24).

Thus, $w_1(\mu - r) + w_2(\mu_f - r) = 0$, and $w_1\sigma + w_2\sigma_f = 0$. This yields the result as in standard diffusion problem, that

$$\frac{\mu - r}{\sigma} = \frac{\mu_f - r}{\sigma_f}. \tag{10.25}$$

From Eqs. (10.22) and (10.23),

$$\begin{aligned}\frac{\mu_f - r}{\sigma_f} &= \frac{\frac{1}{2} f_{SS}\sigma^2 S^2 + f_S(\mu - \lambda[\theta - 1])S + f_t + \lambda E[f(yS) - f(S)] - rf}{f_S \sigma S} \\ &= \frac{\mu - r}{\sigma}.\end{aligned}$$

Thus, the fundamental equation for solution of the MJD option pricing model is obtained as follows

$$\frac{1}{2}\sigma^2 S^2 f_{SS} + (r - \lambda[\theta - 1])Sf_S - f_\tau - rf + \lambda E[f(yS) - f(S)] = 0. \tag{10.26}$$

Theorem 10.2 *The solution*[2] *to the mixed PDE (10.26) is*

$$f(S, \tau) = \sum_{j=0}^{\infty} P_j\, E_j\left[C(V_j, \tau)\right] \tag{10.27}$$

where $P_j = \frac{\exp(-\lambda\tau)\,(\lambda\tau)^j}{j!}$, $V_j = Se^{-\lambda[\theta-1]\tau}\prod_{i=1}^{j} y_i$, *and the RHS conditional expectation is taken w.r.t.* j *jumps as reflected in the subscript in* $E_j(\cdot)$. *The function* $C(S, \tau; r, \sigma, K)$ *is the same Black–Scholes formula we saw in* (10.10). ■

The proof of (10.27) can be done by partial differentiation of RHS of (10.27) w.r.t. S to obtain f_S, f_{SS}, w.r.t. τ to obtain f_τ, and so on, and then assembling all the resulting terms on the LHS of (10.26) to show the sum equals zero on the RHS.

---

[2]The proof here follows that in Merton, RC (1990). *Continuous-Time Finance.* Basil: Blackwell.

Proof.

$$f_S = \sum_{j=0}^{\infty} P_j E_j[C_V \frac{\partial V_j}{\partial S}, \tau]$$

which implies $Sf_S = \sum_{j=0}^{\infty} P_j E_j[C_V V_j, \tau]$.

$$f_{SS} = \sum_{j=0}^{\infty} P_j E_j \left[C_{VV} \left(\frac{\partial V_j}{\partial S} \right)^2, \tau \right]$$

which implies $S^2 f_{SS} = \sum_{j=0}^{\infty} P_j E_j[C_{VV} V_j^2, \tau]$.

$$f_\tau = -\lambda(\theta - 1)Sf_S + \sum_{j=0}^{\infty} P_j E_j[C_\tau, \tau] - \lambda f$$
$$+ \lambda \sum_{j=1}^{\infty} \frac{\exp(-\lambda\tau)\,(\lambda\tau)^{j-1}}{(j-1)!} E_j[C(V_j, \tau)].$$

The last term above can be rearranged as $\lambda \sum_{k=0}^{\infty} P_k E_{k+1}[C(V_{k+1}, \tau)]$.

In (10.26), the expectation term is integral w.r.t. both S and y since y is a RV. From (10.27), thus

$$E[f(yS, \tau)] = E_y \left(\sum_{j=0}^{\infty} P_j E_j[C(yV_j, \tau)] \right)$$
$$= \sum_{j=0}^{\infty} P_j\, E_{j+1}[C(V_{j+1}, \tau)],$$

where $E_y(\cdot)$ denotes integral over the distribution of i.i.d. y. The term $E[f(S, \tau)] = f(S\tau)$.

Adding up all the above components, we have

$$\frac{1}{2}\sigma^2 S^2 f_{SS} + (r - \lambda[\theta - 1])Sf_S - f_\tau - rf + \lambda E[f(yS) - f(S)]$$
$$= \sum_{j=0}^{\infty} P_j \left\{ \frac{1}{2}\sigma^2 E_j[C_{VV} V_j^2] + (r - \lambda[\theta - 1])E_j[C_V V_j] \right.$$
$$+ \lambda(\theta - 1)E_j[C_V V_j] - E_j[C_\tau] - \lambda E_{j+1}[C(V_{j+1}]$$

$$- rE_j[C(V_j)] + \lambda E_{j+1}[C(V_{j+1}]\Big\} + \lambda f - \lambda f$$

$$= \sum_{j=0}^{\infty} P_j E_j \left\{ \frac{1}{2}\sigma^2 V_j^2 C_{VV} + rV_j C_V - rC - C_\tau \right\}.$$

Since the PDE $\frac{1}{2}\sigma^2 V_j^2 C_{VV} + rV_j C_V - rC - C_\tau = 0$ satisfies the Black–Scholes formula for C, the expression sums to zero. ■

As an important special case of (10.27), if $\tilde{y}_i$ jump sizes are i.i.d. lognormal with $\ln y_i \sim N(\ln\theta - \frac{1}{2}\delta^2, \delta^2)$, then the MJD call option with a compound Poisson process has a no-arbitrage price under the risk-neutral measure of

$$f(S,\tau) = \sum_{j=0}^{\infty} \frac{\exp(-\lambda'\tau)(\lambda'\tau)^j}{j!} C(X_j, \tau), \tag{10.28}$$

where $C(\cdot)$ is the familiar Black–Scholes function seen in (10.10), $X_j = S[\prod_{i=1}^{j} y_i]$, and each dX_j follows SDE $dX_j = (r - \lambda[\theta - 1] + j\ln(\theta)/\tau)dt + \sqrt{(\sigma^2 + j\delta^2/\tau)}dW^Q$. Further, $\lambda' = \lambda\theta$. The adjustments include not only the Poisson sum, but also a higher intensity if the jump sizes are on average positive.

Volatility Risk

When volatility σ, as in GBM, is not constant, it can take various forms such as a local volatility function viz. $\sigma(S,t)$, or stochastic volatility.

When we use BS formula to imply out the volatility $\hat{\sigma}$ from a market option price C_t^*, i.e.,

$$C_t^* = S_t N(d_1) - Ke^{-r\tau} N(d_1 - \sigma^* \sqrt{\tau})$$

then σ^* is called an implied volatility (IV). As IV can be obtained using, say, European call option prices at various maturities τ and various strike prices K, we can plot a graph of IV versus τ and K at any point in time t. The earliest papers stated the IV skewness as a graph showing IV versus τ on one side and K on another side of the same axis, but more recent papers have simply dropped the part about τ, and called IV skewness (or "smile" or "smirk") as the graph of σ^* versus K at a point in time t. The term "volatility smile" is more clearly a case where IV is high at OTM and ITM options and lower at ATM options, thus creating a smile curve. This was noticed more visibly after the stock market 1987 crash.

The IV skewness generally shows the graph is not a flat line as posited by the BS GBM assumption of a constant σ. The shape of the option IV skewness curve may vary depending on the underlying security such as whether it is equity, equity index, commodities, or currencies.

The term structure of implied volatility is a graph of ATM implied volatilities versus maturities. Together, the volatility smile or skewness, and the term structure produce a 3D IV surface with K and τ as the two horizontal axes.

Local volatility function (predictable function) models are described by underlying SDE processes in general as

$$dS_t = S_t(\mu(S_t, t)dt + \sigma(S_t, t)dW_t),$$

in using to price the options. An early example of this is the constant elasticity of variance (CEV) model proposed by Fischer Black and others who had observed that local volatility varied inversely with the price level. In the CEV model, the SDE is

$$dS_t = (r - \delta)S_t dt + \sigma S_t^{\alpha} dW_t^Q,$$

where δ is dividend payout, and $\alpha \in (0, 2)$. When $\alpha < 1$, the inverse price-volatility relationship is observed, and the solution of a call and put prices are respectively

$$C(S_t, t) = S_t e^{-\delta\tau} Q_1 - K e^{-r\tau} Q_2$$
$$P(S_t, t) = K e^{-r\tau}(1 - Q_2) - S_t e^{-\delta\tau}(1 - Q_1)$$

Q_1, Q_2 are the complementary cdfs of non-central chi-squares with parameters as functions of $(S_t, r, K, \tau, \sigma, \alpha)$.

More recent attention is focused on the use of stochastic volatility (SV) models driven by underlying processes such as

$$dS_t = S_t(\mu dt + \sqrt{\psi}_t dW_t), \tag{10.29}$$

and

$$d\psi_t = (\alpha - \beta\psi_t)dt + \gamma\sqrt{\psi_t} dZ_t, \tag{10.30}$$

where dW_t and dZ_t are two correlated Wiener processes with cov (dW_t, dZ_t) $= \rho dt$.

The solution to the European option of the underlying SV processes above is called the Heston model.[3] The model can be solved analytically in which the price of volatility risk appears as a non-zero term. Since we can form a portfolio of a stock and another option with similar SDE specifications in conjunction with the option under consideration to diversify away or nullify the W_t and the Z_t noises, a no-arbitrage portfolio can be formed in which risk-neutral measure Q exists. Alternatively, assuming no arbitrage, a risk-neutral Q measure exists. Under this measure, the price of volatility risk is subsumed under the drift term of the variance diffusion process when the risk price is specified to be linear in the instantaneous variance.

First, we form a portfolio price $\Pi = C - XS - YV$ comprising the long call option, price C under SV, short X units of stocks with price S, and short Y units of a third asset price V. (The time subscripts are left out to avoid notational clutterings.)

Second, we determine X^* and Y^* such that the noises are eliminated, viz.

$$\frac{\partial C}{\partial S} - X^* - Y^* \frac{\partial V}{\partial S} = 0, \quad \text{and} \quad \frac{\partial C}{\partial \psi} - Y^* \frac{\partial V}{\partial \psi} = 0.$$

Substituting these values X^* and Y^* in $\Pi = C - X^*S - Y^*V$, and taking its total derivative, the risk-neutral portfolio should yield instantaneous risk-free return rate. Thus, we obtain the fundamental pricing PDE as follows

$$\frac{C_t + \frac{1}{2}\psi S^2 C_{SS} + \rho\gamma\psi S C_{S\psi} + \frac{1}{2}\gamma^2 \psi C_{\psi\psi} + rSC_S - rC}{C_\psi} = -(\alpha - \beta\psi - \phi\psi)$$

where ϕ is the market price of volatility risk.

The LHS of the PDE is to satisfy all securities that are derivative functions of $(t, S, \psi; r, K, \tau, \rho, \gamma, \alpha, \beta)$. The RHS is chosen as a parsimonious natural candidate because first it is only a simple function of ψ, and it contains β and α which have not yet been included on the LHS. More complicated RHS functions involving $(t, S, \psi; r, K, \tau, \rho, \gamma, \alpha, \beta)$ could in principle be used with solutions satisfying the no-arbitrage pricing. However, there may not be tractable solutions. Added complexities sometimes do not buy

[3]This is one of the earliest papers that has received widespread attention and industry application, viz. Steven, HL (1993). A closed-form solution for options with stochastic volatility with applications to bond and currency options. *Review of Financial Studies*, 6(2), 327–343.

better economic interpretations of the empirical data or necessarily produce better forecasts.

Using Fourier transform techniques that we see in Chap. 9, the solution of this European call under SV is

$$C(S,\psi,t) = SP_1 - Ke^{-r\tau}P_2,$$

where for $j = 1$ or $j = 2$,

$$P_j = \frac{1}{2} + \frac{1}{\pi}\int_0^\infty \mathrm{Re}\left[\frac{\exp(C_j(\tau,u) + D_j(\tau,u)\psi + iu\ln[S/K])}{iu}\right] du,$$

$$C_j = rui\tau + \frac{\alpha}{\gamma^2}\left\{(b_j - \rho\gamma ui + d_j)\,\tau - 2\ln\left[\frac{1-ge^{d_j\tau}}{1-g}\right]\right\},$$

$$D_j = \frac{b_j - \rho\gamma ui + d_j}{\gamma^2}\left[\frac{1-e^{d_j\tau}}{1-ge^{d_j\tau}}\right],$$

$$b_j = \beta - (2-j)\rho\gamma + \phi,$$

$$g_j = \frac{b_j - \rho\gamma ui + d_j}{b_j - \rho\gamma ui - d_j},$$

$$d_j = \sqrt{(\rho\gamma ui - b_j)^2 - \gamma^2(2\theta_j ui - u^2)}, \quad \text{and}$$

$$\theta_j = -(-1)^j\frac{1}{2}.$$

SV models and models with local volatility functions have gained popularity because they provide a more realistic capture of the empirical characteristics of the underlying security (e.g., stock prices having higher kurtosis and having skewness) both not captured by the Black–Scholes GBM model. This is additional reason to solving the IV skewness issue.

10.4. American Options

Unlike the European-style options that we have discussed so far, American-style options are different in that the holder or buyer need not wait till maturity, but can exercise the option at any time up till maturity. For example, a European put, strike price \$8, on a stock with current price \$5, cannot be exercised. The holder has to wait till maturity perhaps in another month's time, by which time the stock price may have risen to above the strike so that the put becomes worthless. On the other hand, an American put, with the same strike price \$8 on the same underlying stock, could be exercised now to gain \$(8 − 5) or \$3 if the holder chooses to do that.

Rational Price Bounds

There are some price bounds for American options that are interesting in themselves mathematically, and serve as a guide to whether the option is rationally being priced in the market; or if the market price breaches the bounds, whether what market frictions are present to cause that.

For notations here, let small letters represent European options and corresponding capital letters represent American options. "C" denotes call, and "P" denotes put. Unless otherwise stated, S is the current underying stock or asset price, K is the strike price, $\tau = T - t$ is the time-to-maturity, r is the constant risk-free rate, and σ is the volatility of the underlying stock or asset return. Any subscript attached to prices denotes the time. Subscripts attached to parameters denote different levels or numbers. Any comparisons below between different options are based on the same underlying at the same point in time with all parameters held the same except for those that are considered to differ as expressed in the arguments or otherwise stated.

There are some useful propositions as follows. Their importance lies in the fact that they are not model dependent, and should apply for any rational pricing models.

(3a) $C \geq \max(S - K, 0)$ at all time $t \in (0, T]$.

This is because the holder can always exercise the American call now and realize a gain of $(S - K)^+$. The amount $(S - K)^+$ is called the intrinsic value of the American option. A European holder cannot exercise it now, and thus it has no intrinsic value but just time value, and the expected future value of the call at maturity under Q-measure may be less than $(S - K)^+$. At expiry T, $C_T = (S_T - K)^+$.

For puts, $P \geq \max(K - S, 0)$ at all time $t \in (0, T]$. At expiry T, $P_T = (K - S_T)^+$.

(3b) $C(\tau_1) \geq C(\tau_2)$, for all τ_1 or $(T_1 - t) \geq \tau_2$ or $(T_2 - t)$ where $T_1 \geq T_2$.

This also applies to American puts. Suppose it is not, then one could buy the cheaper longer-maturity option and sell the more expensive shorter-maturity option, and pocket the immediate gains. Then, at any time if the shorter-maturity option is being exercised by the counter-party, one would then also exercise the longer-maturity option as well to nullify any payoff.

(3c) $C(K_1) \geq C(K_2)$ and $P(K_1) \leq P(K_2)$ for $K_1 \leq K_2$.

(3d) $C(S) \leq S$. $P(S) \leq K$.

The first case in (3d) is true as it costs to exercise the call into stock, even if $K \downarrow 0$. The second case holds also because the exercise to realize gain involves buying back the stock at a price ≥ 0.

The above provides for $\lim_{S\uparrow\infty} C(S) = \infty, \lim_{S\downarrow 0} P(S) = K$, $\lim_{S\downarrow 0} C(S) = 0$ and $\lim_{S\uparrow\infty} P(S) = 0$, as boundary conditions for the options. Initial conditions (based on τ) (or called terminal conditions if based on T) are $C(\tau = 0) = \max(S - K)^+$, and $P(\tau = 0) = \max(K - S)^+$. In addition, American options generally have free boundary conditions as well, which we shall see in the next sections.

(3e) c and C are convex functions of strike prices. Thus, for $K_1 < K_2$, for any $\lambda \in (0, 1)$, $\lambda c(K_1) + (1 - \lambda)c(K_2) \geq c(\lambda K_1 + (1 - \lambda)K_2)$, and $\lambda C(K_1) + (1 - \lambda)C(K_2) \geq C(\lambda K_1 + (1 - \lambda)K_2)$. For the European call, at T

$$\lambda \max(S_T - K_1, 0) + (1 - \lambda)\max(S_T - K_2, 0)$$
$$\geq \max(S_T - (\lambda K_1 + (1 - \lambda)K_2), 0).$$

It is easily seen that if $S_T < K_1 < K_2$, then there is equality. If $K_1 < S_T < K_2$, then LHS > RHS. If $S_T > K_2 > K_1$, then LHS again equals RHS. Since the above result holds true at any time before maturity of an American call, in terms of exercised gains, then the prices of c and C must also follow this convexity result.

(3f) $C \geq c$ and $P \geq p$. American options are worth at least the values of the European options. Otherwise, arbitrage profits can be made by selling cheaper American options and buying more expensive European options. At maturity, both will either be exercised or not exercised, and in either case, the payoffs are nullified. More particularly, for American puts and American calls with dividend or other payouts, early exercise may be optimal than wait till maturity. This will become clearer in the next section. For now, it suffices to know that as long as there exists probability 1 that $(S_t - K)e^{r\tau} > S_T - K \geq 0$, then it makes sense to exercise an American call at t. In whatever case, the American option can always be kept till maturity like an European option, and thus should be worth at least as much and not less.

(3g) If the underlying security has no payout till maturity, then $c \geq \max(S - Ke^{-r\tau}, 0)$. This lower bound for a European call is tight. To show this, compare two portfolios A and B. A consists of a European call with strike K and a K number of ZC bonds that mature at \$1 at T. B consists of a stock at price S_t now. At maturity T, if $S_T \leq K$, then A is worth \$ K while B is worth \$ $S_T \leq K$. At T, if $S_T > K$, then A is worth $S_T - K + K = S_T$, while B is worth S_T, the same. Hence, A will have payoffs at least equal to B. Thus to prevent arbitrage, cost of A must be at least as large as cost of B. Thus, $c + Ke^{-r\tau} \geq S$. Since c must ≥ 0, then $c \geq \max(S - Ke^{-r\tau}, 0)$. Hence also $C \geq c \geq \max(S - Ke^{-r\tau}, 0)$. Using the put-call parity, $c - p = S - Ke^{-r\tau}$, we can show also that $P \geq p \geq \max(Ke^{-r\tau} - S, 0)$.

(3h) If the underlying security has no payout till maturity, then the price of an American call is exactly the same as that of a European call, *ceteris paribus*. This is at first sight an intriguing result. In fact, we can derive this result from (3g). There, $C \geq c \geq \max(S - Ke^{-r\tau}, 0)$. If during [0,T], an American call is exercised, its payout is $S - K$. But this is less than $S - Ke^{-r\tau} \leq C$. Hence, it is better to sell the American call at C than to exercise it at a lower gain. This leads to the statement that, "American calls are worth more alive than dead". Since an American call will not be exercised before maturity in this case of no underlying security payout, it is *de facto* equivalent to a European call. Thus $C = c$ in this case.

(3i) Given the European put-call parity, $p = c - S + Ke^{-r\tau}$, where $c - S \leq 0$, then an upper bound for put price is $p \leq Ke^{-r\tau}$.

Without any assumptions on underlying security distributions or on unobserved parameters such as drifts of the underlying security or any assumptions on utility functions, the tightest bounds on the calls and puts can be stated as follows

$$S \geq C \geq c \geq \max(S - Ke^{-r\tau}, 0) \quad \text{and}$$
$$K \geq P \geq p \geq \max(Ke^{-r\tau} - S, 0),$$

and for European put, $Ke^{-r\tau} \geq p$.

(3j) For American options with no dividend payouts, there are two inequality relationships:

$$S - Ke^{-r\tau} \geq C - P \geq S - K$$

We shall leave the proof of this as an exercise.

(3k) If the underlying security issues dividend payouts in $[0, T]$ with a known present value of D, then $c \geq \max(S - Ke^{-r\tau} - D, 0)$. This is because *ceteris paribus*, the stock price's present value will be reduced to $S - D$. The European call holder does not benefit from any receipt of the dividends but is disadvantaged by the price dilution, *ceteris paribus*.

If the amount of D is large enough, then an American call will be worth more than a European call because there is the possibility of early exercise.

The converse of no American call exercise is when $S - Ke^{-r\tau} - D > S - K$, or $K > Ke^{-r\tau} + D$. But when $D > K(1 - e^{-r\tau})$, then there is positive probability of early exercise. Thus, a American call will dominate a European call in terms of payoff, and will be priced higher.

For European put, $p \geq \max(Ke^{-r\tau} - S + D, 0)$. The same reasoning goes that D decreases price which is good for put. Obviously, $C \geq c \geq \max(S - Ke^{-r\tau} - D, 0)$, and $P \geq p \geq \max(Ke^{-r\tau} - S + D, 0)$.

(31) An American put will always be worth more than a European put because of the positive probability of early exercise. First, note that event $\{c < K(1 - e^{-r\tau})\}$ is possible, hence $P(c < K(1 - e^{-r\tau})) > 0$.

Suppose $K - Ke^{-r\tau} > c$. This implies $K - S > c - S + Ke^{-r\tau} = p$. The last equality comes from the European put-call parity. Then, $p < K - S \leq P$. Hence, $P > p$.

There is a link between the bounds and hedging we saw earlier. Suppose an option writer or issuer is exposed to the option price C_0. He or she wants to hedge in a way so that the hedging outcome will cover any loss arising out of the exposure. Let the initial cost of the self-financing option replicating portfolio be Π_0^H, and the outcome for any time $t \in [0, T]$ is such that the payoffs of the replicating portfolio is dominant over the option payoff exposures, i.e., $(\Pi_t^H - \Pi_0^H) \geq (C_t - C_0)$ and thus $\Pi_0^H \geq C_0$. Such self-financing hedging strategies are called superhedging or superreplication. A superhedge's cost also forms an upper bound of the price of the option. Thus, the superhedging produces an upper bound $\Pi_0^H \geq C_0$.

Conversely, if an option holder with an option value C_0 wishes to ensure that the value will not decrease, then he/she can sell a self-financing option replicating portfolio of value Π_0^L so that the combined effect is to produce for any outcome at $t \in [0, T]$ a net value $(\Pi_0^L - \Pi_t^L) - (C_0 - C_t) \geq 0$ and so $C_0 \geq \Pi_0^L$, which produces a lower bound for the option.

The rational bounds in this section are essentially static hedging strategies. Dynamic hedging strategies are only possible when the stochastic processes, hence distributions, of the underlying securities are specified, whether fully (parametrically) or partially (semiparametric), and in incomplete markets as in most cases, suitable choice of martingale measure (remember there is some implicit connection to investor preferences in an equilibrium market situation) to find optimal hedges may be able to yield tighter bounds given the additional assumptions on stochastic processes.[4]

Optimal Stopping and Snell Envelope

The subject of American options is more complicated than European options. European options have boundary conditions such as $c(S,T) = \max(S_T - K, 0)$, $\lim_{S\to\infty} C(S,t) = S$, or $C(0,t) = 0$ (boundaries based on time and also on states that are certain and determinate). However, American options have the added flexibility of early exercise in $t \in [0,T]$, and at any point in time t, and the resulting state S, two simultaneous decisions take place: one is the pricing at t, the other is whether to exercise or not the option. For the latter decision, there would be a critical price S^* where on one side is the exercise or stopping region and on the other is the continuation region (i.e., continue to keep the option alive without exercise). Beforehand, it is not clear if at future $t < T$, which state S_t we would be in, and therefore which region would the price settle. This is called a free boundary problem. If the exercise or stopping region and the other continuation region do not change over time, then the problem is usually easier. If the regions change over time, it is usually called a moving boundary or Stefan problem.

Fortunately, we shall be able to find results in martingale theory (based on Markov processes mostly, of course) that enable us to simplify this free boundary problem of American options into a problem of computing expected values of stopped martingales. Still, closed-form analytical formulae, such as those more readily available in European options, are hard to come by, and most solutions are in terms of numerical methods or some form of analytic approximations.

Suppose X_t is an adapted integrable process to a filtered probability space. Let another adapted integrable process Z_t be defined such that

(4a) $Z_T = X_T$ a.s.

(4b) $Z_t = \max\{X_t, E[Z_{t+1}|\mathcal{F}_t]\}$, for $t = 0, 1, 2, \ldots, T-1$. (We may also express Z_t as the ess sup $\{X_t, E[Z_{t+1}|\mathcal{F}_t]\}$ in the sense that any complement event has zero probability measure, in a more general probability space.)

[4]For further studies in this direction, see Föllmer, H and A Schied, (2002). *Stochastic Finance: An Introduction in Discrete Time.* Berlin: Walter de Gruyter.

Then, the stochastic process $\{Z_t\}$ is called the Snell envelope of the process $\{X_t\}$. Clearly, $Z_t \geq E_t(Z_{t+1})$, so it is a supermartingale. It is also the smallest supermartingale that dominates X_t in the sense that $Z_s \geq X_s$ for all $s \in [0, T]$. At most times, before any stopping event, assume $Z_s > X_s$.

In Chap. 7, we discuss the idea of stopping times as random variables adapted to the filtration $\{\mathcal{F}_t\}$. We shall see how stopping times play an important role in the Snell envelope here. Define the stopping time τ^* as

$$\tau^* = \inf\{t \in [0, T] : Z_t = X_t\}.$$

In other words, this stopping time is the earliest time that Z_t hits X_t, ($X_t = \max(S_t - K, 0)$ for call, and $X_t = \max(K - S_t, 0)$ for put), and that at this stopping time, $X_t > E[Z_{t+1}|\mathcal{F}_t]$. Since we are dealing with $t \in [0, T]$ or finite time, the stopping times are all bounded and we have no issues with having to take expectations at infinite time.

Theorem 10.3 *The stopped $Z(\tau \bigwedge n)$, $n \in [0, T]$, is a martingale.* ∎

Proof. For $n \leq T - 1$, $Z(\tau \bigwedge n) = Z(0) + \sum_{i=1}^{n} 1_{\tau \geq i}(Z(i) - Z(i-1))$. So

$$Z\left(\tau \bigwedge (n+1)\right) - Z\left(\tau \bigwedge n\right) = 1_{\tau \geq n+1}(Z(n+1) - Z(n)). \tag{10.31}$$

For event $\{\tau \geq n+1\}$, $Z(n) > X(n)$, so

$$Z(n) = E(Z(n+1)|\mathcal{F}_n). \tag{10.32}$$

From (10.31) and (10.32),

$$Z\left(\tau \bigwedge (n+1)\right) - Z\left(\tau \bigwedge n\right) = 1_{\tau \geq n+1}(Z(n+1) - E(Z(n+1)|\mathcal{F}_n)).$$

Take conditional expectation on the above, and given that event $\{1_{\tau \geq n+1}\} = \{1^c_{\tau < n}\} \in \mathcal{F}_n$,

$$\begin{aligned} &\mathrm{E}\left[Z\left(\tau \bigwedge (n+1)\right) - Z\left(\tau \bigwedge n\right) |\mathcal{F}_n\right] \\ &= 1_{\tau \geq n+1} E[Z(n+1) - E(Z(n+1)|\mathcal{F}_n)|\mathcal{F}_n] = 0. \end{aligned}$$

∎

Theorem 10.4 (Optional Stopping or Optional Sampling Theorem): *A stopped martingale (supermartingale) is a martingale (supermartingale) at stopping times.* ∎

One implication of this theorem is that at stopping times τ_1, τ_2, etc., $E(Z_{\tau_k}) = E(Z_1)$ where Z_0 is the initial value. Hence, the last two theorems imply that the Snell envelope $Z_t = ess\ sup(X_t, E[Z_{t+1}|\mathcal{F}_t])$ is a martingale, and if it is stopped at stopping times $\{\tau^*\}$, then $Z_t(\tau^*)$ is a martingale.

Now, we come to the meat of the American option solution problem using stopping-time martingale theory.

Theorem 10.5 (Optimal Stopping Theorem): τ^* *is the optimal stopping time in the sense that*

$$Z_0 = E(X(\tau^*)) = ess\ sup[E(X(\zeta)]$$

for all other stopping times $\zeta \in [0, T]$. ■

Proof. By Theorems 10.3 and 10.4, $Z(\tau \bigwedge m)$ is a martingale for $m \in (\tau, T]$. Now

$$\begin{aligned} Z_0 \equiv Z\left(0 \bigwedge \tau^*\right) &= E\Big(Z(\tau^* \bigwedge T)\Big) \\ &= E(Z(\tau^*)) = E(X(\tau^*)). \end{aligned}$$

Under the stopping time τ^* and the fact that Z_t is a Snell envelope which is a supermartingale, then Z_0 dominates $X_t(\zeta)$ for all stopping times ζ. Thus, $Z_0 = \text{ess sup}\{E(X(\zeta)\}$ for all ζ.

Hence $Z_0 = E(Z(\tau^*)) = E(X(\tau^*)) = \text{ess sup}\{E(X(\zeta)\}$ for all ζ.

■

The implication of the above result is that we can compute the American option price by taking expectation of the Snell envelope martingale Z_t. Another implication is that if stopping time τ^* is reached when Z_t equals $X_t = \max(S_t - K, 0)$ for call or $X_t = \max(K - S_t, 0)$ for put, for the first time, then it is optimal to exercise (since $Z(\tau^*) = X(\tau^*)$ would be optimal). This allows the binomial tree (or lattice tree) technique to be used, as shown later in the chapter. The only difference from the European lattice tree is that the nodal values are replaced by the Snell envelope values.

Approximations

The free boundary problem of the fundamental PDE of American options[5] is a difficult problem that is sometimes transformed into a relatively easy

[5]McKean Jr. HP (1965). A free boundary problem for the heat equation arising from a problem in mathematical economics. *Industrial Management Review*, 6, 32–39, was one

ODE problem by way of approximation. We introduce a useful method by Barone-Adesi and Whaley[6] that provides a relatively fast approximation to American calls and puts, and is accurate particularly when the time to maturity is either very short or long.

Recall that the fundamental PDE for derivative $f(S,t)$ with S following GBM is

$$\frac{1}{2}\sigma^2 S^2 f_{SS} + bSf_S - f_\tau - rf = 0, \tag{10.33}$$

where b is the carry cost of holding an underlying asset with price S. This PDE applies to American call price C, European call price c, American put price P, and European put price p. It also applies to $C - c$, and $P - p$, or the early exercise premia. What distinguishes the different prices are the boundary conditions that differ for each derivative.

Suppose the early exercise premium is f and is defined as $f = \Gamma X$. f is driven by the PDE in (10.33). However, X follows PDE:

$$S^2 X_{SS} + NSX_S - MX/\Gamma - (1-\Gamma)MX_\Gamma = 0, \tag{10.34}$$

where $\Gamma = 1 - e^{-r\tau}$, $N = 2b/\sigma^2$, and $M = 2r/\sigma^2$. The approximation to the solution is obtained by ignoring the last term, or approximating it at $(1-\Gamma)MX_\Gamma = 0$. This approximation is good when $\tau \downarrow 0$ so $X_\Gamma \downarrow 0$, or when $\tau \uparrow \infty$ so $(1-\Gamma) \downarrow 0$.

The PDE for X then becomes $S^2X_{SS} + NSX_S - MX/\Gamma = 0$, a second-order ODE where the solution is in power form. The analytic approximate solution for American call is then

$$C = \begin{cases} c + \Gamma a_c S^{q_c}, & S < S^* \\ S - K. & S \geq S^* \end{cases}$$

where c is found in (10.13). $q_c = \left(1 - N + \sqrt{(N-1)^2 + 4M/\Gamma}\right)/2 > 0$. The unknowns a_c and S^* are solved by the smooth-pasting conditions to ensure that the price below S^* and that immediately above S^* are joined at S^*, and that their slopes are equal at the joining point. These lead to

of the earliest researchers who had produced some useful analytical characterizations for the pricing of American options.

[6] Barone-Adesi, G and RE Whaley (1987). Efficient analytic approximation of American option values. *The Journal of Finance*, 42(2), 301–320.

two simultaneous equations to be solved for a_c and S^*:

$$\begin{aligned} S^* - K &= c(S^*, \tau) + \Gamma a_c S^{*q_c} \\ 1 &= e^{(b-r)\tau} N[d_1(S^*)] + \Gamma q_c a_c S^{*q_c - 1}. \end{aligned}$$

As a result, S^* is obtained by solving

$$S^* - K = c(S^*, \tau) + [1 - e^{(b-r)\tau} N[d_1(S^*)]] S^* / q_c.$$

After some algebraic arrangements, the solution is

$$C(S, \tau) = \begin{cases} c(S, \tau) + A_c (S/S^*)^{q_c}, & S < S^* \\ S - K, & S \geq S^* \end{cases}$$

where $A_c = (S^*/q_c)[1 - e^{(b-r)\tau} N[d_1(S^*)]] > 0$.

Employing the same approach, the analytic approximate value of an American put is

$$P(S, \tau) = \begin{cases} p(S, \tau) + A_p (S/S^{**})^{q_p}, & S > S^{**} \\ K - S, & S \leq S^{**} \end{cases}$$

where $A_p = -(S^{**}/q_p)[1 - e^{(b-r)\tau} N[-d_1(S^{**})]] > 0$.

$q_p = (1 - N - \sqrt{(N-1)^2 + 4M/\Gamma})/2 < 0$. And S^{**} is found via solving

$$K - S^{**} = p(S^{**}, \tau) - [1 - e^{(b-r)\tau} N[-d_1(S^{**})]] S^{**} / q_p.$$

Numerical Methods

One of the most common methods that quant analysts typically implement is the recombining lattice tree method. When derivative pricing is path dependent, Monte Carlo methods or quasi-MC methods for improved speed and accuracy, and other techniques such as importance sampling and variance reduction are sometimes used in conjunction to reduce the computational burden and tame the curse of dimensionality. There is usually some tradeoff between faster convergence versus analytical accuracy. We shall continue with the example of the binomial tree as a standard workhorse for pricing American options here. We employ the recombining tree for derivatives that are not path dependent.

Consider an American put option on a stock. The initial stock price is $S_0 = \$6$ and the stock process follows a lognormal diffusion with instantaneous volatility of return as σ. Following the discussion of such

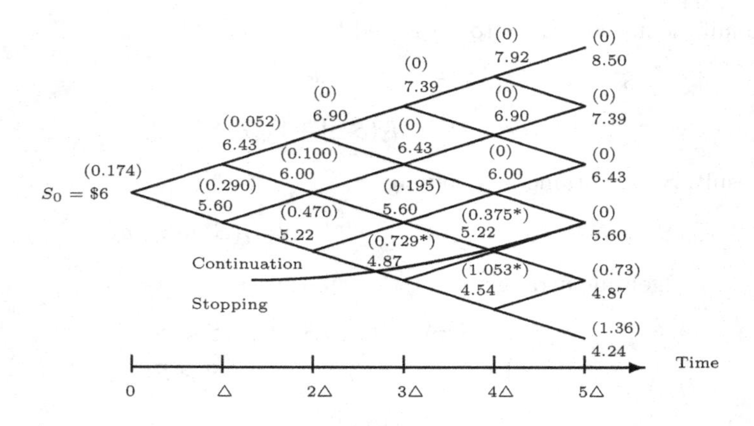

Fig. 10.4: American Option Pricing on Binomial Tree

structures in Chap. 3, the risk-neutral probabilities of stock return factors u and d in the next period are p and $q = 1 - p$, respectively. $u = e^{\sigma\sqrt{\triangle}}$, $d = u^{-1}$, and $p = \frac{1}{2}\left[1 + \left(\frac{r-1/2\sigma^2}{\sigma}\right)\sqrt{\triangle}\right]$ or we can use $p = \frac{e^{r\triangle}-d}{u-d}$. Assume no dividend payouts till maturity, and the time-to-maturity is divided into 5 periods of equal interval $\triangle$ year. The evolution of the underlying stock price and the associated (European put price) at that state are shown on the lattice tree in Fig. 10.4. The strike or exercise price of the American put is \$5.60. In the following example, let $\triangle = 1/52$, $\sigma = 0.5$, $r = 2\%$ p.a., so $u = 1.072$, $d = 0.933$, and $p = 0.486$.

The distinction between an American put and a European put is that the American put can be exercised at any time (at any intervals $j\triangle$, $j = 0, 1, 2, 3, 4, 5$. The exercise gain is $K - S > 0$ depending on the underlying price S at the particular state or node at time $j\triangle$. In this numerical method, the optimal exercise occurs when the condition $(K - S) > c$ is satisfied, i.e., when exercise gain is more than selling the put alive. In this case, it happens at node $(4\triangle, UDDD)$ when the exercise gain is $5.60 - 5.22 = 0.38$ which is higher than selling the put at 0.375, and at node $(4\triangle, DDDD)$ when the exercise gain is $5.60 - 4.54 = 1.06$ which is higher than selling the put at 1.053. It also happens at node $(3\triangle, DDD)$ when the exercise gain is $5.60 - 4.87 = 0.73 > 0.729$. These occurrences are marked with $(*)$ on the prices that are computed as the discounted expected values under risk-neutral probabilities (p, q). The backward computation of the American put prices then replaces the European prices with the higher exercise prices at the nodes with $(*)$. It is seen that for this case, the optimal exercise would occur at the earliest instance when exercise gain exceeds the risk-neutral discounted expected price, i.e., at $t = 3\triangle$ if state DDD would be reached, and it would also occur at $t = 4\triangle$ if state UDDD is reached without going through DDD. At time $t = 0$, this American put is priced at \$0.174. An equivalent European put would be priced at \$0.172 at time $t = 0$. Thus, the early exercise premium of the American put here for the period $[0, 5\triangle]$ is \$0.002.

To illustrate the idea of the free boundary discussed earlier, we also draw an estimated boundary curve separating the stopping or exercise region below (in the put case) and the continuation region above. The curve ends at the strike price 5.60 at expiry T.

10.5. Problem Set 10

1. Show why in order to prevent arbitrage, the following must hold, for $r, \delta > 0$:

$$u > e^{(r-\delta)h} > d > 0.$$

2. The put-call parity of a European call c and a European put p on an underlying non-dividend stock price S is

$$c - p = S - Ke^{-r\tau},$$

where K is the common exercise price of both call and put, and τ is time to maturity of both options. Prove the above parity. Then, show why the delta of the call – delta of the put = 1, and the gamma of the call = gamma of the put. Find the delta of a portfolio of 5 at-the-money European puts, using the put-call parity and the fact that delta of a European call is $N(d_1)$.
3. In Sec. 10.1, find the self-financing portfolio of $(\triangle_1^U, B_1^U)$. How is the initial outlay of the self-financing portfolio at $t = 0$ related to C_{uu}, C_{ud}, C_{dd}, p, and $R = e^{rh}$?
4. In the discrete binomial tree or lattice pricing model of options, suppose at node (time states) (t, UU), option price is f^{uu}, underlying stock price is S^{uu}. At node (t, UD), option price is f^{ud}, underlying stock price is S^{ud}. And at node $(t-1, U)$, option price is f^u, underlying stock price is S^u. How would you estimate the option delta from these information, assuming that the option delta is really an approximation of the continuous-time model delta as the lattice interval shrinks toward zero.

5. Suppose there is a call option 1 whose exposure we want to hedge to make it delta-gamma-vega neutral. Moreover, we use a self-financing hedge. There is the underlying stock with price \$10 and two other call options written on the same stock, call 2 and call 3 in the market, as well as a risk-free bond. The prices of the calls 1,2, and 3 are C_1, C_2, and C_3, respectively. The deltas, gammas, and vegas of the call options are shown in the following table. Show the 4 equations in order to solve for the delta-gamma-vega neutralizing self-financing hedge. Is the delta-gamma-vega hedge always superior to just a delta-gamma hedge in the sense the tracking error or self-financing portfolio deviation at the end of the horizon is smaller?

Option	Delta	Gamma	Vega
Call 1	0.7	0.065	5.3
Call 2	0.5	0.047	6.5
Call 3	0.6	0.053	4.8

6. A continuous-time Poisson process N_t is described by

$$dN_t = \begin{cases} 1 & \text{with probability } \lambda dt \\ 0 & \text{with probability } 1 - \lambda dt \end{cases}.$$

Describe in a similar way the corresponding compensated Poisson process M_t and find its mean.

7. Prove the American options inequalities

$$S - Ke^{-r\tau} \geq C - P \geq S - K.$$

8. A bank sells calls on underlying asset with price $S_t < K$, and at a strike of K. The continuously compounded risk-free rate is r. In an attempt to fully hedge its exposure and to avoid loss if $S_T > K$ at T, the bank devises a stop-loss dynamic strategy in which it buys the underlying stock at market-placed limit order when its price becomes $K + \epsilon$, and then sells it at stop-loss limit order when its price falls back to $K - \epsilon$, where $\epsilon > 0$ is a small number. There is a small exchange cost for each buy or sell order on the underlying. Explain the imperfect hedge and 3 transaction costs that the hedge has to bear. Explain also if there is any trading risk, should the price become very jumpy with large volatility.

Chapter 11

THEORY OF MARKOV CHAINS AND CREDIT MARKETS

This chapter provides some mathematical results on the theory of Markov chains (MCs) which is an important class of stochastic process in its own right. In relation to financial markets, a key application of MCs is in the dynamical adjustment of the credit rating of a firm from period to period. In turn, this credit rating is a major determinant of the pricing of instruments in the credit market, such as risky bonds and credit default swaps as the transition matrix includes a default rating. We also explain the intensity-based approach to pricing defaultable securities. A section on insurance and mortality analysis to show the linkages to default or exit or optimal stopping in credit instruments is also included.

11.1. Insurance and Mortality

A mortality table is a table showing statistics collected from reported deaths in a territory such as a country. A sketch-example is shown in Table 11.1 as follows.

The base is set at 10 million people for each gender born at age 0. It is seen that statistically, only 10,757 of 10 million or 0.1076% males are left, and only 30,698 of 10 million or 0.3070% females are left alive at the age of 100. It should be noted that in going from historical frequencies such as the above to statements of *a priori* or *ex ante* probability, there is the implicit assumption that the probability of demise of a person at age x is the same regardless of whether he or she was borne in year t or year s, $s \neq t$.

The number of deaths at age x is $D(x)$. At the next age $x + 1$, the remaining lives $R(x + 1) = R(x) - D(x)$. We have to be careful to qualify remaining lives at $R(x)$ being the number at the start of x. For clarity, $R(1) = 10,000,000$ is the population at the start of year 1, i.e., at birth in the first year.

Table 11.1: A Mortality Table

	Male			Female		
Age x	Remaining Lives R(x)	Deaths at Age D(x)	Probability of Death at Age	Remaining Lives R(x)	Deaths at Age D(x)	Probability of Death at Age
1	10,000,000	41,800	0.00418	10,000,000	28,900	0.00289
2	9,958,200	10,655	0.00107	9,971,100	8,675	0.00087
3	9,947,545	9,848	0.00099	9,962,425	8,070	0.00081
⋮	⋮	⋮	⋮	⋮	⋮	⋮
⋮	⋮	⋮	⋮	⋮	⋮	⋮
50	9,022,649	56,031	0.00621	9,262,013	42,883	0.00463
51	8,966,618	60,166	0.00671	9,219,130	45,727	0.00496
⋮	⋮	⋮	⋮	⋮	⋮	⋮
⋮	⋮	⋮	⋮	⋮	⋮	⋮
99	31,450	20,693	0.65798	89,200	58,502	0.65585
100	10,757	10,757	1.00000	30,698	30,698	1.00000

The probability of death during age x itself, is $P(x) = D(x)/R(x)$, i.e., this is when the individual survives up to x and dies during the year before $x+1$.

The probability of death at age 100 is set to $P(100) = 1$. Thus it is seen that $D(100) = R(100)$. The probability that an individual will survive for at least one more year as in age x, or survive past age x, is $Q(x) = 1 - P(x) = 1 - D(x)/R(x) = R(x+1)/R(x)$.

The probability that an individual at age x will die within $n+1$ years by end of $x+n$ is $P(x,n) = [R(x) - R(x+n+1)]/R(x) = [R(x) - R(x+1) + R(x+1) - R(x+2) + \cdots + R(x+n) - R(x+n+1)]/R(x) = [D(x) + D(x+1) + \cdots + D(x+n)]/R(x)$.

Thus the probability that an individual age x will survive up to end of $x+n$ is $Q(x,n) = 1 - P(x,n) = R(x+n+1)/R(x)$. Given an individual is age x, the probability he or she will die between start of $m \geq 1$ year(s) and end of $n \geq m$ year(s) into the future is

$$\begin{aligned}
&P_x(x+m, n-m \mid \text{survival from } x \text{ to } x+m-1)\\
&= [1 - P(x, m-1)]\ P(x+m, n-m)\\
&= \frac{R(x+m)}{R(x)}\ \frac{[R(x+m) - R(x+n+1)]}{R(x+m)} = \frac{[R(x+m) - R(x+n+1)]}{R(x)}.
\end{aligned}$$

Notice we make use of conditional probability here. As long as survival up to $x+m-1$ is given, we can simply write the conditional probability as

$P_x(x+m, n-m) > 0$ without any loss of generality. A special case is the probability that an individual at age x would die at age $x+m$, which is $\frac{[R(x+m)-R(x+m+1)]}{R(x)}$. The probability that an individual at age x would live up to including age $x+m$ and then die the following year is the same as the probability that the individual would die specifically at age $x+m+1 = \frac{[R(x+m+1)-R(x+m+2)]}{R(x)}$.

We can also write

$$\begin{aligned}
P(x,n) &= D(x)/R(x) + P_x(x+1,0) + P_x(x+2,0) \\
&\quad + \cdots + P_x(x+n,0) \\
&= D(x)/R(x) + [R(x+1) - R(x+2)]/R(x) \\
&\quad + [R(x+2) - R(x+3)]/R(x) \\
&\quad + \cdots + [R(x+n) - R(x+n+1)]/R(x) \\
&= [D(x) + D(x+1) + \cdots + D(x+n)]/R(x),
\end{aligned}$$

as we have obtained earlier via another method.

A pure endowment is a life insurance policy in which an individual who survives until the contract maturity will receive the face value of the endowment. If the person is dead by the contract maturity, there is no payment. Let $E(x,n)$ be the present value of a dollar to be paid to a surviving individual age x in n years' time when the contract would expire. Then

$$E(x,n) = \frac{1}{(1+r)^n} Q(x,n) \times \$1,$$

where we note that $Q(x,n) = 1 - P(x,n) = R(x+n+1)/R(x)$, and r is the risk-free interest rate per annum (or sometimes taken as the risk-free p.a. deposit rate of the individual). Rearranging the terms,

$$\frac{E(x,n)(1+r)^n}{Q(x,n)} = \frac{E(x,n)(1+r)^n R(x)}{R(x+n+1)} = \$1.$$

Conceptually $E(x,n)$ is also the amount of money each of the $R(x)$ individuals puts into a common pot, where the collection is deposited into a bank to earn interest at rate r for n years, after which the deposit is then withdrawn and distributed amongst the $R(x+n+1)$ surviving individuals such that each surviving individual gets \$1. It should be noted that $E(x,n)$ is an estimated, and not a predetermined, number based on the mortality table. Thus, buying \$10,000 pure endowment for n years at age x actuarially costs (based on mortality table at fair costing) \$10,000 $\times E(x,n)$ at age x.

A whole life annuity purchased at age x is an annuity or constant \$ amount payable to an individual yearly if he is alive, and the yearly payment ceases when the individual dies. Suppose the first payment is at the end of year 1 or age $x+1$. Suppose the whole life annuity pays \$1 each year. What is its current single premium or price \$$A$?

\$$A$ is the sum of the present values of a sequence of pure endowment policies for each future year $x+1$, $x+2$, $\ldots, 99$, or $E(x,1)$, $E(x,2)$, $\ldots$, $E(x, 99-x)$. Thus, $A = \sum_{n=1}^{99-x} E(x,n)$.

A whole life insurance is one where the insurance company pays the face value of the policy when the insured dies. Suppose the face value of the whole life insurance is \$1, then at age x, the insured must pay an actuarially fair premium of \$$L$.

$$\$L = \sum_{n=1}^{99-x} \left[\frac{1}{(1+r)^n}\right] P_x(x+n,0) \tag{11.1}$$

$$= \sum_{n=1}^{99-x} \frac{1}{(1+r)^n} \frac{R(x+n) - R(x+n+1)}{R(x)}. \tag{11.2}$$

Note the term R(101) $= 0$ by definition.

Hence the premium for a face value of \$$Y$ whole life insurance is \$ YL.

A K-year term life insurance is one where the insurance pays the face value, say \$1, when the individual dies before the end of K years in the future. The cover expires after K years, so the premium is

$$\sum_{n=1}^{K} \frac{1}{(1+r)^n} \frac{R(x+n) - R(x+n+1)}{R(x)}.$$

A K-year endowment insurance policy is one which pays face value \$1 after K years if the insured is still alive (like a pure endowment policy). Its premium is $\frac{1}{(1+r)^k}\frac{R(x+k+1)}{R(x)}$. The premium of a K-year term life savings policy with a \$1 savings payment if the insured survives after K years is the sum of the premiums of a K-year endowment policy and a K-year term life insurance policy:

$$\frac{1}{(1+r)^K}\frac{R(x+K+1)}{R(x)} + \sum_{n=1}^{K} \frac{1}{(1+r)^n} \frac{R(x+n) - R(x+n+1)}{R(x)}.$$

For the whole life insurance policy, sometimes the insured does not pay a lump sum $\$L$ for the policy, but instead pays an annual premium over N years that is a predetermined number. This is an annual premium-type policy. If the insured dies before N years, then his estate is no longer obliged to pay for the rest. The insured's beneficiary still obtains \$1 when the insured dies.

Let this annual premium whole life insurance policy have an annual premium of $\$P$ over N years. Face value of whole life policy is \$1. Then, P is found via

$$P\sum_{t=1}^{N}\frac{1}{(1+r)^t}\frac{R(x+t+1)}{R(x)}$$
$$=\sum_{n=1}^{99-x}\frac{1}{(1+r)^n}\frac{R(x+n)-R(x+n+1)}{R(x)}.$$

The probability within the summation on the LHS is specifically seen as the probability that the individual will survive up to age $x+t$, and would thus pay the annual premium P at t in the future, but not pay afterward as he dies the following year.

How does an insurance firm make profits if expected returns under actuarially fair condition are zeros? In practice, even if there is a highly competitive market, which means the premia collected by insurance companies are driven toward those computed above based on fair expectations, there is still some profitability in order that the firm will hope to do well. The insurance company can make a normal profit (in economic terms, this means making profits to cover firm operating costs as well as managerial wages and bonuses) by using a r rate that is slightly lower than the market risk-free rate, or a r rate that is lower than the firm's own funding or borrowing rate. Basically, in all the policies, r is used to discount expected future payments, and acts like a deposit rate given to the insured. As is the case with banks, this deposit rate is made lower than the rate that the bank or firms in turn lend out. In the last case of annual premium-type policy where the insured pays in instalments, making profit is like lending to the insurance firm at below risk-free rate, so the premium charged here at \$P is in fact more than the actuarially fair price. The insurance company can make profit by charging a higher premium payable P than the case where the r's on both LHS and RHS are equal.

In fact, suppose an individual has survived up to age x, his or her probability of dying on that age is $P(x)$ and of surviving onto age $x+1$ or

Table 11.2: Survival Transition Probability

	x	$x+1$	$x+2$	$x+3$	$x+4$	$\cdots$	D
$x-1$	$1-P(x-1)$	0	0	0	0	$\cdots$	$P(x-1)$
x	0	$1-P(x)$	0	0	0	$\cdots$	$P(x)$
$x+1$	0	0	$1-P(x+1)$	0	0	$\cdots$	$P(x+1)$
$x+2$	0	0	0	$1-P(x+2)$	0	$\cdots$	$P(x+2)$
$\vdots$	$\vdots$	$\vdots$	$\vdots$	$\vdots$	$\vdots$	$\cdots$	$\vdots$
100	0	0	0	0	0	$\cdots$	1

longer is $1 - P(x)$. This can be shown on a probability Table 11.2 where D denotes the state of death. "x" denotes the state of survival up to age x.

Thus, the ij^{th} element in the table represents the probability of reaching state j (j^{th} column) from state i (i^{th} row). The state to state transition occurs over 1 year. The sum of probabilities in each row adds to one.

11.2. Corporate Credit Ratings

Three major companies have been offering credit ratings of listed firms and of all kinds of debts and financial instruments, including the now-infamous CDOs and other derivatives. They are Moody's, Standard and Poor's, and Fitch, and they faced questioning by the U.S. Congress on some of the major mistakes they made in over-rating some of these financial products.[1]

Firms rated by Standard & Poor's have grades ranging from high-quality investment-grade AAA, AA and A to medium investment-grade BBB. Below this level of credit-worthiness, firms are said to be of speculative grade, ranging from BB to B, and then CCC to CC (may default), and finally C (failed to pay debt interest and has filed for bankruptcy petition) to D (in default). The probabilities of a firm moving from one credit rating to another or remaining the same in the following year form a one-year credit migration probability matrix, as shown in Table 11.3. The percentages (%) are collected from past histories of yearly rating changes or non-changes over 1981 to 2003. Migration frequencies in the CC and C categories are too few to make good estimates of %, and so were typically left out in tables.

In the credit migration table by Schuermann (2007), the probability of a firm with AAA rating moving to AA rating in the following year is

[1]See their corporate websites at http://www.moodys.com/, http://www2.standardand poors.com, and http://www.fitchratings.com/corporate/index.cfm respectively.

Table 11.3: One-Year Credit Migration Probability Matrix

		Year $t+1$							
		AAA	AA	A	BBB	BB	B	CCC	D
Year t	AAA	92.29	6.96	0.54	0.14	0.06	0.00	0.00	0.000
	AA	0.64	90.75	7.81	0.61	0.07	0.09	0.02	0.010
	A	0.05	2.09	91.38	5.77	0.45	0.17	0.03	0.051
	BBB	0.03	0.20	4.23	89.33	4.74	0.86	0.23	0.376
	BB	0.03	0.08	0.39	5.68	83.10	8.12	1.14	1.464
	B	0.00	0.08	0.26	0.36	5.44	82.33	4.87	6.663
	CCC	0.10	0.00	0.29	0.57	1.52	10.84	52.66	34.030
	D	0.00	0.00	0.00	0.00	0.00	0.00	0.00	100

Source: Schuermann, T (2007). Credit Migration Matrices. In *Encyclopedia of Quantitative Risk Assessment*, E Melnick and B Everitt, (eds.). New York: Wiley.

6.96%, while the probability that it remains in the same AAA rating is 92.99%. Notice that once a firm has defaulted, attaining rating D, it remains defaulted forever.

In credit ratings, we have a countable state space: states "AAA", "AA", "A", "BBB", "BB", "B", "CCC", and "D". The migration probability from one state in time t to another state in time $t+1$, is expressed as a probability of state x_{t+1} at $t+1$ conditional on state x_t at t, i.e., $P(x_{t+1}|x_t)$. Here, the RV X_t takes categorical values, not numerical values. For example, if $X_t =$ "AA", then $P(X_{t+1} =$ "BBB"$|X_t =$ "AA"$) =$ 0.61%. The conditional probabilities $P(X_{t+1}|X_t =$ "AA"$)$ are shown in the second row. All probabilities lying in a row should add to 1. In addition, the most important observation is that X_t has the Markov property $P(x_{t+1}|x_t, x_{t-1}, x_{t-2}, \ldots, x_0) = P(x_{t+1}|x_t)$, hence X_t follows a Markov process. In such a process, given the present state X_t, past states $X_{t-j},\ j > 0$ have no probabilistic influence on future states $X_{t+j},\ j > 0$.

Definition 11.1 We call a Markov process with countable state space X_t a Markov Chain (MC).[2] If the time steps are discrete, it is a discrete time MC. If it is a continuous time process, then it is called a continuous time MC. We call the conditional probability $P(X_{t+1} = y|X_t = z)$ a transition probability of the chain. ■

The credit migration above is an example of a MC. A MC is sometimes represented in "reduced form" by the transition (probability) matrix shown

[2]Some mathematicians prefer to expand the definition of MC to encompass continuous space.

in Table 11.3. Note that all probability numbers in any row of a transition matrix must sum to 1 in order for the MC to be properly defined — in this case the transition matrix is said to be a stochastic matrix. Is the life or death table earlier a MC? The answer is yes — from Table 11.2, we can see that at state x, each row adds up to 1. Moreover, $P(x)$ depends only on x and not on $x-1, x-2, \ldots$, etc.

In the credit migration table, if the probabilities do not depend on the particular year t, but applies equally in any year t, then the transition probability function of the MC is said to be stationary or time-homogeneous.[3]

A MC has a simple dependence structure that makes probabilistic analysis tractable in many instances, but yet is rich enough in its structure (beyond the simpler i.i.d. RVs) to yield interesting and important results for modeling purposes, such as in Markov decision theory, and such as the Markov Chain Monte Carlo method (MCMC) in econometrics.

Let X_t be an MC. Define the (one-step or t to $t+1$) transition probability $P(X_{t+1} = y|X_t = z) \triangleq P(y|z) \geq 0$, and $\sum_y P(y|z) = 1$. As usual, $P(y,z)$ is the joint distribution $P(\{X_{t+1} = y\} \bigcap \{X_t = z\})$. Further, $P(X_t)$ is the marginal probability distribution of X_t. Specifically, $P(X_0)$ is the initial probability distribution of the MC as it starts with RV $\tilde{X}_0$. Unless otherwise stated, we shall assume that we are dealing with a MC having transition probabilities that are time-homogeneous or time independent. A MC is completely defined in this case by its one-step transition probability matrix and the initial probability distribution. The transition matrix for finite state space (i.e., a finite set of N values $a_1, a_2, \ldots, a_N$, that any X_t would take) is assumed to be a $N \times N$ square matrix $P \equiv (P_{ij})$ as in the credit migration matrix. The ij^{th} element of this transition matrix P, P_{ij}, is the transition probability $P(a_j|a_i)$.

11.3. Basic Properties of MC

In this section, we discuss some basic properties of MCs.[4]

---

[3]One should avoid confusing this with the different concept of a stationary (unconditional state) distribution of a MC. Generally, it is not preferable to use a term like "stationary MC", as it more aptly refers to unconditional stationarity. "Time-homogeneity" is a better term.

[4]For details and more specific results and derivations, two intermediate textbooks are valuable, viz. *Introduction to Stochastic Processes*, by Hoel, Port and Stone (1972).

The joint distribution of RVs on a MC is easily expressed as the product of transition probabilities and the initial distribution:

$$\begin{aligned}
P(x_1, x_0) &= P(x_1|x_0)P(x_0), \\
P(x_2, x_1, x_0) &= P(x_2|x_1, x_0)P(x_1, x_0) \\
&= P(x_2|x_1, x_0)P(x_1|x_0)P(x_0) \\
&= P(x_2|x_1)P(x_1|x_0)P(x_0),
\end{aligned}$$

and in general

$$P(x_n, x_{n-1}, \ldots, x_1, x_0) = P(x_n|x_{n-1})P(x_{n-1}|x_{n-2}) \times \cdots \times P(x_1|x_0)P(x_0).$$

Hence, conditional probability can be expressed as follows

$$\begin{aligned}
&P(x_{n+m}, x_{n+m-1}, \ldots, x_{n+2}, x_{n+1}|x_n, x_{n-1}, \ldots, x_1, x_0) \\
&\quad = \frac{P(x_{n+m}, x_{n+m-1}, \ldots, x_1, x_0)}{P(x_n, x_{n-1}, \ldots, x_1, x_0)} \\
&\quad = P(x_{n+m}|x_{n+m-1})P(x_{n+m-1}|x_{n+m-2}) \\
&\qquad \times \cdots \times P(x_{n+2}|x_{n+1})P(x_{n+1}|x_n).
\end{aligned}$$

Define $P^m(y|z)$ as the m-step transition probability function giving the probability of going from $X_t = z$ to $X_{t+m} = y$ in m steps, for any t. Transition probabilities are conditional probabilities. Then,

$$\begin{aligned}
P^2(y|z) &= \sum_k P(y|k)P(k|z), \\
P^3(y|z) &= \sum_j P^2(y|j)P(j|z) \\
&= \sum_j \left(\sum_k P(y|k)P(k|j) \right) P(j|z) \\
&= \sum_j \sum_k P(y|k)P(k|j)P(j|z).
\end{aligned}$$

Houghton Mifflin Co., and *Markov Chains*, by JR Norris (1997). Cambridge University Press.

In general, an m-step transition probability can be expressed in two ways:

$$P^m(y|z) = \sum_j P^{m-r}(y|j)P^r(j|z), \quad \text{for } m > r > 0,$$

called the Chapman–Kolmogorov equation, or more fundamentally as

$$\begin{aligned} P^m(y|z) &= \sum_{a_1}\sum_{a_2}\cdots\sum_{a_{m-1}} P(y|a_{m-1}) \times \cdots \times P(a_2|a_1)P(a_1|z) \\ &= \sum_{a_1}\sum_{a_2}\cdots\sum_{a_{m-1}} P(y, a_{m-1}, \ldots, a_2, a_1|z) = P(X_m = y|X_0 = z). \end{aligned}$$

A special case of this is when $r = 0$, so $P^r(j|z) \equiv P(\tilde{X}_0 = z)$, then $\sum_z P^m(y|z)P(z) = P^m(y)$. The last term is also $P(\tilde{X}_m = y)$. Therefore, we can compute the unconditional probability distribution of X_m in terms of the initial distribution $P(z)$, and the m-step transition probability function, $P^m(y|z)$.

Define $\rho(y|z)$ as the probability that a MC starting at value z will visit value y in finite time. If $\rho(z|z) = 1$, then z is called a recurrent state. If $\rho(z|z) < 1$, then z is called a transient state as there is a non-zero probability that the MC may not return to z after leaving it. If $P(z|z) = 1$, then z is called an absorbing state, i.e., once the MC reaches z, it does not leave it.

Given transition matrix P with $X_t \in S_{N\times 1}$, a finite state space $S = \{a_1, a_2, \ldots, a_N\}$ and suppose $A \subset S$. The hitting time of A is $T_A \triangleq \min(n > 0 : X_n \in A)$, i.e., the first time the MC RV X_t hits A, or takes a value in the set of A.[5]

Let the probability, conditional on initial $X_0 = z$, of event $T_A = m$ be $P(T_A = m|X_0 = z)$. Then

$$P^n(y|z) = \sum_{m=1}^{n} P(T_y = m|X_0 = z)P^{n-m}(y|y).$$

This means the MC starts at z, first hits y at time m, and then leaves y and revisits y (whether second time or third time or whatsoever) at time n.

[5]Sometimes the hitting time when A is a single value or state is called the "first passage time". In other usages, hitting time and first passage time are equivalent.

In addition,

$$P(T_y = 1|X_0 = z) = P(y|z)$$

$$P(T_y = 2|X_0 = z) = \sum_{k \neq y} P(y|k)P(k|z)$$

(Note that this is not the same as $P^2(y|z)$.)

$$P(T_y = n|X_0 = z) = \sum_{k \neq y} P(T_y = n - 1|X_1 = k)P(k|z).$$

Also, $P(T_y < +\infty|X_0 = z) = \rho(y|z) = \sum_{m=1}^{\infty} P(T_y = m|X_0 = z)$. The expected visit time of state y is $\sum_{m=1}^{\infty} mP(T_y = m|X_0 = z)$. $P(T_y < +\infty|X_0 = y)$ is called the return time (back to value y). If every state in a recurrent MC has finite expected return time, then the MC is said to be positive recurrent with positive recurrent states. If a recurrent state has infinite expected return time, then the state is said to be null recurrent and not positive recurrent.

Define $1(X_n = y)$ to be an indicator function yielding 1 if $X_n = y$ and 0 otherwise. If $N(y|X_0 = z)$ denotes the number of times that the chain after leaving $X_0 = z$ visits state y over time $(0, \infty)$, then

$$P[N(y|X_0 = z) \geq 1] = P(T_y < \infty|X_0 = z).$$

$$\begin{aligned} P[N(y|X_0 = z) \geq 2] &= \sum_{r=1}^{\infty} \sum_{s=r+1}^{\infty} P(T_y = r|X_0 = z)P(T_y = s|X_r = y) \\ &= \left(\sum_{r=1}^{\infty} P(T_y = r|X_0 = z) \sum_{s=r+1}^{\infty} P(T_y = s|X_r = y) \right) \\ &= \rho(y|z)\rho(y|y). \end{aligned}$$

Then in general,

$$P[N(y|X_0 = z) \geq m] = \rho(y|z)[\rho(y|y)]^{m-1}.$$

Now, the conditional expectation of the indicator function is found as follows.

$$E[1(X_n = y)|X_0 = z] = P(X_n = y|X_0 = z) = P^n(y|z),$$

and

$$E(N(y)|X_0 = z) = E\left(\sum_{n=1}^{\infty} 1(X_n = y|X_0 = z)\right)$$
$$= \sum_{n=1}^{\infty} E(1(X_n = y|X_0 = z))$$
$$= \sum_{n=1}^{\infty} P^n(y|z).$$

Note that $P(N(y) = \infty|X_0 = z) = P(T_y < \infty|X_0 = z) = \rho(y|z)$.

If the state space S in a MC is partitioned into subsets $S_1, S_2, \ldots, S_L$ such that for any state $z \in S_k$ and any other state $y \in S_k^c$, and if $\rho(y|z) = 0$, i.e., a MC variable will not travel from z to outside of S_k in finite travel, then the part of the MC S_k is said to be closed. A MC that does not have any partitioned subset that is closed is said to be a communicating class and is irreducible. When $S' \in S$ is the only closed subset of the state space, then $P(T_{S'} < \infty|X_0 = z)$ is called the absorption probability. It is the probability that the MC RV X_t will enter into S' and be absorbed and thereafter not communicate with the rest of the chain.

If a MC itself is closed and irreducible, then every state in the MC is recurrent or every state in the MC is transient (e.g., every state is a default state). If every state in a closed irreducible MC is recurrent, it is called a recurrent MC. If every state in a closed irreducible MC is transient, it is called a transient MC. In addition, if a recurrent MC has a finite number of states, then it is a positive recurrent MC.

Suppose P is a time-homogeneous $S \times S$ transition matrix. Let π_t be the $S \times 1$ unconditional probability distribution of the states at time t, i.e., $\pi_t^T = (P(X_t = a_1), P(X_t = a_2), \ldots, P(X_t = a_N))$ where T denotes the transpose. If

$$\pi_t^T P = (P(X_{t+1} = a_1), P(X_{t+1} = a_2), \ldots, P(X_{t+1} = a_N)) = \pi_t^T,$$

or $\pi_t^T = \pi_{t+1}^T, \forall t > 0, \pi$ (dropping the time subscript since it is the same for every t) is called the stationary (or invariant) distribution of the MC. If

$$\lim_{t\to\infty} \pi_t^T P = (P(X_{t+1} = a_1), P(X_{t+1} = a_2), \ldots, P(X_{t+1} = a_N)) = \pi^T,$$

then limiting constant vector π is called the steady-state distribution of the MC. There is a result that states that an irreducible MC is positive recurrent iff it has a stationary distribution.

Applications

As an example, consider the "Gamblers' ruin" MC depicted as follows by the transition matrix in Fig. 11.1. Starting with a MC RV of $X_0 = \$W$, the gambler either wins \$1 with probability p or loses \$1 with probability $q = 1 - p$, in each time-step forward.

	\$0	\$1	\$2	...	\$W-1	\$W	\$W+1	...
\$0	1	0	0	...	0	0	0	...
\$1	q	0	p	...	0	0	0	...
\$2	0	q	0	...	0	0	0	...
⋮	⋮	⋮	⋮	⋮	⋮	⋮	⋮	⋮
\$W-1	0	0	0	...	0	p	0	...
\$W	0	0	0	...	q	0	p	...
\$W+1	0	0	0	...	0	q	0	...
⋮	⋮	⋮	⋮	⋮	⋮	⋮	⋮	⋮

Fig. 11.1: Gamblers' Ruin Markov Chain

It can be characterized by $P_{00} = 1$ (i.e., state \$0 is the absorbing bankrupt state from which the gambler cannot escape into the rest of the chain again), and $P_{i,i-1} = q$, $P_{i,i+1} = p$, for $i = \$1, \$2, \$3, \ldots$ etc.

Let $h_i = P(T_0 < +\infty | X_0 = i)$ or the probability of hitting state 0, the bankruptcy or ruin state, in finite time, starting from state $\$i$. The solution of h_i in the above chain satisfies two equations:

$$h_0 = 1 \tag{11.3}$$

$$h_i = ph_{i+1} + qh_{i-1}, \quad \text{for } i = 1, 2, \ldots \tag{11.4}$$

If $p < q$, the solution to the difference equations (11.3) and (11.4) is, for constants a and b

$$h_i = a + b\left(\frac{q}{p}\right)^i .$$

For $i > 0$ that is very large, if $b > 0$, then there will be $h_i > 1$ which is not plausible since h_i is a probability. If $b < 0$, then there will be $h_i < 0$ which is again not plausible. Hence, $b = 0$. Then, $h_i = a$ where a is a constant. To solve for all $i \geq 0$, including condition (11.3), then $a = 1$.

If $p = q = 1/2$, the solution for constant a and b is

$$h_i = a + bi.$$

For $i > 0$ that is very large, if $b > 0$, then there will be $h_i > 1$ which is not plausible since h_i is a probability. If $b < 0$, then there will be $h_i < 0$ which is again not plausible. Hence, $b = 0$. Thus, $a = 1$.

Therefore, for $p \leq q$ the solution is $h_i = 1$ for all $i \geq 0$. In other words, the probability of hitting state 0, the bankruptcy or ruin state, in finite time, is 1. Thus, the gamblers' ruin is with certainty. In the special case when $q = 0$, then $h_i = 0$ for all $i > 0$. How about the case $0 < q < p$? In this case, a solution is $h_i = (1 - b) + b(\frac{q}{p})^i$ for $b \in (0, 1)$. The value of b can be derived via auxiliary assumptions. If $h_i \to 0$ as $i \to \infty$, then $\lim_{i \to \infty} b = 1$.

As another example, consider the following simplified credit migration matrix with only 3 ratings, A, B, and D.

$$\begin{pmatrix} a & b & 1-a-b \\ c & d & 1-c-d \\ 0 & 0 & 1 \end{pmatrix}.$$

Suppose the steady-state distribution is $(p\ q\ 1-p-q)^T$, then we have

$$\begin{aligned}&(p\ q\ 1-p-q) \quad \begin{pmatrix} a & b & 1-a-b \\ c & d & 1-c-d \\ 0 & 0 & 1 \end{pmatrix} \\ &\quad = (pa + qc \quad pb + qd \quad p(1-a-b) + q(1-c-d) + (1-p-q)).\end{aligned}$$

If we solve $p = pa + qc$, $q = pb + qd$, and $1 - p - q = p(1 - a - b) + q(1 - c - d) + (1 - p - q)$, the solution is $p = q = 0$. The steady-state distribution is $(0 \quad 0 \quad 1)^T$. Hence, the probability of bankruptcy is 1, or bankruptcy is certain in the limit as time proceeds toward infinity in a credit migration MC.

11.4. Strong Markov Property

We shall recall the meaning of a stopping time discussed in Chap. 7. A RV $T : \Omega \longrightarrow \{0, 1, 2, 3, 4, \ldots\}$ is called a stopping time of stochastic process $\{X_n\}$ if for each time n, the occurrence of the event $\{T = n\}$ is determined by $X_0, X_1, X_2, X_3, \ldots, X_n$ for $n = 1, 2, 3, \ldots$. Equivalently, there exists a function f s.t. $1_{\{T=n\}} = f(X_0, \ldots, X_n)$. In other words, with information on the process as it unravels, we will know when the event has occurred as it occurs.

Examples of stopping times are:

(1a) the first passage time $T_j = \min\{n \geq 1 : X_n = j\}$, and

(1b) the first hitting time of event $A \in \sigma(X_0, X_1, \ldots, X_n)$.

However, stopping time does not need to be a hitting time, and can happen many times in a stochastic process. For example, a stopping time where $RV X_t = t$ where t takes odd values consists of $T = 1, 3, 5, \ldots$, etc.

We have discussed what is the Markov property in Chap. 7. We shall use the MC to show what is strong Markov property and its application. Basically, when a stochastic process $\{X_n\}$ has the following property:

$$P(X_{T+1} = j | X_0, X_1, \ldots, X_{T-1}, X_T = i) = P(X_{T+1} = j | X_T = i),$$

for every finite stopping time $T < +\infty$, then the process possesses a strong Markov property. In other words, a process showing Markov property but conditioned on random stopping time is a strong Markov process. The past and the future of the process are conditionally independent given the present stopping time value i. Event $X_T = i$, i.e., X_T at time T taking the value i, is called a stopping event. The stopping need not be defined by the state taking value i, but may generally be defined as some function f indicated earlier, that is adapted to the natural filtration.

We state a strong Markov theorem below for a time-homogeneous MC with finite or bounded probabilities. Such a MC is always a Markov process — this is easily seen by the fact that a one-period transition probability matrix has constant elements that are not dependent on past histories except the current state, and the Chapman–Kolmogorov equation likewise preserves this property for multiperiod transition probability matrices. We show that any discrete-time MC possesses the strong Markov property.

Theorem 11.1 *For any time-homogeneous discrete-time MC with finite probabilities in the transition matrix $\{p_{ij}\}$ where the RV X_n takes values in states defined by i, j, etc., and if T is any finite stopping time such that $P(T < \infty) = 1$, then for every stopping time T*

$$P(X_{T+1} = j | X_0, X_1, \ldots, X_{T-1}, X_T = i) = P(X_{T+1} = j | X_T = i)$$

■

Proof. It is given the MC has finite probabilities, so the usual transition probability $p_{ij} \in (0, 1)$. Given $P(T < \infty) = 1$ or finite stopping time, events such as $X_T = i$ or $j \in A$ are well-defined with usual probabilities $\in$ (0,1). Stopping event $\{T = k\}$ may also be written in long-hand as $\{X_k \in A, T = k\}$ for a defined stopping event or set A. We need not specify A for the proof. Events $\{T = 1\}$, $\{T = 2\}$, $\{T = 3\}$, etc. are obviously disjoint.

Now

$$\begin{aligned}
&P(X_0, \ldots, X_{T-1}, X_T = i, X_{T+1} = j)\\
&= \sum_{k \geq 0} P(X_0, \ldots, X_{T-1}, X_T = i, X_{T+1} = j, T = k)\\
&= \sum_{k \geq 0} P(X_0, \ldots, X_{k-1}, X_k = i, X_{k+1} = j, T = k)\\
&= \sum_{k \geq 0} (P(X_0, \ldots, X_{k-1}, X_k = i, T = k)\\
&\quad \times P(X_{k+1} = j | X_0, \ldots, X_{k-1}, X_k = i, T = k)).
\end{aligned}$$

Since $P(X_{k+1} = j | X_0, \ldots, X_{k-1}, X_k = i, T = k) = p_{ij}$ by the Markov property, hence

$$\begin{aligned}
&P(X_0, \ldots, X_{T-1}, X_T = i, X_{T+1} = j)\\
&\quad = p_{ij} \sum_{k \geq 0} P(X_0, \ldots, X_{k-1}, X_k = i, T = k)\\
&\quad = p_{ij}\, P(X_0, \ldots, X_{T-1}, X_T = i). \qquad (11.5)
\end{aligned}$$

Hence, via Eq. (11.5),

$$P(X_{T+1} = j | X_0, \ldots, X_{T-1}, X_T = i) = p_{ij}. \qquad (11.6)$$

Since $P_{ij} = P(X_{T+1} = j | X_T = i)$, therefore from Eq. (11.6):

$$P(X_{T+1} = j | X_0, \ldots, X_{T-1}, X_T = i) = P(X_{T+1} = j | X_T = i).$$

This statement clearly shows that only $X_T = i$ affects the probability of $P(X_{T+1})$, not any earlier stopping events, such as X_{T-1} in (11.6). Extrapolating, any other earlier stopping events X_{T-k}, for $k \geq 1$ would not change the conditional probability given $X_T = i$, which is p_{ij}. ■

Every discrete MC with finite probabilities as in this case has both the Markov and strong Markov property. A strong Markov process is Markov by indexing the stopping time to ordinary time sequencing. For example, the natural time index or stopping event A such that t is one of values $\{1, 2, 3, \ldots\}$ makes a MC RV X_t stopped at $t = 1$, $t = 2$, $t = 3$, etc. This stopped MC is strongly Markov, but is also the original MC. However, in general, a Markov process that is not a discrete MC is not strongly Markov.

An immediate application of strong Markov process in a MC is that if we are interested in the transition probability of credit ratings originating from stopped events such as a rating hitting below investment grade, we can compute the probability as if the below investment grade state to similar or worse state probability were an MC, and not have to consider intermediate states (other than a jump to default).

11.5. Risky or Defaultable Bonds

Let $P_{S\times S}$ be a credit migration matrix as shown earlier, and assume that the transition probabilities of the MC $P(X_{t+1} = y|X_t = z)$ are time-homogeneous. Each period is a year. A 2-year transition matrix becomes $P \times P = P^2$. In general, a n-year transition matrix is P^n. Let the default state be notated as "D" while the other ratings states are $1, 2, \ldots, S-1$.

The probability of default within n years starting with rating z at the beginning of the first year is p^n_{zD}. The zD-element of P^n is $Q_z(0,n)$. The probability of a firm starting at credit rating z at $t = 0$ and surviving past time $t = n$ years is defined as $\Pi_z(0,n) = 1 - Q_z(0,n)$. Sometimes, the survival probability is also written as $\Pi_z(0,n) = P(\tau > n)$ where τ is the default time of the firm.

Suppose a corporate discount bond with rating z defaults within n years, and upon default, there is some recovery by debt holders of an effective amount assumed to be equal to fraction δ_z of the par value of the bond \$1, to be received at maturity (not at time of default). Generally it is known that δ_z increases with rating quality z.

In order to avoid introducing arbitrary preference parameters, we shall assume there is no arbitrage in this defaultable bond market, and invoke the First Fundamental Theorem of Asset Pricing seen in Chap. 7. Thus there exists risk-neutral default and survival probabilities and we assume the MC transition probabilities are measured in the risk-neutral world.

Suppose maturity is at time T. The risk-neutral expected value at T of the risky corporate discount bond with rating z, is

$$\Pi_z(0,T) \times 1 + Q_z(0,T) \times \delta_z,$$

where $\Pi_z(0,T)$ is the probability of no default, and $Q_z(0,T)$ is the probability of default.

The price of the risky discount bond now at $t = 0$ is to discount the above expectation by the risk-free return of $R(0,T)$ which is equal to $1/B(0,T)$ where $B(0,T)$ is the present market price of a treasury or risk-free discount bond maturing at T with par value of \$1.

Let the time $t = 0$ market price of the rating z risky discount bond be $V_z(0,T)$.

$$\begin{aligned} V_z(0,T) &= B(0,T)[\Pi_z(0,T) + (1 - \Pi_z(0,T))\delta_z] \\ &= \delta_z B(0,T) + (1 - \delta_z) B(0,T)\Pi_z(0,T) \\ &= B(0,T)(\delta_z + (1 - \delta_z)\Pi_z(0,T)). \end{aligned} \tag{11.7}$$

Taking natural logs on Eq. (11.7),

$$-\frac{1}{T}\ln[V_z(0,T)] = -\frac{1}{T}\ln[B(0,T)] - \frac{1}{T}\ln[(\delta_z + (1-\delta_z)\Pi_z(0,T))].$$

Or, the T-year z-credit rating continuously compounded interest rate p.a. is the T-year risk-free interest rate p.a. + credit risk premium of z-rating bond. Thus, the T-year credit risk spread of z rating is

$$\begin{aligned} C_z(T) &= \frac{1}{T}\ln\left[\frac{1}{V_z(0,T)}\right] - \frac{1}{T}\ln\left[\frac{1}{B(0,T)}\right] \\ &= \frac{1}{T}\ln\left[\frac{1}{(\delta_z + (1-\delta_z)\Pi_z(0,T))}\right]. \end{aligned} \tag{11.8}$$

The term structure of credit spread can be derived and is depicted in Fig. 11.2, where the term structure refers to $C_z(T)$ for different $0 < T < N$ for some finite N.

Suppose the year-to-year credit migration transition matrix is not time-homogeneous, but time-dependent as $P_{S\times S}(t, t+1)$ during the year t. Then, for example, $P^2(t, t+1) \neq P(t, t+1)P(t+1, t+2)$. Let the z-row of the non-homogeneous transition matrix $P(t, t+1)$ be $P_z(t, t+1)$. Then, the probability of default within n years starting with rating z at the beginning

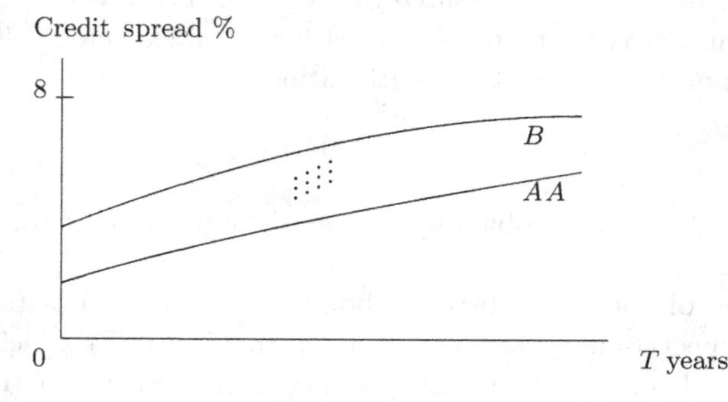

Fig. 11.2: Term Structure of Credit Spread

of the first year is

$$\begin{aligned} P_{zD}^{n} &\equiv Q_z(0,n) \\ &= P_z(0,1)P(1,2)P(2,3) \times \cdots \times P(n-2,n-1)P_D(n-1,n) \end{aligned}$$

where $P_D(n-1,n)$ is the default column of the transition matrix $P(n-1,n)$. Once $Q_z(0,n)$, and hence $\Pi_z(0,n) = 1 - Q_z(0,n)$ is obtained, the rest of the pricing as in Eqs. (11.7) and (11.8) can be done. As long as we do not consider any future migration once a default occurs (i.e., the model then disappears) non-homogeneous transition matrices with given frequencies are still consistent with a Markov process for the credit rating categorical RV X_t.

What is the 1-year conditional probability of default at $n+1$ given survival up to n? Let this be $P(\tau \le n+1|\tau > n)$. On the other hand, the 1-year conditional probability of no default at $n+1$ given survival up to n is $1 - P(\tau \le n+1|\tau > n)$ or $P(\tau > n+1|\tau > n)$.

Using Bayes' formula, the latter is

$$\begin{aligned} P(\tau > n+1|\tau > n) &= \frac{P(\{\tau > n+1\}\bigcap\{\tau > n\})}{P(\tau > n)} \\ &= \frac{P(\tau > n+1)}{P(\tau > n)} \\ &= \frac{\Pi_z(0,n+1)}{\Pi_z(0,n)}. \end{aligned}$$

Hence, the 1-year conditional probability of default at $n+1$ given survival up to n is

$$1 - \frac{\Pi_z(0,n+1)}{\Pi_z(0,n)}.$$

Another possible formulation is $\frac{Q_z(n,n+1)}{1-Q_z(0,n)}$.

The price of a risky discount bond in Eq. (11.7) can be modified in two ways.

(2a) Suppose no recovery takes place, then $\delta_z = 0$, and so $V_z'(0,T) = B(0,T)\Pi_z(0,T)$;

(2b) Suppose $T > 1$, and if default occurs in a year $s < T$ and that recovery is paid not at the maturity T, but earlier, at the end of the year in which default occurs, i.e., in $s < T$, then

$$V_z''(0,T) = B(0,T)[\Pi_z(0,T)] + \sum_{n=1}^{T} B(0,n)\delta_z P(\tau = n)$$

$$= B(0,T)[\Pi_z(0,T)] + \delta_z \sum_{n=1}^{T} B(0,n)[\Pi_z(0,n-1) - \Pi_z(0,n)],$$

where the last term is the unconditional probability of default at time n, and $\Pi_z(0,0) = 1$.

Consider a corporate coupon bond of maturity T years, currently in rating z, where in each year that the firm does not default, it pays a coupon c per par value of \$1 on the bond. The market price of this coupon bond with par value \$1 is

$$\left(c \sum_{n=1}^{T} V_z'(0,n) \right) + 1 \times V_z(0,T). \tag{11.9}$$

Note that Eq. (11.9) assumes no recovery to the coupon part over and above the recovery on the par value at T. Another version in the following equation assumes recovery on the par value takes place at the end of the year in which default happens, not having to wait till maturity T:

$$\left(c \sum_{n=1}^{T} V_z'(0,n) \right) + 1 \times V_z''(0,T). \tag{11.10}$$

Under the risk-neutral measure, it is as if the coupon bond could be divided into T coupons, and the final redemption value and all these could be sold off separately and added up in value to the coupon bond togather. (These formulations have slightly different assumptions on how the recovery is paid since the coupons are meant to be payable over different times.)

11.6. Credit Default Swaps

We shall now price another credit derivative. A credit default swap (CDS) is a contract arrangement between two counterparties, a buyer who wishes to hedge against default in a reference entity or security (e.g., a z-rated bond by an entity firm XYZ which the buyer has a long position), and thus pays a yearly swap premium (much like an insurance premium) or swap rate or swap spread or swap "price" or swap coupon of amount $c(T)$, where T denotes the length in years of the CDS obligation, or the number of yearly payments when default has not occurred, to the seller or insurance provider.

When default occurs, the swap payment by the buyer is terminated (without having to pay for the coupon in the year when default occurs), and the seller pays an amount $0 < (1-\delta) < 1$ times the notional amount of the CDS to the buyer.

We assume the latter payment is made at the end of the year in which default takes place. δ is the recovery fraction of the reference security, so the buyer only gets insured payment for the portion that he or she loses due to the default.

For a T-year notional \$1 CDS on z-rated reference entity, the swap coupon p.a. $c(T)$ is determined by equating the buyer's expected payment for protection (seller's expected receipts) to the buyer's expected receipt for protection (seller's expected payment), all under a risk-neutral probability measure:

$$\sum_{n=1}^{T} c(T)B(0,n)\Pi_z(0,n) = \sum_{n=1}^{T}(1-\delta_z)B(0,n)P(\tau=n)$$

or

$$c(T)\sum_{n=1}^{T} B(0,n)\Pi_z(0,n) = (1-\delta_z)\sum_{n=1}^{T} B(0,n)[\Pi_z(0,n-1)-\Pi_z(0,n)]$$

or

$$c(T)\sum_{n=1}^{T} V_z'(0,n) = (1-\delta_z)\sum_{n=1}^{T} B(0,n)[\Pi_z(0,n-1)-\Pi_z(0,n)]. \quad (11.11)$$

An investor who is long in the risky bond of Eq. (11.10) and who hedges by buying a CDS with the cash inflow on the RHS of Eq. (11.11) will have a combined cash inflow of

$$\left(c\sum_{n=1}^{T} V_z'(0,n)\right) + B(0,T)[\Pi_z(0,T)]$$
$$+\,\delta_z\sum_{n=1}^{T} B(0,n)[\Pi_z(0,n-1)-\Pi_z(0,n)]$$
$$+\,(1-\delta_z)\sum_{n=1}^{T} B(0,n)[\Pi_z(0,n-1)-\Pi_z(0,n)]$$

$$= \left(c \sum_{n=1}^{T} V_z'(0,n) \right) + B(0,T)[\Pi_z(0,T)]$$
$$+ \sum_{n=1}^{T} B(0,n)[\Pi_z(0,n-1) - \Pi_z(0,n)].$$

The combined effect is that when default occurs, future coupons ceased and the bond is redeemed on par value without any loss at default time. The additional cost of this hedge is $c(T) \sum_{n=1}^{T} B(0,n)\Pi_z(0,n)$.

So far, we have employed empirical frequencies on past defaults to construct MC transition probabilities, but have developed the pricing of risky bonds and credit derivatives based on risk-neutral probabilities. One disadvantage of the MC approach in default or credit migration modeling is that the distinction between empirical probabilities and risk-neutral probabilities is often blurred, and attempts have to be made to connect the two sets of probabilities. It is also not convenient for modeling changing non-homogeneous transition matrices.

We shall now show a reduced form intensity-based approach towards such pricing. Yet another type is a structural approach where a firm's total assets and total liabilities are modelled as two distinct stochastic processes, and default is said to occur when the asset value hits the liability value from above. This leads to stopping time computations of the resulting credit derivative prices.

11.7. Intensity-Based Pricing Approach

We shall move from finite-state discrete time MC to finite-state continuous time MC in order to show some key results in Markov processes. Even more general infinite state continuous time MC processes carry most of these results, although the proofs get a lot more technical.

Let continuous time MC process $\{X_t\}$ take values on finite states $S = 1, 2, \ldots, N$. $P_{N\times N}(t)$ is the transition matrix over a period of time length t, i.e.,

$$P_{ij}(t) = P(X_{t+s} = j | X_s = i), \quad \forall s < t, \ \forall i.$$

Assume stationary transition probabilities and that $P(t)$ is continuous in t so there is no jump in the transition probabilities over time. The Chapman–Kolmogorov equation of the MC is now seen as a matrix version in continuous time, $P(t+s) = P(t)P(s)$ $t, s \geq 0$.

For each row $i \in [1, N]$

$$1 = P_{ii}(h) + \sum_{j \neq i}^{N} P_{ij}(h),$$

or

$$\frac{1 - P_{ii}(h)}{h} = \sum_{j \neq i}^{N} \frac{P_{ij}(h)}{h}.$$

Taking limits, let $\lim_{h \downarrow 0} \frac{1-P_{ii}(h)}{h} = q_i < \infty$, and $\lim_{h \downarrow 0} \frac{P_{ij}(h)}{h} = q_{ij} < \infty, i \neq j$. Therefore, $q_i = \sum_{j \neq i}^{N} q_{ij}$. Define $q_{ii} = -q_i = -\sum_{j \neq i}^{N} q_{ij}$, hence $\sum_{i=1}^{N} q_{ij} = 0$.

The q's are the rates of leaving the states in which the process is in. Now, $0 \leq q_i < \infty$, though this could be ∞ in infinite state MC, and $0 \leq q_{ij} < \infty, \forall i, j \neq i$.

A $N \times N$ matrix carrying the q rates

$$A = \begin{bmatrix} -q_1 & q_{12} & \cdots & q_{1N} \\ q_{21} & -q_2 & \cdots & q_{2N} \\ \vdots & \vdots & \vdots & \vdots \\ q_{N1} & q_{N2} & \cdots & -q_N \end{bmatrix}$$

is called the infinitesimal generator of the Markov process $\{X_t\}$. This matrix can be derived as follows:

$$\lim_{h \downarrow 0} \frac{P(h) - I}{h} = A.$$

The elements of A are rates of transition in an infinitesimally small time interval. But

$$\frac{P(t+h) - P(t)}{h} = \frac{P(t)[P(h) - I]}{h} = P(t)\frac{P(h) - I}{h},$$

or

$$\frac{P(t+h) - P(t)}{h} = \frac{[P(h) - I]}{h} P(t).$$

Taking the limit as $h \downarrow 0$, hence

$$P'(t) = P(t)A = AP(t). \tag{11.12}$$

The solution to Eq. (11.12), given initial condition $P(0) = I$ (over zero time, X_t stays in state), is unique. The solution ties in many strands of results and development in modern probability theories. For one, a Poisson process is a common type of such a continuous time MC with finite states or values (we can think of mapping categorical variables to nominal values) whereby:

(3a) it has the Markov property (i.e., it is memoryless), and

(3b) its increments (discrete changes from one state to another) are stationary and independent.

We shall build such a Poisson process from a infinitesimal generator

$$A = \begin{bmatrix} -\lambda & \lambda \\ 0 & 0 \end{bmatrix}.$$

Also, let

$$P(t) = \begin{bmatrix} P_{11} & P_{12} \\ P_{21} & P_{22} \end{bmatrix}.$$

Let the Poisson process represent either the no-default or default condition of a particular firm, hence the two states. The rate λ is the firm's default intensity, and differs from firm to firm according to what the firm's current rating is.

Applying Eq. (11.12), using the Kolmogorov forward equation $P'(t) = P(t)A$ (we could also use the Kolmogorov backward equation $P'(t) = AP(t)$ that would yield the same unique solution), we obtain

$$\begin{aligned} P_{11}'(t) &= -\lambda P_{11}(t) \\ P_{12}'(t) &= \lambda P_{11}(t) \\ P_{21}'(t) &= -\lambda P_{21}(t) \\ P_{22}'(t) &= \lambda P_{21}(t). \end{aligned}$$

Solving using initial condition $P(0) = I$, we obtain

$$P(t) = \begin{bmatrix} e^{-\lambda t} & 1 - e^{-\lambda t} \\ 0 & 1 \end{bmatrix}.$$

It is seen that $P_{11}(t)$ is the probability of an exponentially-distributed RV. In other words, the waiting or holding time in the no-default state or condition of any rated firm is distributed with $P(\tau > t) = e^{-\lambda t}$.

Suppose we add the complication that the instantaneous interest rate and the default intensity is not a constant but is an adapted process itself, i.e., $r(s)$ and $\lambda(s)$, so by the Markov property and Chapman–Kolmogorov relation

$$
\begin{aligned}
P_{11}(t) &= \lim_{n\uparrow\infty}[P_{11}(t/n)]^n \\
&= \lim_{n\uparrow\infty}\left[e^{-\lambda(t/n)}\right]^n \\
&= e^{-\int_0^t \lambda(s)ds},
\end{aligned}
$$

which is the probability of no default up to time t.

The density function of the distribution of default $F(s) = \left(1 - e^{-\int_0^s \lambda(u)\,du}\right)$ is

$$
\frac{d}{ds}\left(1 - e^{-\int_0^s \lambda(u)\,du}\right) = \lambda(s)e^{-\int_0^s \lambda(u)\,du},
$$

which is the probability of default at time s when no default occurred up to time s.

Consider the following three lemmas.[6] The superscripts attached to the expectation operator denotes the integration over those variables under the risk-neutral joint probability measure.

Lemma 11.1 *Consider a contingent claim that pays a random amount X at time T provided default has not occurred, and zero otherwise. The time $t = 0$ value of this claim is, given default time is s,*

$$
\begin{aligned}
&E_0^{r,X,s}\left[\exp\left(-\int_0^T \tilde{r}(u)du\right)\tilde{X}\tilde{1}_{\{s>T\}}\right] \\
&= E_0^{r,X}\left[\exp\left(-\int_0^T \tilde{r}(u)du\right)\tilde{X}P(s>T)\right] \\
&= E_0^{r,X}\left[\exp\left(-\int_0^T \tilde{r}(u)du\right)\tilde{X}\exp\left(-\int_0^T \tilde{\lambda}(u)du\right)\right] \\
&= E_0^{r,X}\left[\exp\left(-\int_0^T \tilde{r}(u)+\tilde{\lambda}(u)du\right)\tilde{X}\right].
\end{aligned}
$$

■

Lemma 11.2 *Consider a security that pays a cash flow $Y(t)$ per unit time at time t provided default has not occurred, and zero otherwise. The time*

[6] Lando, D (1998). On Cox processes and credit risky securities. *Derivatives Research*, (2–3), 99–120.

$t = 0$ *value of this security is, given default time is s,*

$$E_0^{r,Y,s}\left[\int_0^T \exp\left(-\int_0^t \tilde{r}(u)du\right)\tilde{Y}(t)\tilde{1}_{s>t}dt\right]$$
$$= E_0^{r,Y}\left[\int_0^T \exp\left(-\int_0^t \tilde{r}(u)du\right)\tilde{Y}(t)P(s>t)dt\right]$$
$$= E_0^{r,Y}\left[\int_0^T \exp\left(-\int_0^t \tilde{r}(u)du\right)\tilde{Y}(t)\exp\left(-\int_0^t \tilde{\lambda}(u)du\right)dt\right]$$
$$= E_0^{r,Y}\left[\int_0^T \exp\left(-\int_0^t \tilde{r}(u)+\tilde{\lambda}(u)du\right)\tilde{Y}(t)dt\right].$$

■

Lemma 11.3 *Consider a security that pays $Z(s)$ if default occurs at random time s, and zero otherwise. The time $t = 0$ value of the security is*

$$E_0^{r,Z,s}\left[\int_0^T \exp\left(-\int_0^t \tilde{r}(u)du\right)\tilde{Z}(t)\tilde{1}_{s\le t}dt\right]$$
$$= E_0^{r,Z}\left[\int_0^T \exp\left(-\int_0^t \tilde{r}(u)du\right)\tilde{Z}(t)\lambda(t)\exp\left(-\int_0^t \tilde{\lambda}(u)du\right)dt\right]$$
$$= E_0^{r,Z}\left[\int_0^T \exp\left(-\int_0^t \tilde{r}(u)+\tilde{\lambda}(u)du\right)\tilde{Z}(t)\tilde{\lambda}(t)dt\right].$$

■

Now, consider a corporate bond with maturity in T years, and continuous instantaneous coupon rate of c^* per $1 par value. Instantaneous risk-free rate is constant r and default intensity is constant λ. Following the notion of the bond in Eq. (11.10) where coupons are stopped with no recovery when default occurs, and where the par value is redeemed at recovery δ at the point of default instead of at maturity, then the price of the bond is:

$$\left(c^*\int_0^T e^{-(r+\lambda)t}dt\right) + \left(1\times e^{-(r+\lambda)T}\right) + \left(\delta\int_0^T e^{-(r+\lambda)t}\lambda dt\right).$$

The first term uses Lemma 11.2 by putting $Y(t) = c^*, r(u) = r$, and $\lambda(u) = \lambda$. The second term is redemption of par without default, using Lemma 11.1. The third term is the expected recovered value of defaulted bond with possible default at any one time before maturity, and uses Lemma 11.3. Comparing term by term with the discrete MC case in

Eq. (11.10), we have

$$\left(c\sum_{n=1}^{T}\frac{\Pi(0,n)}{(1+R)^n}\right)+\left(1\times\frac{\Pi(0,T)}{(1+R)^T}\right)+\left(\delta\sum_{n=1}^{T}\frac{\Pi(0,n-1)-\Pi(0,n)}{(1+R)^n}\right),$$

where R is the per period simple interest rate. Thus, the price of the defaultable risky bond is:

$$e^{-(r+\lambda)T}+\frac{c^*+\delta\lambda}{r+\lambda}\big(1-e^{-(r+\lambda)T}\big). \tag{11.13}$$

It is seen that the intensity-based approach ended up with pricing models that require a far less number of parameters than those on an MC. Fewer parameters generally allow easier calibration or the estimation of the parameters using current traded market prices of the instruments. The calibrated models can then be used to price other related credit derivatives or used for forecasting future prices. Intensity-based models also allow very flexible modeling of the intensity process itself, e.g., allowing λ to be dependent on a specified exogenous process. One disadvantage of the intensity-based approach could be that it reduces the structure by too much and may induce model errors. The models are also more sensitive to the specifications of the intensity processes employed.

The pricing of derivatives with payoffs dependent on default events was a lucrative business for the Wall Street quants and investment bankers who sold such products up to 2008. CDS remains a sizeable — and to a large extent, useful — derivative instrument to hedge exposures to huge portfolio positions on securities that may face sudden demise. The market for products like cash collaterized debt obligations (cash CDOs with mortgages as collateral) or synthetic CDO (CDOs with CDS as collateral portfolio) or CDO squares (CDOs with other CDOs as collaterals), and similar products such as mortgage-backed securities (MBS) and some asset-backed securities (ABS) have been largely subdued after the 2008 financial crisis, and there have been few, if any, new issuances of CDOs due to its infamous destruction of major corporations like Lehman Brothers, Bear Stearns, Merrill Lynch, Wachovia, and many other smaller banks that had invested heavily in such instruments due to their ostensible profitability in those times. It was touted as a clever piece of financial engineering to redistribute debt portfolios into different tranches of varying levels of risk for investors with different risk appetites. At the same time, they made huge profits for the SPVs (special purpose vehicles of investment banks or commercial banks) that engineered it, more profits for the real estate brokerage firms that solicited buyers lacking credit worthiness, more profits for building firms, and more profits for rating agencies that provided the stamp of credit worthiness on the securities. It also led to more profits for both the institutional and retail investors who could presumably obtain higher returns in a AAA-rated senior tranche of the CDO than in an AAA corporate bond, more profits for commericial banks to lend huge sums of money for what seemed like good business loans, and even profits for the jobless, who could then buy a home and wait to cash out when housing prices went up again. It was a perfect Cinderella story until the clock struck and the walls came tumbling down.

The major problems by now are well understood, and one could point fingers at the easy Federal Reserve monetary policy for keeping interest rates too low ever since the uptake after the tech bubble episode in 2001–2002, or the greed and unbelievably high leverage banks are willing to lend to investment firms and institutional borrowers, the fault of incompetent rating agencies and many financial analysts, and the quants who loved their models too much and ignored reality checks.[7]

11.8. Problem Set 11

1. From the mortality tables, find the probability that a male, a female at age 50 will survive for at least one year? Will survive for at least 49 years?
2. The risk-free deposit rate is 3% p.a. For males, $R(52) = 8{,}841{,}435$, and $R(65) = 7{,}329{,}740$. If a male who is of age 52 now wishes to buy a pure endowment that pays $100,000 at his age 65, provided he survives till then, how much must he fork out now for this policy?
3. Given $Q(56{,}1) = 0.965$, $Q(56{,}2) = 0.963, Q(56{,}3) = 0.960$, $Q(56{,}4) = 0.957$, $Q(56{,}5) = 0.952$, and $Q(61,1) = 0.935$, $Q(61,2) = 0.930$, $Q(61,3) = 0.924$, $Q(61,4) = 0.920$, $Q(61,5) = 0.915$. Find the annual premium payable over 5 annual instalments at age 55 of a term

[7]For how CDOs and MBS have been priced, see Liu ZY, GZ Fan and KG Lim (2009). Extreme events and the copula pricing of commercial mortgage-backed securities. *Journal of Real Estate Finance and Economics*, 38(3); and also, Cao LJ, JQ Zhang, LK Guan and Z Zhao (2008). An empirical study of pricing and hedging collaterized debt obligation (CDO). *Advances in Econometrics*, 22, 15–54. For the method of copula used in modeling joint default probabilities of a basket of securities, see Cherubini, U, E Luciano and W Vecchiato (2004). *Copula Methods in Finance*. New York: John Wiley.

annuity of $20,000 starting at age 61 till 65. Risk-free interest rate is 3% p.a. Suppose risk-free interest rate has a term structure so that the annual rate over horizon of 10 years start at 3% but increases by 0.1% per additional year, what is the annual premium? (Assume first annual premium payment is at the end of the first year.)

4. Consider an independent Bernoulli RVs $X_i, \forall i$,

$$X_i = \begin{cases} +1, & \text{with probability } p_i = \frac{2}{3} \\ -1, & \text{with probability } 1 - p_i = \frac{1}{3} \end{cases}$$

and $S_0 = 0$. Is the asymmetric random walk $S_n = S_0 + \sum_{i=1}^{n} X_i$ an MC? If so, show the transition matrix.

5. Consider the Cox–Ross–Rubinstein binomial model of security price S_0 at $t = 0$. In each subsequent period, the price either goes up by factor $u > 1$ with probability p or goes down by factor $u^{-1} < 1$ with probability $1 - p$. Show how the price process or transition is a MC for an N-period model.

6. In the credit migration matrix, is there a closed subset? Is the credit rating MC an irreducible MC? Is it a recurrent or transient MC?

7. Find the invariant distribution of a two-state MC $\begin{pmatrix} 1-a & a \\ b & 1-b \end{pmatrix}$, where a, b are not 0 or 1.

8. Given $P(T_y = 2|X_0 = z)$ and $P^2(y|z)$, which is larger?

9. From Eq. (11.8), explain how given a 1-year transition matrix P and rates δ_z for every $z \in S$, the term structure of credit risk spread can be obtained for every rating z.

   Assume that an A-rated firm's probability of survival after 2 years is 0.95, and if it defaults, recovery fraction is 0.1. Find its credit risk premium. The Treasury 2-year risk-free rate is 2%. Hence find the price of a 2-year A-rated discount bond with a par value of $100. (You may assume that the Treasury rate is obtained using the continuous compounding formula of $\frac{1}{T} \ln \left[\frac{1}{B(0,T)} \right]$.)

10. Given that the one-year conditional probability of default at $n + 1$ given survival up to n is $1 - \frac{\Pi_z(0,n+1)}{\Pi_z(0,n)}$ for every $n > 0$, what is the unconditional probability of default at time $n + 1$? What is the unconditional probability of default by or up to time $n + 1$?

Chapter 12

INTEREST RATE MODELING AND DERIVATIVES

Classical interest rate theory is concerned with macroeconomic issues of price stability and the business cycle. Fisher Irving is best known for his Fisher's effect which says that an increase in expected inflation will cause an approximately similar increase in the nominal interest rate of the country, assuming the real rate is constant. Or, nominal interest equals expected real interest multiplied by the expected inflation factor. Interest rate theory before the advent of continuous-time mathematics and the option pricing literature from the late 1970s onward, has mostly been about the determination of the general level of interest based on factors affecting the money supply and the demand for money.

Modern interest rate theory uses much of the no-arbitrage condition as a workhorse. At the same time, the availability of many new market instruments on interest rates also lends fuel to new models and empirical validation of these models.

12.1. Interest Rates

Simple interest rate r p.a. means that \$1 today will yield $\$(1+r)$ at the end of the year. The compound amount at simple interest after t years is \$ $(1+r)^t$. Interest rates, unless otherwise stated, are usually quoted in the market on a p.a. basis.

For compound interest over subintervals of a year, e.g., monthly compounding of rate r, then \$1 earns total return (interest plus capital repayment) of \$ $(1+\frac{r}{12})^{12}$ at the end of the year, and $(1+\frac{r}{12})^{12t}$ at the end of t years.

For continously compounded interest rate r, \$1 earns total return of \$ e^{rt} at the end of t years. This is obtained as a limit, $\lim_{n\to\infty}(1+\frac{r}{n})^{nt}$. There are many ways of getting the exponent; one is to put $y_n = (1+\frac{r}{n})^{nt}$, take the natural logarithms of both sides, and apply L'Hôpital's rule to obtain $\lim_{n\to\infty} \ln y_n = rt$.

A spot interest rate is an interest rate charged on a loan or an interest rate received on a deposit or earned on a bond for holding period from the current time till some future time. For example, a discount bond (without coupons) bought today at price $B(t,T)$ for redemption at par \$1 at maturity time $T > t$ in the future (time-to-maturity $T-t$ in terms of number of years) has an effective return rate (nonannualized) of $1/B(t,T) - 1$. The discrete-time spot rate or spot yield p.a. (or yield-to-maturity for a discount- or zero-coupon (ZC) bond) is

$$y(t,T) = \left(\frac{1}{B(t,T)}\right)^{1/(T-t)} - 1.$$

In most of the advanced analyses of interest rates, due to the heavy use of modeling in continuous-time, it is often convenient to employ continuous-time compounding.

Hence, the continuously compounded spot rate is

$$y(t,T) = \frac{1}{T-t} \ln\left(\frac{1}{B(t,T)}\right). \tag{12.1}$$

The corresponding effective spot yield factor or return is

$$Y(t,T) = \exp(y(t,T) \times (T-t)).$$

At any time t, assume the market is trading a continuous spectrum or series of discount bonds with different maturities T at prices $B(t,T)$, respectively. We can then compute from their prices the corresponding spot rates via (12.1). The spot rate $y(t,T)$ for a particular T provides information on the investment return over $[t,T]$. If we plot the graph of $y(t,T)$ against time T, the graph is called a spot rate (spot yield) curve, or simply the yield (yield-to-maturity) curve of the discount bonds. A yield curve is also called the term structure of the yields.

A special case of the continuously compounded spot rate is when the holding period is infinitesimally small. In this case, it is an instantaneous spot rate, defined by

$$y(t) \triangleq \lim_{\triangle \downarrow 0} y(t, t+\triangle),$$

which may also be denoted, although more clumsily, as $y(t,t)$ or $y(t,t_+)$. This instantaneous spot rate is also called the short rate (spot rate at the zero-intercept short-end of the spot rate curve or term structure).

Henceforth, we shall use the continuously compounded version (12.1) of p.a. spot rate unless otherwise indicated. The difference between the annualized discrete spot rate and annualized continuously compounded spot rate is typically very small and of the order $O(y^2)$. Actual market practice uses different (daycount) conventions such that effective return is the quoted interest rate, r, multiplied by day-count factor $n/360$ for money market instruments less than a year to maturity, or multiplied by $n/365$ for bond market instruments with more than a year to maturity, where n is day count to maturity. Thus, we may set theoretical spot rate $y(t,T) = \frac{1}{T-t}\ln(1 + r[n/360])$ for a quoted market rate $r(t,T)$.

A p.a. forward interest $f(t,T_1,T_2)$ is an interest rate contracted at time t for a risk-free loan over time period $[T_1,T_2]$. For example, if $f(0,1,2)$ is 5% p.a., then a loan contracted now, receiving \$1 in 1 years' time, will require payment of principal plus interest of \$1.05 in 2 years' time. In general $f(t,T_1,T_2)$ is the p.a. forward interest rate contracted at t (or predetermined at t) for a loan or deposit over period $[T_1,T_2]$ at the rate. If \$1 loaned at T_1 under this forward contract should return $\$1 + F$ at $\mathbb{T}_2$, then F is the effective forward return or forward interest factor:

$$F(t,T_1,T_2) \triangleq \exp(f(t,T_1,T_2) \times (T_2 - T_1)). \tag{12.2}$$

Spot-Forward-Bond Price Relationships

Spot rates, forward rates, and discount bond prices are related as follows. Implicitly we rule out default. Thus, these rates are *de facto* risk-free rates or treasury rates (assuming no sovereign default risk for a major developed country). We shall use the time notations: $t < T_1 < T_2$.

(1a) $y(t,t+\triangle) = f(t,t,t+\triangle)$. In other words, the forward rate at t and effecting at t is also the spot rate at t.

(1b) $\exp(y(t,T_2)(T_2-t)) = \exp(y(t,T_1)(T_1-t))\ F(t,T_1,T_2)$. This is a no-arbitrage condition. Putting \$1 in a deposit account at T_2-maturity spot rate $y(t,T_2)$ should yield the same return at time T_2 compared to putting the \$1 at T_1-maturity spot rate $y(t,T_1)$ and then roll over at T_1 into the forward deposit at interest factor $F(t,T_1,T_2)$.

(1c) $\frac{B(t,T_1)}{B(t,T_2)} = F(t,T_1,T_2)$. This relates the forward interest rate to the prices of two discount bonds.

Relationship (1c) can be obtained from (1b) by using the definition of spot yield in (12.1) which implies that $\exp(y(t,T)(T-t)) = 1/B(t,T)$. Therefore, the LHS of (1b) is $1/B(t,T_2)$ while the RHS of (1b) is $F(t,T_1,T_2)/B(t,T_1)$

Taking natural logarithms on both sides on (1c), and dividing by $T_2 - T_1$, we obtain

$$\frac{1}{T_2 - T_1} \ln \left[\frac{B(t, T_1)}{B(t, T_2)} \right] = \frac{1}{T_2 - T_1} \ln[F(t, T_1, T_2)].$$

Using the definition of $F(t, T_1, T_2)$ in (12.2), therefore

$$\frac{1}{T_2 - T_1} [\ln B(t, T_1) - \ln B(t, T_2)] = f(t, T_1, T_2). \tag{12.3}$$

Just as there is an instantaneous spot rate $y(t)$, there is an instantaneous forward rate $f(t, T)$ which is the forward rate over an infinitesimal interval $[T, T + dt]$. The last Eq. (12.3) yields

$$f(t, T, T + \triangle) = -\frac{1}{\triangle}[(\ln B(t, T + \triangle) - \ln B(t, T))].$$

By taking limits on both sides as we let $\triangle \downarrow 0$, we obtain the instantaneous forward rate:

$$f(t, T) = \lim_{\triangle \downarrow 0} - \left[\frac{\ln B(t, T + \triangle) - \ln B(t, T)}{\triangle} \right] = -\frac{\partial \ln B(t, T)}{\partial T}. \tag{12.4}$$

The forward rate curve in Fig. 12.1 shows the instantaneous forward rates $f(t, T)$ for various maturities $T > t' > t$.

Effective forward returns can be combined to form more forward return factors:

$$F(t, t', t' + \triangle)\, F(t, t' + \triangle, t' + 2\triangle) = F(t, t', t' + 2\triangle).$$

Hence, it is sufficient to know $f(t, t', t' + ds)$ for every t', and we can then construct any arbitrage-free forward rate $f(t, T_1, T_2)$, via

$$F(t, T_1, T_2) = \prod_{j=0}^{N-1} F(t, T_1 + j\triangle, T_1 + (j + 1)\triangle),$$

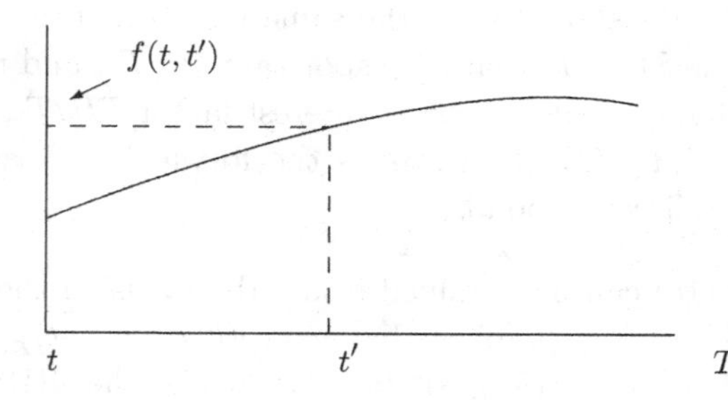

Fig. 12.1: Forward Rate Curve

where $N\triangle = T_2 - T_1$. If we take natural logarithm on both sides and shrink $\triangle \downarrow 0$, we obtain

$$\begin{aligned}
\ln F(t, T_1, T_2) &= \lim_{\triangle \downarrow 0} \ln \left[\prod_{j=0}^{N-1} F(t, T_1 + j\triangle, T_1 + (j+1)\triangle) \right] \\
&= \lim_{\triangle \downarrow 0} \sum_{j=0}^{N-1} \ln F(t, T_1 + j\triangle, T_1 + (j+1)\triangle) \\
&= \lim_{\triangle \downarrow 0} \sum_{j=0}^{N-1} f(t, T_1 + j\triangle, T_1 + (j+1)\triangle)\triangle \\
&\qquad \text{(via Eq. (12.2))} \\
&\rightarrow \int_{T_1}^{T_2} f(t, u)du.
\end{aligned}$$

Or,

$$F(t, T_1, T_2) = \exp\left(\int_{T_1}^{T_2} f(t, u)du \right). \tag{12.5}$$

From Eq. (12.5) and (1c), hence

$$\frac{B(t,t)}{B(t,T)} = \frac{1}{B(t,T)} = F(t, t, T) = \exp\left(\int_{t}^{T} f(t, u)du \right).$$

Thus, discount bond price can be related to the aggregation of instantaneous forward rates as follows

$$B(t, T) = e^{-\int_t^T f(t,u)du}. \tag{12.6}$$

Note that we have now a relationship involving discount bond prices and instantaneous forward rates in (12.6). We also have a relationship between spot rate (over a finite term, i.e., not instantaneous) and discount bond price in (12.1), and a relationship between spot rates over a finite term and the corresponding forward rate in (1b). The latter two relationships are basically definitional. It is interesting to note at this point there is no apparent relationship between the instantaneous spot rate (or short rate) and discount bond price. We shall explore this, among other concepts, later in this chapter. One intuition why this is so goes as follows. At time t, discount bond price $B(t, T)$ is known for maturity T. Forward rates $f(t, u)$ for u up to T are also contractable, i.e., the forward rate curve is known.

Thus, it is possible to relate two known quantitites in the market. Spot rates with finite terms up to T are also known. However, the short rates $y(s)$ for $s > t$ are random variables that will happen only in the future, and are not known now at time t. Hence, it is not trivial to try to connect these future short rates with current discount bond prices.

For most of what follows, we shall mostly use the entities of instantaneous spot rate $y(t)$, instantaneous forward rate $f(t,T)$, and discount bond price $B(t,T)$. These form the basic building blocks of most interest rate (or term structure) models. They are also interesting random variables. As we do not consider default in the modeling here, we shall interpret these rates as risk-free and use the common notation $r(t)$ to denote the short rate $y(t)$. The stochastic evolutions of $r(t)$ and $f(t,T)$ are generally referred to as interest rate processes. The process of $B(t,T)$ is called discount or ZC bond price process.

We shall first consider how no-arbitrage applies in a dynamic world of spot rates, forward rates, and discount bond prices. This is an essential condition that will be used to derive many of the recent interest rate models.

12.2. No-Arbitrage Dynamics of Yield Curve

For this section, assume that the world operates in discrete time periods, $t = 0, 1, 2, \ldots$, etc., and that spot rates $r(t)$ for each period are determined at the start of the period. Furthermore, like the CRR binomial option pricing model, we assume the spot rates follow a binomial process where in the next period, the 1-period spot rate realizes one of two values either in an up-state (spot rate increases) or a down-state (spot rate decreases). We may jump ahead and visualize how this 1-period spot rate can become a short rate when the period interval is shrunk towards zero.

For now, we shall start the modeling using the evolution of this (discrete) spot rate r_t^j where subscript t refers to the time when the spot rate is realized and which applies over the interval $[t, t+1)$, and superscript j refers to the state, u or d. Since the states follow the binomial tree, at $t = 1$, there are two states u and d, and at $t = 2$, there are four states, uu, ud, du, and dd.

Associated with the evolution of the spot rates is the evolution of the risk-free ZC (or discount) bond prices. Assume there are 3 different maturity bonds at the start, $t = 0$. See the term structure lattice tree in Fig. 12.2.

Term Stucture

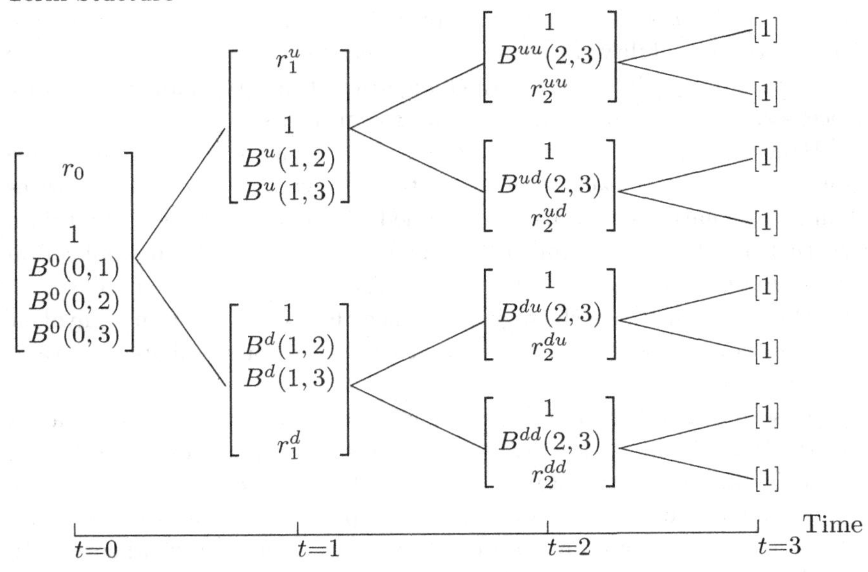

Fig. 12.2: Interest Rate Lattice Time

At $t = 0$, the ZC bond with 1 period to maturity is priced at $B^0(0,1) = \frac{1}{1+r_0}$. At $t = 1$, state u, the ZC bond with 1 period to maturity is priced at $B^u(1,2) = \frac{1}{1+r_1^u}$. At $t = 1$, state d, the ZC bond with 1 period to maturity is priced at $B^d(1,2) = \frac{1}{1+r_1^d}$.

At $t = 2$, state j, the ZC bond with 1 period to maturity is priced at $B^j(2,3) = \frac{1}{1+r_2^j}$.

Now, consider bonds with two periods till maturity. Here, we have to work backwards just as in binomial option pricing. At $t = 1$, there is a bond with 2 periods till maturity.

At $t = 1$, state u

$$B^u(1,3) = \frac{pB^{uu}(2,3) + (1-p)B^{ud}(2,3)}{1 + r_1^u} \tag{12.7}$$

At $t = 1$, state d

$$B^d(1,3) = \frac{pB^{du}(2,3) + (1-p)B^{dd}(2,3)}{1 + r_1^d} \tag{12.8}$$

Thus, bond prices are priced in a no-arbitrage martingale method, where p is the u-state risk-neutral probability, and $1 - p$ is the d-state risk-neutral probability. In the more complicated stochastic process of the spot rates, the probabilities may change over time. In simpler spot rate processes, the probabilities may remain as constants.

Hence, we see that given the spot rate process — how they take values over time in each possible state — a vector of ZC bond prices of different maturities is also determined via a no-arbitrage equilibrium (whether markets are complete or not). This follows from application of the First Fundamental Theorem of Asset Pricing seen in Chap. 7 whereby we assume no-arbitrage and hence the existence of an equivalent martingale or risk-neutral probability measure with state probabilities p and $1 - p$ here.

As seen in Eq. (12.1) and the binomial tree above, each time-t state-j vector of ZC bond prices gives rise to a set of spot rates of different maturities at time t and state j. Each of these is a contingent spot rate curve. Thus, given the 1-period spot rate process r_t^j, we see how a no-arbitrage condition can imply a rich stochastic evolution of yield curves. In actual physical situation or reality, the market can only observe a particular realized state j at each time t. Thus, at each time t, there is only a realized (*ex post*) term structure or spot rate curve, although *ex ante* there are contingent spot rate curves each likely to happen with some risk-neutral probabilities according to the evolution of the binomial tree.

Given a particular spot rate model or process, the no-arbitrage yield curve that is theoretically derived at $t = 0$ should ideally be consistent with the observed current market yield curve. To do this, we calibrate or estimate the parameters of the spot rate process so that the theoretical curve and the empirically observed curve coincide as closely as possible.

It is instructive to review the pricing of the 2-period ZC bond at time $t = 1$ for Eqs. (12.7) and (12.8), respectively in states u and d. They can be simplified as

$$B^u(1,3) = p\left[\frac{1}{(1+r_1^u)(1+r_2^{uu})}\right] + (1-p)\left[\frac{1}{(1+r_1^u)(1+r_2^{ud})}\right]$$

and

$$B^d(1,3) = p\left[\frac{1}{(1+r_1^d)(1+r_2^{du})}\right] + (1-p)\left[\frac{1}{(1+r_1^d)(1+r_2^{dd})}\right].$$

Then, at current time $t = 0$, a 3-period ZC bond has price

$$
\begin{aligned}
B^0(0,3) &= \frac{pB^u(1,3) + (1-p)B^d(1,3)}{1+r_0} \\
&= p^2\left[\frac{1}{(1+r_0)(1+r_1^u)(1+r_2^{uu})}\right] \\
&\quad + p(1-p)\left[\frac{1}{(1+r_0)(1+r_1^u)(1+r_2^{ud})}\right] \\
&\quad + (1-p)p\left[\frac{1}{(1+r_0)(1+r_1^d)(1+r_2^{du})}\right] \\
&\quad + (1-p)^2\left[\frac{1}{(1+r_0)(1+r_1^d)(1+r_2^{dd})}\right].
\end{aligned}
$$

In general, let $\tilde{Y}_{0,T} = (1+r_0)(1+\tilde{r}_{\triangle})(1+\tilde{r}_{2\triangle}) \times \cdots \times (1+\tilde{r}_{N\triangle})$, where $N\triangle = T$. Then,

$$\ln \tilde{Y}_{0,T} = \sum_{i=0}^{N} \ln(1+\tilde{r}_{i\triangle}).$$

Let $\tilde{r}_{i\triangle}^{C} = \ln(1+\tilde{r}_{i\triangle})$. The LHS is the continuously compounded spot rate, and is the short rate r_u when $\triangle \downarrow 0$. Then,

$$\ln \tilde{Y}_{0,T} = \sum_{i=0}^{N} \tilde{r}_{i\triangle}^{C} \to \int_0^T \tilde{r}_u du,$$

as $\triangle \downarrow 0$ and $N = \frac{T}{\triangle} \uparrow \infty$, and where we drop the superscript C for notational convenience.

Hence, in continuous-time,

$$\tilde{Y}_{0,T} = \exp\left(\int_0^T r_u du\right).$$

Then, it is seen that

$$B^0(0,T) = E_Q\left[\frac{1}{\tilde{Y}_{0,T}}\right] = E_Q(e^{-\int_0^T r_u du}).$$

More generally,

$$B(t,T) = E_t^Q(e^{-\int_t^T r_u du}). \tag{12.9}$$

We can see that under no-arbitrage condition, ZC bond prices are indeed related to the expectation (under the risk-neutral Q-measure) of future par bond values discounted by future random spot rates. If we use the physical measure, then we generally require to know the risk premium associated with the covariation of r_s with market return.

This completes the missing link we query in the last section between current known ZC bond price $B(t, T)$, current known forward rates in Eq. (12.6), $f(t, u)$, and expectation of a random discount function involving future random short rates as in Eq. (12.9).

As another feature of the risk-neutral or Q-measure short rate tree in Fig. 12.2, at any particular time t, it can be shown that the next period rate of return to a portfolio of a long position of one T-maturity ZC bond at \$ $B(0, T)$ and return to a short position of one $T+1$-maturity ZC bond at \$ $B(0, T+1)$ (and unwinding them the next period) is the same for any T, including buying one ZC bond with 1-period to maturity at \$ $B(0, 1)$. In other words, under risk-neutral measure, the expected return to holding any T-maturity ZC bond for that period t, is constant at $1 + r_0$. This is sometimes called the local expectations hypothesis (and should not be confused with the unbiased expectations hypothesis of the forward rate on future spot rate).

Continuous-Time Stochastic Models

We shall now examine a few continuous-time interest rate models. An interest rate model basically consists of interest rate processes (short rates need not be the starting point; it could be instantaneous forward rates as starting points, or even some discrete period interest rates like LIBOR) that are well specified, and hopefully well-calibrated (i.e., agree with empirical data) such that the model can provide positive theory about interest rate derivative prices including first of all plain vanilla bonds and, also desirably, simple derivatives such as interest rate swaps and interest rate options.

In the early 1980s, banks and financial institutions were hunting for good models to price bonds with embedded options such as callable bonds, puttable bonds, and convertible bonds. Good models should be a result of equilibrium condition (no-arbitrage condition is an equilibrium market condition that is preference-free) and should also at the start agree or fit with the initially observed market yield curve data, as indicated earlier.[1]

[1]Ho, TSY and SB Lee (1986). Term structure movements and pricing interest rate contingent claims. *Journal of Finance*, 41, 1011–1029, was an important paper showing

The earliest models start with the modeling of short rates, and solving for bond prices under no-arbitrage equilibrium. Calibrating to existing yield curve as another condition led to a second wave of models. We consider two equilibrium models in the later part of this section, the Vasicek model[2] and the Cox–Ingersoll–Ross interest rate model.[3]

Equations (12.6) and (12.9) are characterizations of the relationships between discount bond price and the instantaneous forward and instantaneous spot rates or short rates. However, by themselves, they do not provide an analytical solution of the bond price because we have not specified the stochastic process underlying either the forward rates or the short rates. In particular, from Eq. (12.9), it is seen that an expectation needs to be taken at time t (under the risk-neutral or EMM measure), but this requires that the short rate probability distribution be known. We shall now more formally show how analytical bond prices could be derived from Eqs. (12.6) and (12.9) once the underlying interest rate processes are specified. This is analogous to the pricing of derivatives $f(S_t)$ that we saw in Chapters 9 and 10, where the stochastic process of the underlying S_t (e.g., its SDE) needs to be specified. Given the SDE specification, its associated probability distribution can be solved. Then the risk-neutral expectations can be computed to obtain the price f via the martingale method. The Feynman–Kac formula or lattice tree convergence method can be used. Another approach is to derive a fundamental no-arbitrage PDE and to solve it directly given the initial boundary conditions, using either the Fourier transform method, Kolmogorov equations, or the numerical finite difference method. These methods are applied after suitable Girsanov transforms. Hence, specification of the continuous-time stochastic process (or referred to as "model" since the rest of the methodology to derive bond and related derivative prices are typically standard technologies once the process is defined) of interest rate is a critical first step.

We first consider short rate process as the primitive driver of prices. Later, we see how forward rate process and discrete spot rate process such as LIBOR can also be used as driver, sometimes for specifically different interest derivatives.

how to calibrate a short rate process to existing yield curve in a way very similar to our exposition in the last section.

[2]Vasicek, OA (1977). An equilibrium characterization of the term structure. *Journal of Financial Economics*, 5(2), 177–188.

[3]Cox, J, J Ingersoll and S Ross (1985). A theory of the term structure of interest rates. *Econometrica*, 53, 385–407.

We also consider single or one-factor interest rate model here whereby the interest rate process is driven by a single stochastic process dW^P under physical measure. Suppose the short rate follows an Itô process:

$$dr = \mu(r,t)dt + \sigma(r,t)dW^P. \tag{12.10}$$

Since the discount bond price is a function of the short rate process $\{r_t\}$, we can write the function as $B(t,T,r)$.

By Itô's lemma, therefore

$$\begin{aligned} dB &= B_t dt + \frac{1}{2}B_{rr}\sigma(r,t)^2 dt + B_r dr \\ &= \left(\frac{1}{2}\sigma(r,t)^2 B_{rr} + \mu(r,t)B_r + B_t\right) dt + \sigma(r,t)B_r dW \\ &= B\ (\alpha(r,t)dt + \beta(r,t)dW) \end{aligned} \tag{12.11}$$

where

$$\alpha(r,t) = \left(\frac{1}{2}\sigma(r,t)^2 B_{rr} + \mu(r,t)B_r + B_t\right) \Big/ B \tag{12.12}$$

$$\beta(r,t) = \sigma(r,t)B_r/B. \tag{12.13}$$

Consider two discount bonds of different maturities, $T_1 < T_2$, both driven by the same innovation dW, so their instantaneous returns, from Eq. (12.11), are perfectly correlated as follows.

$$\frac{dB_1}{B_1} = \alpha_1(r,t)dt + \beta_1(r,t)dW \tag{12.14}$$

$$\frac{dB_2}{B_2} = \alpha_2(r,t)dt + \beta_2(r,t)dW. \tag{12.15}$$

Recall Lemma 9.1 in Chap. 9 under the identical situation of a single factor, where it was shown that the Sharpe ratios of all securities that are thus perfectly correlated are the same. Here, r_t is random but adapted at time t

$$\frac{\alpha_1 - r_t}{\beta_1} = \frac{\alpha_2 - r_t}{\beta_2} = \lambda(r,t), \tag{12.16}$$

where "price of risk" function $\lambda(r_t)$ is the same for all bonds with different maturities, and hence must be independent of any discount bond's maturity T_j. We anticipated this and so the terms $\alpha_j(r,t)$ and $\beta_j(r,t)$ did not contain the maturity terms T_j.

Hence, for any bond j

$$\alpha_j - r_t = \beta_j \lambda(r_t) = \lambda(r_t)\sigma(r,t)\frac{B_r}{B_j}.$$

In the above, the term $(B_r)/(B_j)$ acts like a bond modified duration measure, which is the percentage change in bond price due to a percentage change in the short rate. The duration measure in fixed income is like the beta in CAPM, a measure of systematic risk. It is multiplied by a constant $\sigma(r,t)$. Hence, $\lambda(r,t)$ which is also constant in the economy here, acts like the market price of risk or a risk premium (being function of at most r_t and t, and could also be a constant itself) per unit volatility $\times$ modified duration to all the bonds.

Substituting the expressions for α in Eq. (12.12) and β in Eq. (12.13) into $\alpha - r = \beta\,\lambda$, we obtain

$$\left(\frac{1}{2}\sigma(r,t)^2 B_{rr} + \mu(r,t)B_r + B_t\right) \Big/ B - r_t = \lambda(r_t)\sigma(r,t)B_r/B$$

or

$$\frac{1}{2}\sigma(r,t)^2 B_{rr} + [\mu(r,t) - \lambda(r_t)\sigma(r,t)]B_r - r_t B + B_t = 0, \tag{12.17}$$

with terminal condition $B(T,T) = 1$. Note that the solution to Eq. (12.17) involves B, B_r, B_{rr} and B_t that are functions of r, t, and T.

Equation (12.17) can also be obtained using the direct method of forming a portfolio of two bonds, maturities T_1 and T_2, with respective numbers of bonds n_1, n_2, and then seeking to eliminate n_1, n_2 by ensuring that the portfolio's risk (or the coefficient of the term dW, since there is only a single factor) is zero, and at the same time, the risk-free portfolio expected return equals portfolio value multiplied by $r\,dt$, the instantaneous risk-free return. The solution to this, after using Eqs. (12.12) and (12.13), is exactly the PDE in Eq. (12.17). Yet another approach is to model the dynamics of the stochastic discount factor to arrive at the PDE in Eq. (12.17).[4]

We can apply the Feynman–Kac result, Theorem 9.4 of Chap. 9, to the PDE Eq. (12.17) which is a PDE of ZC bond prices $B(t,T,r)$ under no-arbitrage equilibrium. PDE Eq. (12.17) under Q-measure implies a solution to be

$$B(t,T,r) = E_t^Q[e^{-\int_t^T r_u du}], \tag{12.18}$$

[4]See, for example, Cochrane, JH (2001). *Asset Pricing*, Chapter 19. Princeton: Princeton University Press.

where terminal condition $B(T,T,r) = 1$ regardless of the short rate process. At the same time, the fundamental PDE was derived partly based on the underlying SDE or stochastic process of r_t. Thus, the Q-measure PDE in Eq. (12.17) must now be related to a transformed SDE under Q-measure so that the coefficient of dt in Eq. (12.10) should equal the coefficient of B_r in Eq. (12.17). We apply the Girsanov Theorem 9.2 to adjust the SDE Eq. (12.10). Let the Radon-Nikodým derivative be $\frac{dQ}{dP}$ (for Q equivalent to P in measure), and $W_t^Q = W_t^P + \int_0^t \lambda(r_u)du$. Then, Eq. (12.10) becomes

$$dr = \mu(r,t)dt + \sigma(r,t)(dW_t^Q - \lambda(r_t)dt),$$

or

$$dr = [\mu(r,t) - \lambda(r_t)\sigma(r,t)]dt + \sigma(r,t)dW_t^Q. \tag{12.19}$$

We see that the drift of the SDE under measure P is now changed from $\mu(r,t)$ to a new Q-drift drift $\mu(r,t) - \lambda(r_t)\sigma(r,t)$ under measure Q, which is the coefficient of B_r in Eq. (12.17). (Sometimes "risk-neutral measure" only refers to one involving an SDE with a drift of $r\ dt$.) In addition the probability transformation uses stochastic exponential, i.e., $\frac{dQ}{dP} = \exp\left(-\frac{1}{2}\int_0^t \lambda(r_u)du - \int_0^t \lambda(r_u)dW_u\right)$, which is adapted to $\mathcal{F}_t$.

Now, we are in business. Solving Eq. (12.18) under the distribution of r_u given by Eq. (12.19) [not Eq. (12.10)] gives the required no-arbitrage equilibrium bond price $B(t,T,r)$ at time t. Of course the solution of Eq. (12.19), an SDE — which should be fully specified, i.e., a model such as Vasicek, which we shall see soon, is itself a task.

Several remarks are in order with references to results in Chaps. 7 and 9.

(2a) In practice, sometimes the physical measure is not mentioned, and Chap. 7's fundamental asset pricing result is invoked, directly allowing the modeller to work with the EMM martingale probability measure Q and its associated SDE with noise or factor W^Q to find the no-arbitrage price of a derivative. However, it is usually good practice to start with a physical measure whenever possible and then work toward an EMM measure for no-arbitrage pricing. Starting with a specification of SDE of the underlying asset price under physical or empirical measure allows verification by actual empirical observations, e.g., GBM.

(2b) Depending on whether the SDE is a one-factor (one noise, i.e., only dW_t) or two-factor model (two noises i.e., dW_t and dZ_t) or even more factors, we find 2 or 3 or more asset prices following the SDE respectively, though each with different drift and diffusion coefficients (and different jump coefficients if it is a mixed jump-diffusion, or the more general Lévy process that is RCLL with stationary independent increments, including both GBM, Poisson process, and mixed JD processes). Combine them in a specific portfolio to eliminate the noise and then equate the instantaneous return of this risk-free portfolio to the risk-free return. In the diffusion cases, since only drift and volatility are involved as coefficients, it naturally leads to a condition on the Sharpe ratio (a quotient of the excess instantaneous return over the instantaneous volatility) being a constant function across all assets for given states at a particular time, and which includes the case of being a constant. This Sharpe ratio can be interpreted as the price of the state risk or the price premium per unit of volatility. In the Black–Scholes GBM case, the Sharpe ratio is a constant.

(2c) The formed SDE in Eq. (12.19) already is under the EMM Q-measure as it admits no arbitrage. Thus, under the fundamental asset pricing theorem, the bond prices arising out of the short rate process in Eq. (12.19) are martingales. That the SDE is already under Q-measure and not P-measure can be seen by applying the Girsanov theorem where $dW^Q = dW^P + \lambda dt$, and where $\frac{dQ}{dP} = \exp\left(-\frac{1}{2}\int_0^t \lambda(r_t)^2 ds - \int_0^t \lambda(r_t) dW_s\right)$. A key point to note is that there are many Q-measures (nonunique when the market is incomplete, and each Q-measure is associated with a distinct price of risk function $\lambda(r_u)$.

For example, if $\lambda = \frac{\mu - r_u}{\sigma_u}$, the market-wide Sharpe ratio, then we obtain a risk-neutral Q-measure in that the SDE is $dr_u = r_u dt + \sigma_u dW_t^{Q^1}$. If we apply say $\lambda = \sigma_T$, then SDE is $dr_u = (\mu - \sigma_T \sigma_u) dt + \sigma_u dW_t^{Q^2}$, where the superscripts to Q denotes different equivalent martingale measures (EMMs).

(2d) An important point about using whichever Q-measure is so that we can find a suitable numéraire or normalization such as to make the derivative price we want to find a martingale for easy computation. The particular Q-measure is then used to find the probability distribution in order to analytically find the expectation of the martingale. Of course, the choice of a particular Q-measure or its associated price of risk function (λ in the bond case here) would in general lead to different prices. Therefore, it is good if the price function $\lambda(\cdot)$ can be sufficiently parameterized so that implied parameters can allow the resulting model to fit with actual observations.

As the classic example of Black–Scholes model that we have elaborated all along, Girsanov's theorem is applied to transform the original SDE under physical P-measure to one under the Q-measure via the Radon–Nikodým derivative $\frac{dQ}{dP}$, whereby the transformed SDE has the risk-free rate r as the drift, so that taking expection on $dS/S = r\,dt + \sigma dW$ gives $E(dS/S) = rdt$ or risk-free rate (hence risk-neutral probability under measure Q). It should be observed that the Q-measure SDE now has a drift term that should appear as the coefficient of X_S or X_r in the fundamental PDE if we were pricing derivative $X(S,t)$ or $X(r,t)$. This accords with the remarks under the Feynman–Kac Theorem 9.4 that both the SDE and PDE are to be under the same Q-measure eventually so that the expectation is also taken w.r.t. the Q-measure in order to obtain the derivative price as an expected value under the Q-measure. Use of the money account numéraire completes the picture in the solution so that now $E_t(f(S_T/M_T)) = f_t/M_t$, a martingale, where $f_t(\cdot) \in C^2$ is a function of S_t and $M_t = e^{rt}$.

(2e) We can sometimes try to solve directly the Q-measure PDE with the appropriate initial and boundary conditions. Or we can try to find a numéraire M_t such that $(X_t)/(M_t)$ is a driftless diffusion process, and thus a martingale under the Q-measure. (This reaffirms by the fundamental theorem(s) of asset pricing in Chap. 9 that we obtain a no-arbitrage equilibrium price of derivative X under expectation on measure Q.) The exact analytical solution of course (is easier as it is a martingale) depends on the Q-measure SDE of the underlying asset price process. This can be solved as a probability distribution using numerical method (lattice tree or finite difference methods) or by Itô integral if it is of a simple form, or by Monte Carlo simulation.

(2f) In Sec. 9.3, we transformed a normalized X/M under SDE(W^P) measure, specifically GBM with drift $(\mu - r)dt$ to a driftless GBM under risk-neutral Q^1 measure using Girsanov transform $W^{Q^1}(t) = W^P(t) + \frac{\mu - r}{\sigma}t$. This may be said to be done in a single step.

Then, in Sec. 9.4, the same outcome is obtained by transforming underlying primitive process X under SDE(W^P), GBM with drift μdt, to an SDE(W^{Q^2}) with drift rdt using Girsanov transform $W^{Q^2}(t) = W^P(t) + \frac{\mu - r}{\sigma}t$. Note the Q^2-measure here is different from the Q^1-measure earlier. However, for X under SDE(W^{Q^2}), $E_s^{Q^2}(X_s) = \frac{M_s}{M_t}X_t$, for $t < s$, and assuming constant r for simplicity here. Then, $E_s^{Q^2}(\frac{X_s}{M_s}) = \frac{X_t}{M_t}$ when transformation is done in a second step with appropriate choice of numéraire.

Thus, some textbooks remark that, "an EMM is always associated with a numéraire," should be interpreted appropriately that an EMM is associated with a transform of the underlying and its derivative prices to martingales (so martingale pricing can be obtained under the EMM probability distribution) and that any numéraire that helps to make the martingale possible whether in a single step or in two steps should itself also be consistently measurable under the same EMM measure.

(2g) In (2b), sometimes the derivative to be solved is based on underlying assets with more than one state variable or noise that is not a tradeable asset, such as short rate or volatility (I had made an earlier comment that VIX may be a close surrogate but not a perfect traded asset for volatility.), then the fundamental equation such as Eq. (12.17) will involve the price of risk term λ that is a function of the non-traded state variable. We also saw this in the Heston volatility option model in Chap. 10. Additional or auxiliary specification of such a price risk function, such as the affine class, will be required to come up with an analytical solution where possible, if not a numerical solution.

Finally, we have two remarks on the usefulness of knowing both the P-measure and the Q-measure, and their relationship.

(2h) The relationship between the physical probability measure or density and the EMM probability measure or density function can be recovered via the Radon–Nikodým derivative, i.e., $\frac{dQ}{dP} = \exp\big(- \frac{1}{2}\int_0^t \gamma_s^2 ds - \int_0^t \gamma_s dW_s \big)$, so that we can use empirical time series data of the underlying to estimate the physical measure parameters and also backed out implied parameters of the EMM measures from derivative prices, and then attempt to test if the transformation from P-measure to Q-measure is empirically supported. If it is not supported, it could imply an incorrect physical SDE specification to start with, or else nonexistence of no-arbitrage equilibrium derivative prices due to imperfect markets or illiquid traded prices that are observed with errors or with wrong time stamps.

(2i) Working solely with and directly with EMM probability measures and the associated SDE(W^Q) and backing out implied parameters from some observed derivative prices — and using the same implied parameters perhaps for pricing related derivatives based on the same underlying EMM SDE(W^Q) — usually does not give much insight into the actual physical process driving the underlying security prices, and this can be a disadvantage when the market is incomplete and choices of which EMM measures need to be made (although this is usually done by some criterion such as minimum least squares of actual empirically observed prices from model prices).

Vasicek Model

In the single-factor Vasicek model, the short rate process is described by

$$dr = a(b - r)dt + \sigma dW, \tag{12.20}$$

where a, b, and σ are positive constants (the above is also generally called an Ornstein–Uhlenbeck process). b is interpreted as the long-run mean towards which short interest rate will revert, and a is the speed of such reversion. A higher (lower) speed of a implies that the short rate will move closer to the long-run mean at a faster (slower) rate. The solution to the mean-reversion process of Eq. (12.20) may be construed to be in the form of a stochastic process $y_t = r_t\, e^{at}$. Since y_t, which follows the Ornstein–Uhlenbeck stochastic process, is a function of r_t, we can employ Itô's lemma, where

$$\begin{aligned} dy_s &= ar_s e^{as} ds + e^{as} dr_s \\ &= e^{as}\{ar_s + a(b - r_s)\}ds + e^{as}\sigma dW_s \\ &= [ab\ e^{as}]ds + e^{as}\sigma dW_s. \end{aligned} \tag{12.21}$$

From Eq. (12.21), we integrate the LHS and RHS over $[t, T]$ to obtain

$$y_T - y_t = b[e^{as}]_t^T + \int_t^T e^{as}\sigma dW_s.$$

Thus

$$r_T e^{aT} = r_t e^{at} + b(e^{aT} - e^{at}) + \sigma \int_t^T e^{as} dW_s.$$

Or

$$r_T = b + (r_t - b)e^{-a(T-t)} + \sigma e^{-aT} \int_t^T e^{as} dW_s. \tag{12.22}$$

Taking unconditional expectation, the mean of the spot rate at T is

$$E(r_T) = b + (E(r_t) - b)e^{-a(T-t)}.$$

From Eq. (12.22), for $s > t$, deviation from mean is

$$r_s - E(r_s) = (r_t - E(r_t))e^{-a(s-t)} + \sigma e^{-as}\int_t^s e^{au}dW_u. \tag{12.23}$$

Hence

$$E[r_s - E(r_s)]^2 = e^{-2a(s-t)}E\left[r_t - E(r_t)\right]^2 + \sigma^2 e^{-2as}\, E\left[\int_t^s e^{au}dW_u\right]^2.$$

The last term can be evaluated as

$$\sigma^2 e^{-2as}\int_t^s e^{2au}du = \sigma^2\frac{e^{-2as}}{2a}\left(e^{2as} - e^{2at}\right) = \frac{\sigma^2}{2a}(1 - e^{-2a(s-t)}).$$

Therefore,

$$E[r_T - E(r_T)]^2 = e^{-2a(T-t)}E\left[r_t - E(r_t)\right]^2 + \frac{\sigma^2}{2a}(1 - e^{-2a(T-t)}).$$

Or

$$\mathrm{var}(r_T) = e^{-2a(T-t)}\mathrm{var}(r_t) + \frac{\sigma^2}{2a}(1 - e^{-2a(T-t)}).$$

The solution to the Vasicek model in Eq. (12.22) is a Gaussian process if r_t is Gaussian. Extrapolating, it may be said that r_T for different T follows a Gaussian process if initial r_0 is normally distributed. This is because the deterministic sum of normally distributed dW_s over $[0, T]$ and r_0 is again normally distributed for any $T > 0$. If $r_0 \sim N(0, \sigma_0^2)$, then r_T is distributed as $N\big(b(1-e^{-aT}),\ \frac{\sigma^2}{2a} + e^{-2aT}(\sigma_0^2 - \frac{\sigma^2}{2a})\big)$. When $T \uparrow \infty$, then the distribution converges to $N(b, \frac{\sigma^2}{2a})$.

However, since in finance we are usually working with given information at time of decision, t, or given its $\mathcal{F}_t$, then the conditional distribution of r_T, given r_t (treated as a constant now), is normal with conditional mean and conditional variance as follows.

$$E(r_T) = b + (r_t - b)e^{-a(T-t)}. \tag{12.24}$$

$$\mathrm{var}(r_T) = \frac{\sigma^2}{2a}(1 - e^{-2a(T-t)}). \tag{12.25}$$

From Eq. (12.22), we can derive

$$\begin{aligned} E\left[-\int_t^T r_u du\right] &= -\int_t^T E(r_u)du \\ &= -\int_t^T b + (r_t - b)e^{-a(T-u)}du \end{aligned}$$

$$= -\left[b(T-t) + (r_t - b)\int_t^T e^{-a(T-u)}\,du\right]$$

$$= -b(T-t) - \frac{(r_t - b)}{a}[1 - e^{a(T-t)}]. \qquad (12.26)$$

From Eq. (12.22), we can derive:

$$r_s - E_t(r_s) = \sigma e^{-as}\int_t^s e^{au}\,dW_u \quad s > t.$$

Or

$$\int_t^T (r_s - E_t(r_s))\,ds = \sigma\int_t^T e^{-as}\left[\int_t^s e^{au}\,dW_u\right]ds$$

$$= \sigma\int_t^T\left[\int_u^T e^{-a(s-u)}\,ds\right]dW_u.$$

Hence

$$\text{var}\left(\int_t^T r_u\,du\right) = E\left[\int_t^T (r_s - E_t(r_s))\,ds\right]^2$$

$$= \sigma^2\int_t^T E\left[\int_u^T e^{-a(s-u)}\,ds\right]^2 du$$

(since the last term $dW_u^2 = du$, we have implicitly used the Itô isometry relationship $E\left[\int H_s\,dW_s\right]^2 = E\left[\int H_s^2\,ds\right]$ for integrable H_s)

$$= \sigma^2\int_t^T\left[\int_u^T\int_u^T e^{-a(v-u)}e^{-a(w-u)}\,dvdw\right]du$$

$$= \sigma^2\int_t^T e^{2au}\left[\int_u^T\int_u^T e^{-a(v+w)}\,dvdw\right]du$$

$$= \frac{\sigma^2}{a^2}\int_t^T e^{2au}\left[e^{-2aT} - 2e^{-a(u+T)} + e^{-2au}\right]du$$

$$= \frac{\sigma^2}{2a^3}(2a(T-t) - 3 + 4e^{-a(T-t)} - e^{-2a(T-t)}). \qquad (12.27)$$

Using Eqs. (12.18), (12.26), and (12.27), and the fact that the Vasicek model short rate distribution is normal, and assuming the specification in Eq. (12.20) is under Q-measure (this can be done by comparing the Q-measure general specification in Eq. (12.19) and, for example, restricting $a = \sigma$, $b = \mu/\sigma$, and $\lambda = r$; or it could be $\lambda = ar/\sigma$, and $\mu = ab$), price of the ZC bond is

$$\begin{aligned} B(t,T,r) &= E_t^Q(e^{-\int_t^T r_u du}) \\ &= \exp\left[E\left(-\int_t^T r_u du\right) + \frac{1}{2}\text{var}\left(\int_t^T r_u du\right)\right] \\ &= \exp\left[-b(T-t) - \frac{(r_t - b)}{a}(1 - e^{a(T-t)})\right. \\ &\quad \left. + \frac{\sigma^2}{4a^3}(2a(T-t) - 3 + 4e^{-a(T-t)} - e^{-2a(T-t)})\right]. \end{aligned} \quad (12.28)$$

Therefore, it is seen that the no-arbitrage bond price can be expressed in the form $\exp[A(t,T)\, r_t + Z(t,T)]$, where in this case

$$A(t,T) = -\frac{1 - e^{-a(T-t)}}{a},$$

and

$$Z(t,T) = -\left(b - \frac{\sigma^2}{2a^2}\right)[A(t,T) + (T-t)] - \frac{\sigma^2 A(t,T)^2}{4a}.$$

The Markov nature of the short rate process in this case allows the pricing feature shown in Eq. (12.28) where the bond price is a function only of the lastest state r_t and not of its history.

We have thus solved for a ZC bond price under the Vasicek or Ornstein–Uhlenbeck short rate process. The issue with this model is that the short (real) rate can take negative values which is not possible in the real world.

Cox, Ingersoll, and Ross Model

We next consider another one-factor short rate model, that of the Cox–Ingersoll–Ross mean-reverting square-root diffusion process, again assuming it is under an equivalent martingale measure:

$$dr = \kappa(\mu - r)dt + \sigma\sqrt{r}dW, \quad (12.29)$$

where $\kappa, \mu > 0$.

Similarly, the solution to the mean-reversion square-root process of Eq. (12.29) may be construed in the form of a stochastic process $y_t = r_t e^{\kappa t}$. Employing Itô's lemma,

$$\begin{aligned} dy_s &= \kappa r_s e^{\kappa s} ds + e^{\kappa s} dr_s \\ &= e^{\kappa s}\{\kappa r_s + \kappa(\mu - r_s)\} ds + e^{\kappa s}\sigma\sqrt{r_s} dW_s \\ &= \kappa\mu e^{\kappa s} ds + \sigma e^{\kappa s/2}\sqrt{y_s} dW_s. \end{aligned} \tag{12.30}$$

Integrating the LHS and RHS over $[t, T]$, we obtain

$$y_T - y_t = \mu[e^{\kappa s}]_t^T + \sigma \int_t^T e^{\kappa s/2}\sqrt{y_s}\, dW_s.$$

Thus

$$r_T e^{\kappa T} = r_t e^{\kappa t} + \mu(e^{\kappa T} - e^{\kappa t}) + \sigma \int_t^T e^{\kappa s}\sqrt{r_s}\, dW_s.$$

Or

$$r_T = \mu + (r_t - \mu)e^{-\kappa(T-t)} + \sigma e^{-\kappa T} \int_t^T e^{\kappa s}\sqrt{r_s}\, dW_s. \tag{12.31}$$

However, Eq. (12.31) is not a Gaussian process since the integrand contains the product of normally distributed dW_s and $\sqrt{r_s}$. For $\kappa, \mu > 0$, however, this process has the interesting property that, unlike the Ornstein–Uhlenbeck process, r_T will not become negative. As soon as r_s touches zero, volatility goes to zero, and there is a deterministic positive drift $\kappa(\mu - r_s) > 0$. Thus, the CIR process is more appropriate in describing non-negative nominal interest rates. Then

$$E(r_T) = \mu + (r_t - \mu)e^{-\kappa(T-t)}.$$

This expectation is of the same form as in the Vasicek model.

The unconditional variance can also be obtained as

$$\text{var}(r_T) = \frac{\sigma^2}{\kappa}(1 - e^{-\kappa T})\left[r_t e^{-\kappa T} + \frac{\mu}{2}(1 - e^{-\kappa T})\right].$$

In particular, the transition probability density function, for $s > t$,

$$p(r_s, s; r_t, t) = C_0 e^{-(u+v)}\left(\frac{v}{u}\right)^{q/2} I_q(2\sqrt{uv})$$

where $C_0 = 2\kappa/[\sigma^2(1 - e^{-\kappa(s-t)})]$, $u = C_0 r_t e^{-\kappa(s-t)}$, $v = C_0 r_s$, and $I_q(\cdot)$ is modified Bessel function of the first kind of order q, $q = \frac{2\mu\kappa}{\sigma^2} - 1$. The modified Bessel function of the first kind is

$$I_q(z) = \sum_{j=0}^{\infty} \frac{(z/2)^{2j+q}}{j!\,\Gamma(j+q+1)},$$

and $\Gamma(w)$ is a gamma function.

The solution to the ZC bond under the CIR process (12.29) is: $B(t,T,r) = \exp\left[A(t,T)r_t + Z(t,T)\right]$, where

$$A(t,T) = \frac{2(1-e^{\gamma(T-t)})}{(\gamma+\kappa)(e^{\gamma(T-t)}-1)+2\gamma},$$

$$Z(t,T) = \frac{2\kappa\mu}{\sigma^2}\ln\left[\frac{2\gamma e^{(\kappa+\gamma)(T-t)/2}}{(\gamma+\kappa)(e^{\gamma(T-t)}-1)+2\gamma}\right],$$

and

$$\gamma = \sqrt{\kappa^2 + 2\sigma^2}.$$

The solution satisfies the terminal condition of $B(T,T,r) = 1$. Note that if we start with P-measure in Eq. (12.29), then a non-zero price risk will appear.

Exponential Affine Class of Short-Rate Models

Duffie and Kan[5] introduced a class of short-rate models called affine term structure models. The main advantage of this class of models is tractability, and it is characterized by 3 key properties as follows. We shall illustrate with a one-state model. Let X_t be a state variable.

(3a) The short rate is an affine function (in the most general multivariate form, it refers to a displacement vector plus a matrix transformation — it suffices here to consider it as a time t constant plus a scalar multiple) of the state variable X_t, so $r_t = w_0(t) + w_1(t)X_t$. (Note restrictions of the coefficients to constants may sometimes not allow existence of solution.)

$$dX_t = \mu(X_t)dt + \sigma(X_t)dW_t.$$

Then, in a similar way we obtain a fundamental bond pricing equation in terms of the state variable X_t, as in Eq. (12.17):

$$\frac{1}{2}\sigma(X_t)^2 B_{XX} + [\mu(X_t) - \lambda(X_t)\sigma(X_t)]B_X - r_t B + B_t = 0, \qquad (12.32)$$

with terminal condition $B(T,T,X) = 1$.

[5]Duffie, D and R Kan (1996). A yield factor model of interest rates. *Mathematical Finance*, 6, 379–406.

(3b) $\lambda(X_t) = \Lambda\sigma(X_t)$ where Λ is a constant. μ is assumed constant or just a function of τ.

(3c) $\sigma(X_t)^2 = a(\tau) + b(\tau)X_t$, i.e., diffusion is affine in X_t.

Moreover, for affine structures, Duffie and Kan has shown that the general solution of a ZC bond price is $B(t,T,X) = \exp(A(t,T)X_t + Z(t,T))$. So, we obtain

$$B_X = AB, \quad B_{XX} = A^2B, \quad \text{and} \quad B_t = -B_\tau = -\left[\frac{dA}{d\tau}X_t + \frac{dZ}{d\tau}\right]B,$$

where $\tau = T - t$.

Then, the interest rate fundamental Eq. (12.32) can be expressed in full as

$$\begin{aligned}
&\frac{1}{2}(a + bX_t)A^2B \\
&\quad + \left[\mu(X_t) - \Lambda(a + bX_t)\right]AB - (w_0 + w_1X_t)B \\
&\quad - \left[\frac{dA}{d\tau}X_t + \frac{dZ}{d\tau}\right]B = 0. \qquad (12.33)
\end{aligned}$$

Collecting terms in X_tB, and in B, we obtain 2 ODEs to solve

(4a) $\frac{1}{2}b(\tau)A^2 - \Lambda b(\tau)A - w_1 = \frac{dA}{d\tau}$

(4b) $\frac{1}{2}a(\tau)A^2 + [\mu - \Lambda a(\tau)A] - w_0 = \frac{dZ}{d\tau}$.

Solving the 2 ODEs simultaneously, we obtain the ZC bond price in terms of r_t by replacing $X_t = \frac{r_t - w_0}{w_1}$. The equations in (4a) and (4b) are called Ricatti type equations and are complicated ODE equations that carry quadratic terms with non-zero coefficients. The ZC bond price must satisfy the terminal or boundary condition of $B(T,T,r) = 1$.

If (3a) holds, $r_t = w_0(t) + w_1(t)X_t$, and where the SDE of X_t is of the CIR form, i.e.,$dX_t = \kappa(\mu(t) - X_t)dt + \sigma(t)\sqrt{X_t}dW_t$, then the model is sometimes called a CIR++ model which is an extended affine form of the original CIR model.

Other Popular Short-Rate Models

Many interest rate models have been written in the course of the last three decades, mainly due to a burgeoning industry in interest rate bonds and derivatives due to increased volatility of interest rates as well as global lending and borrowing. Three of the more popular short-rate models, as well as multi-factor versions of some the other models, are as follows.[6]

The models are basically distinguished by their Q-measure short-rate process specification, as the solutions for the ZC bond and other interest derivatives based on the Q-measure expectation are standard procedures.

Hull and White[7] (extended Vasicek model) uses

$$dr_t = (\theta(t) - \alpha(t)r_t)dt + \sigma(t)dW_t.$$

Black–Derman–Toy[8] uses

$$d\ln r_t = \theta(t)dt + \sigma dW_t,$$

where the short-rate solution is $r_t = r_u \exp(\int_u^t \theta(s)\,ds + \sigma(W_t - W_u))$.

Black–Karasinski[9] uses

$$d\ln r_t = [\alpha(t) - \beta(t)\ln r_t]\,dt + \sigma dW_t,$$

where the short-rate solution is

$$r_t = \exp\left(\ln r_u e^{-\beta(t-u)} + \int_u^t e^{-\beta(t-s)}\alpha(s)\,ds \right.$$
$$\left. + \int_u^t \sigma e^{-\beta(t-s)}\,dW_s\right).$$

for some function $K(\cdot)$.

The Heath–Jarrow–Morton Model

Other important models in modern interest rate theory include the Heath–Jarrow–Morton (HJM) model[10] of forward interest rate dynamics.

So far, interest rate modeling has centered on the use of short rates. To the extent this solves for a ZC bond, then any forward rate could be

---

[6]See Brigo, D and F Mercurio (2006). *Interest Rate Models: Theory and Practice.* Berlin: Springer Verlag, for more details.

[7]Hull, J and A White (1990). Pricing interest-rate derivative securities. *The Review of Financial Studies*, 3(4), 573–592.

[8]Black, F, E Derman and W Toy (1990). A one-factor model of interest rates and its applications to treasury bond options. *Financial Analysts Journal*, 33–39.

[9]Black, F and P Karasinski (1991). Bond and option pricing when short-rates are lognormal. *Financial Analysts Journal*, 46(1), 52–59.

[10]Heath, D, RA Jarrow and A Morton (1992). Bond pricing and the term structure of interest rates: A new methodology for contingent claims valuation. *Econometrica*, 60(1), 77–105.

obtained using Eq. (12.2) with two ZC bonds of different maturities. In addition, we also saw how an instantaneous forward rate at t is actually $f(t,t) = r(t)$, and any instantaneous forward rate in the future at time T is $f(t,T)$, which is $\lim_{\triangle \downarrow 0} f(t,T,T+\triangle)$. Clearly there must also be a converse way of modeling the continuous-time process of instantaneous forward rate to arrive at ZC bond prices and spot rates. We look at this problem here.

Consider the risk-neutral Q-measure process for bond price $B(t,T)$, such that LEH holds, so $E_t^Q(dB/B) = rdt$.

$$dB(t,T) = r_t B(t,T)dt + \sigma(t,T,\mathcal{F}_t)B(t,T)dW_t^Q, \tag{12.34}$$

where $\mathcal{F}_t$ denotes information at t used to determine the volatility of the bond price $B(t,T)$ at time t.

One important restriction on the volatility function of the bond price is that

$$\sigma(t,t,\mathcal{F}_t) = 0, \quad \forall t \tag{12.35}$$

since over a zero time interval, there should be no volatility in a diffusion process. This includes the case $\sigma(T,T,\mathcal{F}_t) = 0$ at the ZC bond's maturity since the price at maturity is fixed at 1 (no credit risk involved here), and thus there is no more randomness at T.

From Eq. (12.3), we have

$$f(t,T_1,T_2) = \frac{1}{T_2 - T_1}[\ln B(t,T_1) - \ln B(t,T_2)]. \tag{12.36}$$

Applying Itô's lemma to Eq. (12.34)

$$d\ln B(t,T_1) = \left[r_t - \frac{1}{2}\sigma(t,T_1,\mathcal{F}_t)^2\right]dt + \sigma(t,T_1,\mathcal{F}_t)dW_t^Q,$$

and

$$d\ln B(t,T_2) = \left[r_t - \frac{1}{2}\sigma(t,T_2,\mathcal{F}_t)^2\right]dt + \sigma(t,T_2,\mathcal{F}_t)dW_t^Q,$$

from which Eq. (12.36) leads to

$$df(t,T_1,T_2) = \frac{\sigma(t,T_2,\mathcal{F}_t)^2 - \sigma(t,T_1,\mathcal{F}_t)^2}{2(T_2 - T_1)}dt + \frac{\sigma(t,T_1,\mathcal{F}_t) - \sigma(t,T_2,\mathcal{F}_t)}{T_2 - T_1}dW_t^Q. \tag{12.37}$$

Put $T_1 = T$ and $T_2 = T + \triangle$, for $\triangle > 0$, then limit of the coefficient of first term on the RHS of Eq. (12.37) is

$$\lim_{\triangle \downarrow 0} \frac{\sigma(t, T+\triangle, \mathcal{F}_t)^2 - \sigma(t, T, \mathcal{F}_t)^2}{2(T+\triangle - T)}$$

$$= \frac{1}{2} \frac{\partial[\sigma(t, T, \mathcal{F}_t)^2]}{\partial T}$$

$$= \sigma(t, T, \mathcal{F}_t) \, \frac{\partial[\sigma(t, T, \mathcal{F}_t)]}{\partial T}.$$

The limit of the coefficient of the second term on the RHS is

$$\lim_{\triangle \downarrow 0} \frac{\sigma(t, T, \mathcal{F}_t) - \sigma(t, T+\triangle, \mathcal{F}_t)}{T+\triangle - T}$$

$$= -\frac{\partial[\sigma(t, T, \mathcal{F}_t)]}{\partial T}.$$

Therefore, we obtain in the limit as $\triangle \downarrow 0$

$$df(t,T) = \left[\sigma(t, T, \mathcal{F}_t) \, \frac{\partial[\sigma(t, T, \mathcal{F}_t)]}{\partial T}\right] dt - \left[\frac{\partial[\sigma(t, T, \mathcal{F}_t)]}{\partial T}\right] dW_t^Q. \qquad (12.38)$$

Equation (12.38) under the Q-measure or no-arbitrage condition shows that the drift and the diffusion coefficient (volatility) of the instantaneous forward rate process are closely related, as first pointed out by Heath *et al.* (1992). Specifically, the drift of a risk-neutral instantaneous forward rate process is completely specified by the volatility function.

First, note that $\int_t^T \frac{\partial \sigma(t,s,\mathcal{F}_t)}{\partial T} ds = \frac{\partial}{\partial T} \int_t^T \sigma(t, s, \mathcal{F}_t) ds = \sigma(t, T, \mathcal{F}_t)$. Then, if $v(t,s) \equiv -\frac{\partial \sigma(t,s,\mathcal{F}_t)}{\partial T}$, for $t < s$, we have $\int_t^T v(t,s) ds = -\sigma(t, T, \mathcal{F}_t)$ as shown above. From Eq. (12.38), we can rewrite it as

$$df(t,T) = \left[v(t,T) \int_t^T v(t,s) ds\right] dt + [v(t,T)] dW_t^Q, \qquad (12.39)$$

where we have left out the argument term $\mathcal{F}_t$ with the understanding that all the functions $v(\cdot)$ and $\sigma(\cdot)$ are well-defined w.r.t. the proper filtrations.

The HJM specification of, or model of instantaneous forward rates with diffusion coefficient $v(t,T)$ and hence also the restricted Q-measure drift function, in principle can be consistent with some short rate process when one can take the expectation on the RHS in Eq. (12.9) to become an exponential of terms say in a function $g(t, r_t, \mu(r,t), \sigma(r,t), \lambda(r_t))$ and equate this with the exponent of terms involving integral of $f(t,u)$ in Eq. (12.6).

To recollect, ZC bond price $B(t,T) = e^{-\int_t^T f(t,u)du}$, so $\frac{\partial B(t,T)}{\partial T} = -f(t,T)B(t,T)$. The latter is negative because we are increasing the maturity T of the ZC bond, so its price must decrease. However, if we take the derivative w.r.t. t, then $\frac{\partial B(t,T)}{\partial t} = f(t,t)B(t,T) = r(t)B(t,T)$. Here, when maturity T is fixed, carrying the same ZC bond will yield return rate r_t over the next instance. Rearranging, $\frac{dB}{B} = r_t dt$ where we keep T and r_t constant at the instance of time t.

At time t, a consistent short-rate process $\{r_s\}$ would produce ZC bonds of different maturities $B(t,T)$ so that at the same time forward rates $f(t,u,u+\triangle) = B(t,u)/B(t,u+\triangle)$ are obtainable for $u \in [t,T]$. This then allows the same bond price to be obtainable from $B(t,T) = e^{\int_t^T f(t,u)}du$. Can we obtain the bond price $B(t,T)$ directly from the instantaneous forward rate process, e.g., Eq. (12.39)? Yes, but cumbersome.

In the last paragraph, we see how each term $f(t,u,u+\triangle)$ or in the limit when $\triangle \downarrow 0$, then the term $f(t,u)$, for $u \in [t,T]$, contributes to the bond price formation in $B(t,T) = e^{\int_t^T f(t,u)}du$. Equation (12.39) is an SDE on changes in $f(t,u)$ (no loss of generality to think of u instead of the special case of T) over infinitesimal $\triangle$ to $f(t+\triangle,u)$ while keeping u constant. Hence sometimes, the notation $df_t(t,u)$ is used on the LHS as in Eq. (12.39). Each $\triangle$-time forward, we get a whole new set of terms $f(t+\triangle,u)$ for a fixed u, $u \in [t,T]$. $f(t+\triangle,u)$ are not adapted to $\mathcal{F}_t$, and hence are random variables at time t. In theory, thinking of the lattice numerical method, we can construct lattice tree for the evolution of $f(t+\triangle,u)$ over time t in order to derive the start point value of $f(t,u)$ for each u, and with the terminal condition for each u, of $f(u,u) = r_u$, $u > t$. However, immediately we see a difficulty. Since we require terms $f(t,u)$ for all $u \in [t,T]$ in order to obtain bond price $B(t,T)$, it means we have to construct (even in discretization) a huge number of lattice trees each ending in time $u = t+\triangle, t+2\triangle, \ldots, T$.

From Eq. (12.39), $E_t^Q[df(t,T)] = [v(t,T) \int_t^T v(t,s)ds]dt$. Clearly, each infinitesimal expected forward rate forward over dt is non-Markovian as it depends not just on a function only at t, but also on the path history having the term being integral over $[t,T]$ of $v(t,s)$. Thus, unlike most short rate models where we can construct numerical methods using recombining lattice trees for fast computation of the price, using HJM instantaneous forward rate approach will require non-recombing lattices which is computationally taxing. The solution would typically require Monte Carlo simulations in order to arrive at the bond price.

T-Forward Measure

We have seen in Chap. 7 how the money market account serves as a numéraire so that the normalized stock price and hence also derivative price on the underlying stock price can become martingales under the Q-measure.

For interest rate derivatives, however, the money account is usually not useful as a numéraire. Consider a derivative price $V(S_t, t)$, and money account process $M(T) = e^{rT}$. Under the GBM risk-neutral measure of V_t where

$$V_T = V_t e^{(r-\frac{1}{2}\sigma^2)(T-t)+\frac{1}{2}\sigma^2(W_T-W_t)},$$

so $E_t^Q(V_T) = V_t e^{r(T-t)}$, then $E_t^Q(\frac{V_T}{M(T)}) = \frac{V_t}{M(t)}$, a martingale process ensuring no-arbitrage equilibrium option pricing via the fundamental theorem of asset pricing.

Now, suppose the derivative is an interest rate derivative $V(r_t, t)$, then

$$E_t^Q\left(\frac{V_T(r_t, t)}{M(T)}\right)$$

would now pose a problem as both the numerator V_T and the denominator $M(T)$ are now functions of the stochastic short rates (or other interest rates) $r_s \in [t, T]$. In this case, $M(T) = e^{\int_0^T r_s ds}$. Taking the Q-risk neutral expecation would involve covariances between the numerator and denominator and it is troublesome to rearrange it into a martingale process. This calls for a new EMM probability measure to do the trick of converting the key process of normalized V_T into a martingale.

First, consider the Radon–Nikodým derivative (RV) $\frac{dQ_T^*}{dQ} = \frac{D(T)}{B(0,T)}$. Note that $E_0^{Q^*}(\frac{dQ}{dQ_T^*}) = 1$. Also, $D(0, T) = e^{-\int_0^T r_u du} = 1/M(T)$.

Let V_T be a future random security price at time T. Suppose its spot price is V_t at current time t, and there exists a T-maturity forward contract in the market, price $F(V_t, T)$, that promises a selling price of $\frac{V_t}{B(t,T)}$ for delivering the security at time T. By the no-arbitrage cost-of-carry model in finance, one could buy the security now at spot price V_t, and hold it with interest cost of $1/B(t, T)$ till t, and then sell it at the forward price V_T that was contracted at time t. If the forward price V_T at t were $>$ $(<)\frac{V_t}{B(t,T)}$, arbitrage profit could be made by selling (buying) the forward contract and buying (short-selling) the spot security with carrying cost.

At time T, prices $F(V_T, T) \equiv S_T$ as is the case in forward (or futures) contract at maturity T.

Under the Q-risk neutral measure mentioned earlier, we have

$$E_t^Q(D(T)V_T) = D(t)V_t$$

$$\Rightarrow \frac{E_t^Q(D(T)V_T)}{D(t)B(t,T)} = \frac{D(t)V_t}{D(t)B(t,T)}.$$

Under the Q^*-risk neutral measure, we have

$$\frac{E_t^{Q*}\left(\frac{dQ}{dQ^*}D(T)V_T\right)}{D(t)B(t,T)} = \frac{D(t)V_t}{D(t)B(t,T)}.$$

Or

$$\frac{E_t^{Q*}\left(\frac{B(0,T)}{D(T)}D(T)V_T\right)}{D(t)B(t,T)} = \frac{D(t)V_t}{D(t)B(t,T)}$$

$$\Rightarrow D(t)^{-1}E_t^{Q*}\left(\frac{B(0,T)}{B(t,T)}V_T\right) = \frac{V_t}{B(t,T)},$$

$$\Rightarrow E_t^{Q*}(V_T) = \frac{V_t}{B(t,T)} = F(V_t,T).$$

Recollect that $F(V_t,T)$ is the forward price deliverable at T of the contract at t based on underlying V_t. Now, we can similarly show

$$E_s^{Q*}(\tilde{V}_T) = \frac{V_s}{B(s,T)} = \tilde{F}(V_s,T), \quad s > t$$

where at time t, $F(V_s,T)$, $s > t$ is a RV, not adapted to $\mathcal{F}_t$. If we take conditional expectation of the above based on information set at t, then

$$E_t^{Q*}[E_s^{Q*}(\tilde{V}_T)] = E_t^{Q*}\left[\frac{V_s}{B(s,T)}\right] = \frac{V_t}{B(t,T)}, \tag{12.40}$$

by the law of iterated expectations. Hence

$$E_t^{Q*}F(V_s,T) = F(V_t,T),$$

for $s > t$. Thus, the forward contract prices $F(V_s,T)$ with fixed maturity at T is a martingale under the Q^*-measure which is called the T-forward measure as it uses the discount or ZC bond $B(s,T)$ as the numéraire and these are discount bond prices fixed at s till maturity T when the price V_s is discounted. Thus, there is no correlation between the numerator and denominator in this case, and taking expectation is convenient. In fact, one can use any discount bond $B(t,T')$ as numéraire for $T' > T$. $F(V_s,T')$ is a martingale process for $t < s < T'$.

The above result is convenient as it holds for securities with prices $V_s(s, T, r)$ that are functions of future interest rates, but yet are independent of the numéraire $B(s, T)$ at time s. However, we still require to find the new probability measure Q^* in more detail to be able to perform the expectation or integration of V_t.

The LIBOR Market Model

Short-rate models have one serious drawback when it comes to pricing not ZC bond or ZC bond option, but pricing the large market of derivative products involving LIBOR p.a. rates (usually 3-month or 6-month London Interbank Borrowing Rates for favored interbank institutional customers). This is because while the (term) LIBOR rates are easily available minute-by-minute and observable, this is not the case with the short rates which are not only not observable, but are actually not physically traded. Also, there is no exact instantaneous interest rate product.

To provide an enhanced solution to this class of actively traded LIBOR derivatives, the Brace–Gatarek–Musiela (BGM) model[11] and others have introduced market LIBOR rate as the underlying stochastic process. The resulting interest derivative models based on the 3-month or 6-month (discrete holding period) LIBOR rates are called LIBOR market models.

Suppose the spot 3-month LIBOR rate today at $t = 0$ is $L_0(0, 3)$ (or $L_t(0, 3)$). A little explanation of the notation is useful here. $L_s(n, m)$ means an annualized (p.a.) LIBOR interest rate that is contracted at time s (could be in the future where $s > t$, in which case it is treated as a RV at time t now) to take effect at time $n > s$ and for which the actual payment is done at time $m > n$ where in this case, it is a $(m - n)$-month LIBOR rate. Note the difference between commodity forward contracts where typically the transaction and payment is done at maturity n whereas for interest rate forwards, the payment is done afterward at the end of the tenor, i.e., at time m. Typically, for the usual 3-month, 6-month, 9-month, 1-year LIBOR rates traded in the market, the respective notations are $L_s(3, 6)$, $s \in [0, 3]$, $L_s(6, 12)$, $s \in [0, 6]$, $L_s(9, 18)$, $s \in [0, 9]$, and so on. We shall study only the 3-month LIBOR case as the theory applies similarly to the others.

Investing \$1 in a spot LIBOR today, time t, will pay out \$ $1 + \delta_3 L_0(0, 3)$ at the end of 3 months where LIBOR $L_0(0, 3)$, e.g., 4% is quoted on p.a.

[11] Brace, A, D Gatarek and M Musiela (1997). The market model of interest rate dynamics. *Mathematical Finance*, 4, 127–155.

basis, δ_3 is the accrual or daycount factor which in this case is 1/4 for 3 months. Hence, the effective return is \$ $1 + 1/4 \times 4\% = 1.01$. $L_0(0,6)$ and $L_0(0,9)$ and so on are the spot 6-month and spot 9-month LIBOR rates at $t = 0$. As it is possible to fix and make forward risk-free rate contracts under ZC risk-free bonds of different maturities at time t, it is similarly possible to fix and make forward LIBOR forward contract rates when the spot LIBOR yield curve is given. In practice, LIBOR are interbank rates and carry some counter-party risks in case one of the OTC counter-party banks defaults. Thus, the LIBOR rates are typically some basis points above treasury rates or risk-free rates calculated in treasury discount bills or equivalent disocunt bonds (stripped bonds). However, we shall not deal with credit risk premium in this context here.

LIBOR rates are related as follows. We continue to use the definition $B(3,6)$ to denote a price of a ZC bond starting at time $t = 3$ and maturing at $t = 6$. This is of course a RV at time $t = 0$ since $1/B(3,6)$ is a 3-month term interest rate (or 3-month spot rate) occurring in the future. Spot ZC bonds would be $B(0,6)$ that matures in 6 months with an effective interest rate of $1/B(0,6) - 1$.

$$B(0,3) = (1 + \delta_3 L_0(3,6))\, B(0,6).$$

Another way to look at this is

$$1/B(0,6) = 1/B(0,3) \times (1 + \delta_3 L_0(3,6)),$$

where the LHS is investing \$1 in a ZC bond and holding till 6 months with return $1/B(0,6)$, and RHS is investing \$1 in a 3-month ZC bond and then simultaneously committing to forward roll this into the next 3 months' forward contract with a total eventual return at the end of 6 months of $1/B(0,3) \times (1+\delta_3 L_0(3,6))$. It is important to note that the LIBOR forward contract with rate $L_0(3,6)$ is contracted now at $t = 0$ so that it is not a RV.

The 3-month LIBOR forward rate at the end of 3 months, specifically at $t = 3$, but contracted now at $t = 0$ is (similar to the forward rate formula seen earlier as in (1c) except there it is the forward return or one plus the forward rate):

$$L_0(3,6) = \frac{1}{\delta_3}\left(\frac{B(0,3) - B(0,6)}{B(0,6)}\right).$$

The interest payment is

$$\delta_3 L_0(3,6) = \left(\frac{B(0,3) - B(0,6)}{B(0,6)}\right).$$

In general, for time now at t, and $t < s < u < T$

$$\delta_3 L_t(u,T) = \left(\frac{B(t,u) - B(t,T)}{B(t,T)}\right), \tag{12.41}$$

where a tradeable security price at $B(t,u) - B(t,T) > 0$ (which is $[F(t,u,T) - 1] \times B(t,T)$) or kind of "spread" between a $(u-t)$ period-to-maturity ZC bond and a cheaper $(T-t)$ period-to-maturity ZC bond, is expressed in terms of the denominator $B(t,T)$ as numéraire. For 3-month forward LIBOR, $T - u$ is fixed at 3 months.

Recollecting from the previous section, a special case of Eq. (12.40) is, by fixing $V_s = B(s,u)$ where $0 < t < s < u < T$, then we obtain

$$E_t^{Q*}\left[\frac{B(s,u)}{B(s,T)}\right] = \frac{B(t,u)}{B(t,T)},$$

for any $t < s < u < T$. But $\frac{B(s,u)}{B(s,T)} = F(s,u,T)$, a forward contract random return as at time $t < s$. Hence under Q^*, forward interest rates $F(s,u,T)$, for $s \in [t,u]$ are martingales. In the LIBOR application here, fix $\frac{B(s,u)}{B(s,T)} \equiv F(s,u,T) = 1+\delta_3 L_s(u,T)$ and $\frac{B(t,u)}{B(t,T)} = 1+\delta_3 L_t(u,T)$. So, under T-forward measure Q^*

$$E_t^{Q*}[1 + \delta_3 L_s(u,T)] = 1 + \delta_3 L_t(u,T),$$

and hence, we obtain

$$E_t^{Q*}\left(L_s(u,T)\right) = L_t(u,T), \quad \text{for } t < s. \tag{12.42}$$

Under Q^*-measure, we see that forward LIBOR rates, $L_s(u,T)$ are a martingale, for $s \in [t,T]$, where current time is t.

Thus, in the same way, an interest rate derivative on LIBOR such as a call option on LIBOR with price $C(L_t(u,T))$ expressed with $B(t,T)$ as numéraire, is a martingale, and can be found via the martingale method of option pricing at maturity u as

$$E_t\left[\frac{C(L_u(u,T))}{B(u,T)}\right] = \frac{C(L_t(u,T))}{B(t,T)},$$

or

$$B(t,u)E_t\left[\max\left(C(L_u(u,T)) - K, 0\right)\right] = C(L_t(u,T)),$$

where tenor or the LIBOR term is $T - u$, i.e., with maturity or reset at time u and actual payment at time $T > u > t$, and we cancel out the term constant accrual factor δ_3 on both sides. Note a caplet payment based on 3-month LIBOR is also made 3 months after the LIBOR is realized at reset.

$B(0,6)$ serves as the numéraire that makes the 3-month forward LIBOR payment $\delta_3 L_s(3,6)$, $s \in [0,3]$, a martingale process, i.e., $E_t^{Q*}\left[\delta_3 L_s(3,6)\right] = \delta_3 L_t(3,6)$, just as $E_t^{Q*}[F(s,3,6)] = F(t,3,6)$, for $0 \le t < s \le 6$. As accrual or daycount factor $\delta_3 = 1/4$ is a constant, the 3-month forward LIBOR rate itself, $L_s(3,6)$, is also a martingale under the Q^* or T-forward probability measure (where $T = 6$ months here).

As we move through time from $t = 0$ till $t = 3$, the 3-month forward LIBOR rate, $L_s(3,6)$ to take effect at $t = 3$ and with payment at $t = 6$, changes randomly as a martingale. The continuous-time stochastic representation of this is:

$$dL_t(3,6) = \zeta_t(3,6)L_t(3,6)\,dW_t^{Q*},$$

or, in general

$$dL_t(u,T) = \zeta_t(u,T)L_t(u,T)\,dW_t^{Q*}. \tag{12.43}$$

Solving Eq. (12.43), we obtain

$$L_s = L_t \exp\left(-\frac{1}{2}\int_t^s \zeta_a^2 da + \int_t^s \zeta_a dW_a^{Q*}\right),$$

where we drop the notations "(3,6)" with the understanding that we are dealing with a 3-month forward LIBOR at reset time $t = 3$.

In Eq. (12.43), we had directly employed an EMM (martingale) T-forward measure Q^* under the discount bond price $B(t,T)$ as numéraire. This allows for a simple lognormal diffusion form of forward LIBOR of $L_s(3,6)$, $0 < s < 3$, and enables a call option pricing formula similar to Black's futures option formula on commodity (including stock) futures.

Pricing Caps

The reset times are an important contract specification in an interest rate cap (derivative contract). The buyer of a 2-year cap with quarterly resets basically receives every 3-months an insurance payment — which could be zero — from the seller of the cap. However, the actual payment is done 3 months later for a 3-month LIBOR cap. Thus, payments are made first at the end of 6 months (based on first reset 3-month LIBOR rate at $t = 3$), then at the end of 9 months (based on the second 3-month LIBOR rate at $t = 6$), and so on until the last payment made at the end of 2 years and 3 months (based on the last reset 3-month LIBOR at $t = 24$). (There are some variations where the first payment is made in 3 months based on the current LIBOR, but we shall not worry about these, as the model can be adjusted without difficulty for such minor variations.)

At each reset, the actual payment received by the buyer (holder of cap) from the seller, in 3 months' time at the payment date, is determined by the cap strike rate $K\%$ p.a. that is fixed at contract time, and which obviously determines the price premium of the cap that the buyer has to pay for the interest rate hike insurance. In fact, each reset payment is *de facto* a call option called a caplet. Thus, for a 2-year cap on 3-month LIBOR resets, there are altogether 8 caplets.

At each reset, the payment is determined by the formula

$$G \times \delta_3 \times \max(L_i(i, i+1) - K, 0),$$

where the ith reset is at time $t = 3$ months, $i+1$ is $t = 6$ months, and so on, and G is a notional amount. Not surprisingly, using the same Black–Scholes technology given that the forward LIBOR process in Eq. (12.43) is basically lognormal diffusion, then the price of the ith caplet now at t which is before the reset time i, is

$$C_i = \delta_3 B(t, i+1)[L_t(i, i+1)N(d_1) - KN(d_2)], \tag{12.44}$$

where $d_1 = \frac{\ln(L_t(i,i+1)/K)+\frac{1}{2}\int_t^i \zeta_u^2 du}{\sqrt{\int_t^i \zeta_u^2 du}}$, and $d_2 = d_1 - \sqrt{\int_t^i \zeta_u^2 du}$.

This looks like the Black's formula on futures option since the payment is actually 3 months after reset. Thus, under the usual BGM type process for forward LIBOR, a caplet or a single call option on a forward LIBOR rate can be easily priced using Eq. (12.44). It would appear simple enough to price a cap or the series of caplets by simply adding up the costs of all the caplets in the contract using a similar formula Eq. (12.44). However, it is not so simple as each caplet is priced based on a different Q-measure w.r.t. a different numéraire, and thus has a different probability distribution even if the distributions are of the same form.

The different measures pose a problem since the next reset LIBOR rate is a stochastic evolution from the previous LIBOR rate, and if the two rates are under two different probability measures, then it is inappropriate to simply take expecations of them based on different measures and sum things up. The solution to a cap on a series of caplets is to find a common equivalent martingale probability measure for all caplets. Under this common EMM, each forward LIBOR $L_t(i, i+1)$, $L_t(i+1, i+2)$, $L_t(i+2, i+3)$, etc., associated with each caplet will have different drifts.

We shall provide a theorem that will help in the construction of a terminal forward-martingale measure for all caplets $i = 1, 2, \ldots, 8$ in a 2-year quarterly reset cap.

Theorem 12.1 *For all caplets $j = 1, 2, \ldots, N-1$, there is a (rolling) terminal forward equivalent martingale measure Q^N with underlying LIBOR SDE:*

$$\frac{dL_t(j, j+1)}{L_t(j, j+1)} = -\sum_{k=j+1}^{N-1} \frac{\zeta_k \delta_3 L_t(k, k+1)}{1 + \delta_3 L_t(k, k+1)} \zeta_j dt + \zeta_j dW^{Q^N}. \tag{12.45}$$

■

Proof. From Eqs. (12.14) to (12.16), a no-arbitrage SDE of a bond price is

$$\frac{dB_u}{B_u} = (r + \lambda \sigma_u) dt + \sigma_u dW^Q,$$

where λ is the price of bond systematic risk (same across all bonds under the one-factor model), and we use the short-form notation B_u for the ZC bond price $B(t, u)$, and $\sigma_u = \sigma(t, u)$, $t < u$. In a risk-neutral measure, the zero systematic risk implies price $\lambda = 0$. However, a different price of the risk will lead to different equivalent martingale measures.

With two bonds of maturities u and T, where $u < T$, we have

$$d \ln B_u = (r + \lambda \sigma_u - \frac{1}{2} \sigma_u^2) dt + \sigma_u dW^Q,$$

and

$$d \ln B_T = (r + \lambda \sigma_T - \frac{1}{2} \sigma_T^2) dt + \sigma_T dW^Q.$$

Taking the difference

$$d \ln \left(\frac{B_u}{B_T} \right) = \left[\lambda (\sigma_u - \sigma_T) - \frac{1}{2} (\sigma_u^2 - \sigma_T^2) \right] dt + (\sigma_u - \sigma_T) dW^Q.$$

By Itô's lemma,

$$d\left(\frac{B_u}{B_T}\right) = \frac{B_u}{B_T}\left[\lambda(\sigma_u - \sigma_T) - \frac{1}{2}(\sigma_u^2 - \sigma_T^2) + \frac{1}{2}(\sigma_u - \sigma_T)^2\right] dt + \frac{B_u}{B_T}(\sigma_u - \sigma_T)\, dW^Q.$$

From Eq. (12.41), $1 + \delta_3 L_t(u,T) = \frac{B_u}{B_T}$. Hence,

$$\begin{aligned} &d(1 + \delta_3 L_t(u,T)) \\ &\quad = (1 + \delta_3 L_t(u,T))\left[\lambda(\sigma_u - \sigma_T) - \frac{1}{2}(\sigma_u^2 - \sigma_T^2) + \frac{1}{2}(\sigma_u - \sigma_T)^2\right] dt \\ &\qquad + (1 + \delta_3 L_t(u,T))(\sigma_u - \sigma_T)\, dW^Q. \end{aligned}$$

For the T-forward measure, fix $\lambda = \sigma_T$, then the coefficient of the term in dt becomes zero, thus

$$\delta_3\, dL_t(i, i+1) = [1 + \delta_3 L_t(i, i+1)](\sigma_i - \sigma_{i+1}) dW^{Q^*},$$

where σ_i refers to the volatility function $\sigma(t,u)$ of dB_u/B_u.

Equating with (12.43), noting that $u \equiv i$(or $t = 3$), and $T \equiv i + 1$(or $t = 6$),

$$\sigma_i - \sigma_{i+1} = \frac{\delta_3 \zeta_i L_t(i, i+1)}{1 + \delta_3 L_t(i, i+1)}, \qquad (12.46)$$

where likewise ζ_i is function $\zeta_t(u,T)$ or simply the volatility of the 3-month LIBOR rate at time i(or $t = 3$) known at time t.

Thus in general, for any caplet i, when we fix $\lambda = \sigma_{i+1}$, we should be able to obtain Eq. (12.46). This gives rise to measure Q^*, which we now use a more convenient notation Q^{i+1} to denote its association with the risk price $\lambda = \sigma_{i+1}$.

Let $\gamma_i = \sigma_i - \sigma_{i+1} = \frac{\delta_3 \zeta_i L_t(i,i+1)}{1+\delta_3 L_t(i,i+1)}$. Apply Girsanov theorem so that for all $i = 1, 2, \ldots, N$, (Note $N = 8$ in the case of the 2-year cap.)

$$dW^{Q^i} = dW^{Q^{i+1}} - \gamma_{i+1}\, dt,$$

where $\frac{dQ^{i+1}}{dQ^i} = \exp\left(-\frac{1}{2}\int_i^{i+1} \gamma_i^2\, ds - \int_i^{i+1} \gamma_i\, dW_s\right)$.

From Eq. (12.43), a LIBOR rate from a previous quarter and measure Q^i can always be expressed in the measure associated with LIBOR of the next quarter Q^{i+1}, using the Girsanov theorem above. Thus

$$\begin{aligned}\frac{dL_t(i,i+1)}{L_t(i,i+1)} &= \zeta_i dW^{Q^i} \\ &= \zeta_i(dW^{Q^{i+1}} - \gamma_{i+1}dt) \\ &= -\frac{\zeta_{i+1}\,\delta_3 L_t(i+1,i+2)}{1+\delta_3 L_t(i+1,i+2)}\zeta_i\,dt + \zeta_i dW^{Q^{i+1}}.\end{aligned}$$

If we roll over the measure from Q^{i-1} to Q^i to Q^{i+1} and so on, we get

$$dW^{Q^{i-1}} = dW^{Q^i} - \gamma_i dt = [dW^{Q^{i+1}} - \gamma_{i+1}dt] - \gamma_i dt.$$

Then

$$\begin{aligned}\frac{dL_t(i-1,i)}{L_t(i-1,i)} &= \zeta_{i-1}dW^{Q^{i-1}} \\ &= \zeta_{i-1}(dW^{Q^{i+1}} - (\gamma_i+\gamma_{i+1})dt) \\ &= -\left(\frac{\zeta_i\delta_3 L_t(i,i+1)}{1+\delta_3 L_t(i,i+1)}\zeta_{i-1} + \frac{\zeta_{i+1}\delta_3 L_t(i+1,i+2)}{1+\delta_3 L_t(i+1,i+2)}\zeta_{i-1}\right)dt \\ &\quad + \zeta_{i-1}dW^{Q^{i+1}}.\end{aligned}$$

If we fix the terminal reset as $i+1 = N$, then for $j = 1,2,\ldots,i$ (or $i = N-1$), by extrapolation we have

$$\frac{dL_t(j,j+1)}{L_t(j,j+1)} = -\sum_{k=j+1}^{N-1}\frac{\zeta_k\delta_3 L_t(k,k+1)}{1+\delta_3 L_t(k,k+1)}\zeta_j\,dt + \zeta_j dW^{Q^N}, \tag{12.47}$$

where Q^N is the terminal forward EMM. ■

Apart from the SDE of terminal forward LIBOR rate $L_t(N,N+1)$ which is a martingale as seen in Eq. (12.43), the other non-terminal LIBOR rates are no longer martingales under the consistent Q^N measure common to all. They carry drift terms. Under the terminal forward EMM Q^N, and stochastic processes of various LIBOR rates $L_t(j,j+1)$, for $j = 1,2,\ldots,$ $N-1$ generated by the SDEs in Eq. (12.47), one may then consistently evaluate, typically by numerical methods such as Monte Carlo method, the prices of all the caplets and then add them up to find the price of the cap. As the formula in Eq. (12.47) involves the volatilities ζ_k of different forward quarterly LIBORs, these need to be determined before any cap can be priced. One pricing approach is to use actual cap prices of different maturities and thus different resets to imply out the forward volatility term structure, and then use this for prediction and pricing of other caps with different strike prices.

Interest Rate Swaps

The interest rate swap contract is one of the most commonly traded derivatives between two counter-parties in the OTC market. A buyer of a payer swap enters into a predetermined N-year Y-monthly (quarterly resets or 6-monthly resets are the most prevalent types) contract with a seller. The payer swap buyer pays a fixed contracted predetermined interest rate called the swap rate or swap premium or swap strike. If the rate is say 8% p.a., then a 6-monthly payment will be effectively 4% of the notional amount, say $\$G$. In return, the buyer of payer swap receives a floating rate payment that is determined at each reset date and collected at the next reset, or 6 months later in a 6-monthly reset. The contract is typically prespecified to base the floating rate on, say, a 6-monthly LIBOR rate determined at reset dates. There are two types of swaps. One is when the first floating rate reset is at the start of contract, and the other is when the first reset is at the second period instead. We shall consider the first type. The analysis for the second type is a simple extension of the first.

The payer swap buyer is also called a receiver swap seller. The receiver swap buyer receives fixed rate and pays floating rate, so the seller pays fixed rate and receives floating, which is exactly the definition of a payer swap buyer.

Figure 12.3 below illustrates the transactions involved in an interest rate swap. Let N = 2-year swap. Let Y equals 6 monthly reset of 6-monthly LIBOR rate. Let the swap strike rate be $K\%$. The accrual or daycount factor is 6 months or 0.5.

The payoff to a payer swap is the floating leg consisting of 4 separate receipts $G \times 0.5(L_j - K)$ for $j = 1, 2, 3, 4$ at the time of 2nd, 3rd, 4th resets and final payment at the end of 2 years. If we start with \$1 at the current time $t = 0$, then it could be put into a LIBOR account to earn

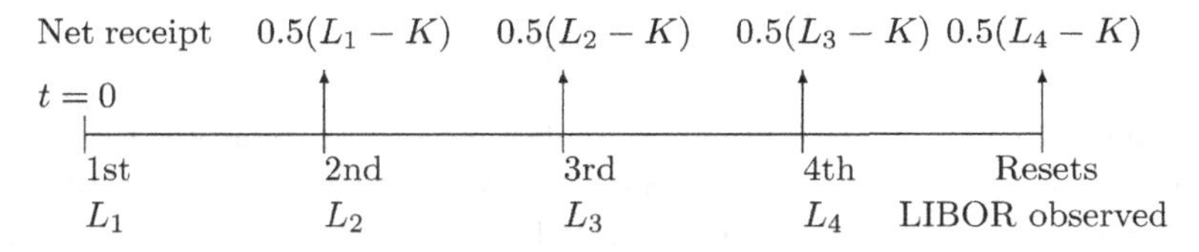

Fig. 12.3: Payoffs to Buyer of Payer Swap, Multiplied by Notional \$G

LIBOR return $0.5L_1$ at 2nd reset, rolled into the next LIBOR account at the 2nd reset rate L_2 to earn return $0.5L_2$ at 3rd reset, rolled into rate L_3 at 3rd reset to collect $0.5L_3$ at 4th reset, and then finally collect the final floating payment of $0.5L_4$ at final reset in 2 years. Since this is a risk-free strategy, the present value of collecting the 4 receipts is the current \$1 less the present value of \$1 at the end of 2 years. In other words, the value of this floating leg of 4 reset LIBOR-based payment is like a floating rate bond except there is no principal repayment at the end, hence the latter's present value has to be substracted from the valuation.

The value of this floating rate leg is $\$G \times (1 - P(0,2))$ where $P(0,2)$ is the discount bond value using the LIBOR spot rate curve where $P(0,2) = \frac{1}{(1+L_0(0,2))^2}$, and $L_0(0,2)$ is the current 2-year spot rate (p.a.) on the LIBOR rate term structure. This LIBOR term structure should be very close to, and possibly a few basis point above similar term treasury spot curve based on ZC bond $B(0,t) = \frac{1}{(1+R_0(0,t))^t}$ where $R_0(0,t)$ denotes the risk-free treasury spot rate over term t.

As the treasury spot rates are known at $t = 0$ for various terms, the present value of the fixed leg payments at 2nd reset, 3rd reset, 4th reset and finally at the end of 2 years can be evaluated as $\$G \times 0.5K\ (B(0,0.5) + B(0,1) + B(0,1.5) + B(0,2))$.

Thus, the present value (PV) of the 2-year payer swap is PV of floating leg less PV of fixed leg. It is

$$G \times \left[(1 - P(0,2)) - 0.5K \sum_{t=1}^{4} B(0,0.5t) \right].$$

Usually at the start of issue and trading of a new floating-to-fixed interest rate swap, the swap strike is set to a par swap rate such that the initial market value (PV) of the swap is zero. Thus, at $t = 0$, the par swap rate for the above 2-year swap and the given LIBOR and treasury spot rate curves, $P(0,t)$ and $B(0,t)$ respectively, should be

$$K^* = \frac{1 - P(0,2)}{0.5 \sum_{t=1}^{4} B(0,0.5t)}.$$

The par swap rate K^* for a zero-value N-years Y-monthly reset swap obviously changes continuously over time as the LIBOR and the treasury spot rate curves change over time and the timings to the next resets also change. If time $u < 0.5$ (in terms of year as unit) elapses, and the present time u par swap rate is now K^{**}, then the value of the buyer swap which originated at a swap strike K^* (or par swap rate at the originating time) is now

$$(K^{**} - K^*) \sum_{t=1}^{4} B(0, 0.5t - u).$$

Swaption

A payer (receiver) swaption is an option, typically American style, that the holder can exercise into a payer (receiver) swap. For example, the underlying is the same 2-year swap seen in the previous section. A payer swaption on that swap, is traded now with a maturity in 6 months and a strike price of K^S. Within the next 6 months, the swaption holder can choose to exercise into the money or gain if the par swap rate of a 2-year swap is sufficiently high at $K^{**} > K^S$. In this case, the gain is

$$(K^{**} - K^S) \sum_{t=1}^{4} B(0, 0.5t - u)$$

at the time of exercise $u < 0.5$ year.

If the par swap rate K_t^{**} process is assumed to follow a lognormal martingale under some Q^S measure, then the usual Black–Scholes technology can be applied to derive the price of the payer swaption at t as

$$\sum_{s=1}^{4} B(t, t + 0.5s) \left(K_t^{**} N(d_1) - K^S N(d_2)\right),$$

where $d_1 = \frac{\ln[K_t^{**}/K^S] + \frac{1}{2}\Sigma_{t,t+2}^2}{\Sigma_{t,t+2}}$, $d_2 = d_1 - \Sigma_{t,t+2}$, and $\Sigma_{t,t+2} = \int_t^{t+2} \sigma_u^2 \, ds$, where $dK_t^{**} = \sigma_t K_t^{**} dW^{Q^S}$.

In the case of many swaptions in the market each with a different maturity, a similar problem of finding a consistent EMM for all swaptions occurs, as in the case of the cap. Taking the swaption with the longest maturity, some terminal EMM measure can be constructed, although the analytics become quite complicated due to the fact that swap pricing involves a kind of averaging across many forward LIBORs. There is also another practical issue which presents some difficulties for quants, which is the calibration simultaneously not only of the volatility term structure of caps, but now also of swaptions under the Black models.

In the literature, unless one uses complicated multifactor models, satisfactory and useful calibrations of the two volatility term structures is not possible. Often times approximations are used, or different measures are applied to each type of derivatives.[12]

12.3. Problem Set 12

1. Show using L'Hôpital's (or Bernoulli) rule that $\lim_{n\to\infty}(1+\frac{r}{n})^{nt} = e^{rt}$.
2. Show that $F(t,s,u) \times F(t,u,w) = F(t,s,w)$.
3. In the 3-period spot rate model in Sec. 2, suppose the spot rate factors at different time-states are: $r_0 = 1.02$, $r_1^u = 1.017$, $r_1^d = 1.022$, $r_2^{uu} = 1.016$, $r_2^{ud} = 1.020$, $r_2^{du} = 1.019$, $r_2^{dd} = 1.025$. $p = 0.75$.

   (a) Find all the ZC bond prices at $t = 0$. Assume par value is $1.
   (b) Compute and show the term structure of spot rates at $t = 0$.
   (c) Find the forward rates $F(0,1,2)$, $F(0,2,3)$.

4. Using the same data in Question 3, suppose there is a discount bond with 3 periods to go till maturity. Suppose there is also a European call option on this discount bond, but which expires in two periods. The call strike price is 0.98. Thus, payoff to the call is $\max(B^j(2,3) - 0.98, 0)$ at $t = 2$. What is the call price now at $t = 0$? If a hedger buys such a call, is he or she trying to reduce the risk exposure to rising or falling (spot) interest rate?
5. Use the same data in Question 3. Suppose there is a risky coupon bond with 3 periods to maturity, paying a 5% coupon at the end of each period, i.e., at $t = 1$, $t = 2$, and $t = 3$. Par value of the bond is $100. Find the present $t = 0$ no-arbitrage equilibrium price of this coupon bond using the zero prices at $t = 0$. (Assume no default occurs.)
6. In the Vasicek short rate model of Eq. (12.20), show that for $v > s > t$, the covariance of r_v and r_s is

$$e^{-a(v+s-2t)}\mathrm{var}(r_t) + \frac{\sigma^2}{2a}\, e^{-a(v+s)}[e^{2a\min(v,s)} - e^{2at}].$$

   Moreover, if $t = 0$, then

$$\mathrm{cov}(r_v, r_s) = e^{-a(v+s)}\mathrm{var}(r_t) + \frac{\sigma^2}{2a}\, e^{-a(v+s)}[e^{2a\min(v,s)} - 1].$$

[12]For more discussion of these issues, see Hull, J (2006), or later editions, *Options, Futures, and Other Derivatives.* New Jersey: Pearson-Prentice Hall, and others such as Pelsser, A (2000). *Efficient Methods for Valuing Interest Rate Derivatives.* London: Springer Verlag.

7. Show that if underlying security GBM price process X_t when expressed in money market account dollars as numéraire can be suitably transformed from physical measure P into a martingale under the risk-neutral Q-measure, then likewise any derivative price C_t on X_t, $C(X_t, t) \in C^2$ (twice continuously differentiable) via Itô's lemma can also be transformed into a martingale under the risk-neutral measure. Thus, we can apply the fundamental asset pricing theorem to find no-arbitrage equilibrium prices by taking Q-measure expectations.
8. In a 2-year 6-monthly reset payer swap, suppose the first cash fixing is based not on the initial $t = 0$ reset LIBOR rate but on the LIBOR rate 6 months after the start of contract. This also means that the last payment is at the end of 2.5 years. If notional principal is \$G, LIBOR ZC bond prices and Treasury ZC bond prices are $P(0, s)$ and $B(0, s)$ respectively. For s taking values 0.5, 1, 1.5, 2, 2.5 years and so on, find the present value of this payer swap. You may assume the swap rate is K in the formula.

APPENDIX

Answers to Problem Set 1

[1] (1c): If $E_i \in \mathcal{F}$ and $E_j \in \mathcal{F}$, then $E_i \cup E_j \in \mathcal{F}$. This condition includes cases where E_i and E_j are disjoint. Let $E_k \in \mathcal{F}$ be another event such that it is disjoint from $E_i \cup E_j \in \mathcal{F}$. By (1c), then $E_k \cup (E_i \cup E_j) = (E_k \cup E_i \cup E_j) \in \mathcal{F}$. Continuing, we can see that for a sequence of disjoint events $E_n \in \mathcal{F}$, for $n = 1, 2, \ldots, N$, $\cup_{n=1}^{N} E_n \in \mathcal{F}$. Then, we can find a probability measure on $\mathcal{F}$ such that $P(\cup_{n=1}^{N} E_n) = \sum_{n=1}^{N} P(E_n)$. This is the finite version of (2c).

[2] $\sigma^2(X) = E(X-\mu)^2 = E(X^2 - 2\mu X + \mu^2) = E(X^2) - \mu^2$.

[3]

$$E\left(\sum_i^K Y_i\right)^2 = E\left(\sum_i^K Y_i^2 + \sum_i^K \sum_{j \neq i}^K Y_i Y_j\right)$$

$$= \sum_i^K E(Y_i^2) + 2\sum_i^K \sum_{j>i}^K E(Y_i Y_j)$$

$$= \sum_i^K E(Y_i^2) + 2 \sum_{1 \leq i < j \leq K} E(Y_i Y_j).$$

Put $Y_i = (X_i - \mu)$ to obtain the result.

[4] Discontinuity at x_1 implies $P(x_1) > 0$. Continuity at x_2 implies that $P(x_2)$=0. Hence, $P(x_1) > P(x_2)$.

[5] No. For example, $X = cZ$ where c is a constant.

[6] For some constants a and b, and RVs X', Y', $E(aX' - bY')^2 = a^2E(X'^2) + b^2E(Y'^2) - 2abE(X'Y') \geq 0$ always. Hence

$$2abE(X'Y') \leq a^2E(X'^2) + b^2E(Y'^2)$$

Put $a = \sqrt{E(Y'^2)} > 0$ and $b = \sqrt{E(X'^2)} > 0$, then

$$2\sqrt{E(Y'^2)E(X'^2)}E(X'Y') \leq E(Y'^2)E(X'^2) + E(X'^2)E(Y'^2)$$

or

$$\sqrt{E(Y'^2)E(X'^2)}E(X'Y') \leq E(X'^2)E(Y'^2)$$

or

$$E(X'Y') \leq \sqrt{E(Y'^2)E(X'^2)}.$$

Since this holds when we put Y' as $-Y'$ as well, then, we can use absolute values, i.e.,

$$|E(X'Y')| \leq |\sqrt{E(Y'^2)E(X'^2)}|$$

Put $X' = X - \mu_X$ and $Y' = Y - \mu_Y$, so we obtain

$$|\sigma_{XY}| \leq |\sigma_X \sigma_Y|$$

This is a version of the Schwarz Inequality. Then, also

$$\left|\frac{\sigma_{XY}}{\sigma_X \sigma_Y}\right| \leq 1$$

and hence $-1 \leq \rho \leq +1$.

[7] $aX + bY = \sum_{ij}(ax_i + by_j)[1_{A_i \cap B_j} + 1_{(A_i \cap B_j)^c}]$.

$$\begin{aligned} E(aX + bY) &= \sum_{ij}(ax_i + by_j)[E(1_{A_i \cap B_j}) + E(1_{A_i^c \cup B_j^c})] \\ &= \sum_{ij}(ax_i + by_j)[P(A_i \cap B_j) + P(A_i \cap B_j)^c] \\ &= a\sum_{ij} x_i P(A_i, B_j) + b\sum_{ij} y_j P(A_i, B_j) \\ &= a\sum_{i} x_i P(A_i) + b\sum_{j} y_j P(B_j), \end{aligned}$$

using law of total probability,

$$= aE(X) + bE(Y)$$

[8] Prob of an even no. is 1/2.
$P(N = 5) + P(N = 4) + P(N = 3) = {}^4C_2(1/2)^2(1/2)^2(1/2) + {}^3C_2(1/2)^2(1/2)(1/2) + (1/2)^3 = 1/2$ where ${}^4C_2$ indicates number of ways in which the first two even numbers are positioned among the first 4 throws, and ${}^3C_2$ indicates the number of ways in which the first two even numbers are positioned among the first three throws.

Expected number of throws $E[N] = 3 \times P(N = 3) + 4 \times P(\text{N=4}) + 5 \times P(N = 5) + 6 \times P(N = 6) + \cdots\cdots$

$$
\begin{aligned}
E[N] &= \sum_{k=3}^{\infty} k \times P(N = k) \\
&= \sum_{k=3}^{\infty} k\ {}^{k-1}C_2 \left(\frac{1}{2}\right)^k \\
&= \sum_{k=3}^{\infty} \frac{k(k-1)(k-2)}{2^{k+1}} \\
&= \left(\frac{1}{2^4}\right) \left(\frac{1 \cdot 2 \cdot 3}{2^0} + \frac{2 \cdot 3 \cdot 4}{2^1} + \frac{3 \cdot 4 \cdot 5}{2^2} + \frac{4 \cdot 5 \cdot 6}{2^3} + \cdots\cdots \right) \\
&= \frac{1}{16} \sum_{n=1}^{\infty} \frac{n(n+1)(n+2)}{2^{n-1}}.
\end{aligned}
$$

Let $S = \sum_{n=1}^{\infty} \frac{n(n+1)(n+2)}{2^{n-1}} = 6 + \sum_{n=1}^{\infty} \frac{(n+1)(n+2)(n+3)}{2^n}$.
Now $\frac{1}{2}S = \sum_{n=1}^{\infty} \frac{n(n+1)(n+2)}{2^n}$. Doing subtraction, so

$$
\begin{aligned}
S - \frac{1}{2}S = \frac{1}{2}S \\
&= 6 + \sum_{n=1}^{\infty} \frac{(n+1)(n+2)(n+3)}{2^n} - \sum_{n=1}^{\infty} \frac{n(n+1)(n+2)}{2^n} \\
&= 6 + \sum_{n=1}^{\infty} \frac{(n+1)(n+2)[(n+3) - n]}{2^n} \\
&= 6 + 3 \sum_{n=1}^{\infty} \frac{(n+1)(n+2)}{2^n}.
\end{aligned}
$$

Therefore, $S = 12 + 6 \sum_{n=1}^{\infty} \frac{(n+1)(n+2)}{2^n}$.
Let $T = \sum_{n=1}^{\infty} \frac{(n+1)(n+2)}{2^n} = 3 + \sum_{n=1}^{\infty} \frac{(n+2)(n+3)}{2^{n+1}}$.
Now $\frac{1}{2}T = \sum_{n=1}^{\infty} \frac{(n+1)(n+2)}{2^{n+1}}$. Doing subtraction, so

$$
\begin{aligned}
T - \frac{1}{2}T = \frac{1}{2}T &= 3 + \sum_{n=1}^{\infty} \frac{(n+2)(n+3)}{2^{n+1}} - \sum_{n=1}^{\infty} \frac{(n+1)(n+2)}{2^{n+1}} \\
&= 3 + \sum_{n=1}^{\infty} \frac{(n+2)[(n+3) - (n+1)]}{2^{n+1}} \\
&= 3 + 2 \sum_{n=1}^{\infty} \frac{(n+2)}{2^{n+1}}.
\end{aligned}
$$

Therefore, $T = 6 + 4\sum_{n=1}^{\infty}\frac{(n+2)}{2^{n+1}}$.
Let $U = \sum_{n=1}^{\infty}\frac{(n+2)}{2^{n+1}} = \frac{3}{4} + \sum_{n=1}^{\infty}\frac{(n+3)}{2^{n+2}}$.
Now $\frac{1}{2}U = \sum_{n=1}^{\infty}\frac{(n+2)}{2^{n+2}}$. Doing subtraction, so

$$\begin{aligned} U - \frac{1}{2}U = \frac{1}{2}U &= \frac{3}{4} + \sum_{n=1}^{\infty}\frac{[(n+3)-(n+2)]}{2^{n+2}} \\ &= \frac{3}{4} + \sum_{n=1}^{\infty}\frac{1}{2^{n+2}} \\ &= \frac{3}{4} + \frac{1/8}{1-1/2} = 1. \end{aligned}$$

Therefore, $U = 2$. $T = 6 + 4U = 14$. $S = 12 + 6T = 96$.
And $E[N] = \frac{1}{16}S = 6$.

[9] After 19 games, John expects to lose $1. Therefore, he expects to lose $19 after $19 \times 19 = 361$ games.

[10] In a Don't pass bet, probability of losing is when the shooter or roller wins, i.e., $\frac{8}{36} + 2 \times (\frac{1}{36} + \frac{4}{90} + \frac{25}{396})$. Probability of tie but not winning when a "12" is rolled is $1/36$. Hence, probability of winning is

$$1 - \frac{9}{36} - 2 \times \left(\frac{1}{36} + \frac{4}{90} + \frac{25}{396}\right) = 0.479293$$

[11] Kasparov walks away with $1 million if he wins three more games with probability $(1/2)^3 = 1/8$ or two more wins and a loss with probability ${}^3C_2(1/2)^3 = 3/8$. He draws with one more win and two losses with a probability of ${}^3C_1(1/2)^3 = 3/8$. He loses if he loses the next 3 games with probability $(1/2)^3 = 1/8$. His expected collection is $(1/8 + 3/8) \times \$1\text{million} + (3/8) \times \$0.5\,\text{million} = \$\frac{11}{16}$ million.

[12] Let X and Y be independent Poisson RVs. Let $Z = X + Y$. We want to show that Z is also a Poisson RV. The MGF of Z is $E[e^{\theta Z}] = E[e^{\theta(X+Y)}] = E[e^{\theta X}]\,E[e^{\theta Y}] = \exp\{\lambda_X(e^\theta - 1)\}\,\exp\{\lambda_Y(e^\theta - 1)\} = \exp\{[\lambda_X + \lambda_Y](e^\theta - 1)\}$. The resulting MGF is clearly that of another Poisson RV with intensity parameter $\lambda_X + \lambda_Y$.

[13] The easier method is actually to use MGF again:

$$E[e^{\theta Z}] = E[e^{\theta X}]\,E[e^{\theta Y}] = \exp\left(\frac{1}{2}\theta^2\right)\,\exp\left(\frac{1}{2}\theta^2\right) = \exp(\theta^2)$$

Hence, clearly Z is another normal RV. Now, $M'_Z(0) = [2\theta M_Z]|_{\theta=0} = 0$, hence mean of Z is zero. $M''_Z(0) = [2M_Z + 4\theta^2 M'_Z]|_{\theta=0} = 2$. Hence, variance of Z is $M''_Z(0) - (M'_Z(0))^2 = 2 - 0 = 2$.
The harder method suggested by the problem is called the convolution method, and is a more general method worth knowing:

$$\begin{aligned} F(z) &= \int_{-\infty}^{\infty}\int_{-\infty}^{z-x} \frac{1}{2\pi}\exp\left[-1/2(x^2+y^2)\right]dy\,dx \\ &= \int_{-\infty}^{\infty}\int_{-\infty}^{z} \frac{1}{2\pi}\exp\left[-1/2(x^2+(u-x)^2)\right]du\,dx \\ &\quad \text{(putting } y = u - x\text{)} \\ &= \int_{-\infty}^{z}\left(\int_{-\infty}^{\infty} \frac{1}{2\pi}\exp\left[-1/2(x^2+(u-x)^2)\right]dx\right)du. \end{aligned}$$

Hence, the PDF of RV Z is, via Leibniz's rule,

$$\begin{aligned} f(z) &= \frac{dF(z)}{dz} \\ &= \int_{-\infty}^{\infty} \frac{1}{2\pi}\exp\left[-1/2(x^2+(z-x)^2)\right]dx \\ &= \frac{1}{\sqrt{2\pi}}\int_{-\infty}^{\infty} \frac{1}{\sqrt{2\pi}}\exp\left[-1/2\left(\sqrt{2}x - \frac{z}{\sqrt{2}}\right)^2 - \frac{z^2}{4}\right]dx \\ &= \frac{1}{\sqrt{2\pi}}\,e^{-1/4z^2}\,\frac{1}{\sqrt{2}}\int_{-\infty}^{\infty}\frac{1}{\sqrt{2\pi}}\,e^{-1/2u^2}\,du \\ &\quad \left(\text{put } u = \sqrt{2}x - \frac{z}{\sqrt{2}}\right) \\ &= \frac{1}{\sqrt{2\pi}}\,e^{-1/4z^2}\,\frac{1}{\sqrt{2}} \\ &= \frac{1}{\sqrt{2}\,\sqrt{2\pi}}\exp\left(-\frac{1}{2}\left[\frac{z}{\sqrt{2}}\right]^2\right). \end{aligned}$$

Clearly $f(z)$ is PDF of a normal RV $N(0,2)$.

[14] $M(\theta) = \exp\left(\mu\theta + \frac{\sigma^2\theta^2}{2}\right)$. Skewness and kurtosis of a normal RV X are 0 and 3, respectively.

[15]

$$\begin{aligned}
E[t^2] &= \int_0^\infty t^2\,\lambda\exp(-\lambda t)\,dt \\
&= -\int_0^\infty t^2\,d(e^{-\lambda t}) \\
&= -\left[t^2e^{-\lambda t}\right]_0^\infty + 2\int_0^\infty t\exp(-\lambda t)\,dt \\
&= 0 + \frac{2}{\lambda}\int_0^\infty t\,\lambda\exp(-\lambda t)\,dt = \frac{2}{\lambda^2}
\end{aligned}$$

Thus, $\text{var}[t] = \frac{2}{\lambda^2} - \left(\frac{1}{\lambda}\right)^2 = \frac{1}{\lambda^2}$.

[16] Put $\lambda = 2$. Probability of x arrival per minute is computed as $P(X = x) = e^{-2}\frac{2^x}{x!}$. Probabilities of number of arrivals at ATM, $X = 0, 1, 2, 3, 4, 5, 6$ are 0.135, 0.271, 0.271, 0.180, 0.090, 0.036, and 0.012 respectively. Thus, $P(X \geq 2) = 1 - 0.135 - 0.271 = 0.594$, and $P(X \geq 3) = 0.323$. If one ATM is installed, probability of queueing forming is $0.594 > \frac{1}{3}$ which is unacceptable. With 2 ATMs, probability of queueing forming is $0.323 < \frac{1}{3}$ which is acceptable.

Answers to Problem Set 2

[1]

$$\begin{aligned}
q &= \frac{1}{1-\rho^2}\left\{\left(\left[\frac{y-\mu_y}{\sigma_y}\right] - \rho\left[\frac{x-\mu_x}{\sigma_x}\right]\right)^2 + (1-\rho^2)\left(\frac{x-\mu_x}{\sigma_x}\right)^2\right\} \\
&= \frac{1}{1-\rho^2}\left\{\left(\frac{y-[\mu_y+\rho(\sigma_y/\sigma_x)(x-\mu_x)]}{\sigma_y}\right)^2 + (1-\rho^2)\left(\frac{x-\mu_x}{\sigma_x}\right)^2\right\} \\
&= \frac{1}{1-\rho^2}\left(\frac{y-[\mu_y+\rho(\sigma_y/\sigma_x)(x-\mu_x)]}{\sigma_y}\right)^2 + \left(\frac{x-\mu_x}{\sigma_x}\right)^2
\end{aligned}$$

Thus

$$\begin{aligned}
f(x,y) &= \frac{1}{\sqrt{2\pi}\sigma_x}\exp\left\{-\frac{1}{2}\left(\frac{x-\mu_x}{\sigma_x}\right)^2\right\} \\
&\times \frac{1}{\sqrt{2\pi}\sigma_y\sqrt{1-\rho^2}}\exp\left\{-\frac{1}{2}\left(\frac{y-[\mu_y+\rho(\sigma_y/\sigma_x)(x-\mu_x)]}{\sqrt{1-\rho^2}\sigma_y}\right)^2\right\}
\end{aligned}$$

Then, $f(y|x) = \frac{1}{\sqrt{2\pi}\left(\sqrt{1-\rho^2}\sigma_y\right)} \exp\left\{-\frac{1}{2}\left(\frac{y-[\mu_y+\rho(\sigma_y/\sigma_x)(x-\mu_x)]}{\left(\sqrt{1-\rho^2}\sigma_y\right)}\right)^2\right\}$ which is the PDF of a normal RV with mean

$$E(Y|x) = \mu_Y + \rho(\sigma_y/\sigma_x)(x - \mu_X).$$

[2] The conditional distribution of Y given $X = 5$ is normally distributed with mean $E(Y|x = 5) = 10 + \rho\left(\frac{5}{1}\right)(5 - 5) = 10$, and variance $\text{var}(Y|\text{x=5}) = 25(1 - \rho^2)$. Thus

$$\begin{aligned} 0.95 &= P\left(Y \in (4, 16)|x = 5\right) \\ &= \int_4^{16} \frac{1}{5\sqrt{(1-\rho^2)}} \frac{1}{\sqrt{2\pi}} \exp\left(-\frac{1}{50(1-\rho^2)}\left[y - 10\right]^2\right) dy \end{aligned}$$

Or, for standard normal CDF function $\Phi(\cdot)$

$$\Phi\left(\frac{(16-10)}{5\sqrt{(1-\rho^2)}}\right) - \Phi\left(\frac{(4-10)}{5\sqrt{(1-\rho^2)}}\right) = \Phi(z) - \Phi(-z) = 0.95,$$

where

$$z = \left(\frac{6}{5\sqrt{(1-\rho^2)}}\right)$$

The value z is 1.96 since $\Phi(1.96) = 0.975$ and $\Phi(-1.96) = 0.025$. Solving $1.96 = \left(\frac{6}{5\sqrt{(1-\rho^2)}}\right)$ yields $\rho = 0.79$.

[3] $X \sim N(\mu, \sigma^2) < v$, and $v' = \frac{v-\mu}{\sigma}$. To show $\mu - \frac{\sigma^2\phi(v')}{\Phi(v')} < v$, we require to use the property from a standard normal PDF $\phi(z)$, viz.

$$\int_{-\infty}^{v'} z\,\phi(z)dz = -\phi(v').$$

This can easily be verified since differentiating w.r.t. v' on LHS gives, using Leibniz's rule, $v'\phi(v')$. Differentiating w.r.t. v' on the RHS gives

$$\begin{aligned} \frac{d}{dv'}\left(-\frac{1}{\sqrt{2\pi}}e^{-1/2\,v'^2}\right) &= -\left(-v'\frac{1}{\sqrt{2\pi}}e^{-1/2\,v'^2}\right) \\ &= v'\phi(v'). \end{aligned}$$

Therefore

$$
\begin{aligned}
-\frac{\phi(v')}{\Phi(v')} &= \int_{-\infty}^{v'} z\, \frac{\phi(z)}{\Phi(v')}\, dz \\
&< \int_{-\infty}^{v'} v'\, \frac{\phi(z)}{\Phi(v')}\, dz \\
&= v' \int_{-\infty}^{v'} \frac{\phi(z)}{\Phi(v')}\, dz \\
&= v' = \frac{v-\mu}{\sigma}.
\end{aligned}
$$

Hence, $\mu - \sigma\frac{\phi(v')}{\Phi(v')} < v$.

[4] From Table 2.4 in Chapter 2, we have $E(X|\omega_1 \in \mathcal{F}_b) = X_1(\omega_1)$, $E(X|\omega_2 \in \mathcal{F}_b) = X_2(\omega_2)$, $E(X|\omega_3 \in \mathcal{F}_b) = X_3(\omega_3)$. So $X(\omega)$ is the RV with exactly the same distribution as $E(X|\omega \in \mathcal{F}_b)$. Using the above result for substitution, hence, $E(E(X|\mathcal{F}_b)|\mathcal{F}_a) = E(X|\mathcal{F}_a)$.

[5] $E(R) = E(S) - 100,000 + E(e) = 100,000$ since $E(e) = 0$. But $E(R|S = 250,000) = 250,000 - 100,000 + E(e|S) = 150,000$. Mean square forecast error of conditional mean is $E[R - E(R|S)]^2 = E[e^2] = \sigma_e^2$. Mean square forecast error of unconditional mean is $E[R - E(R)]^2 = E[S - E(S) + e]^2 = \sigma_S^2 + \sigma_e^2$. Obviously, the conditional mean is a better forecast.

[6] $g(y)$. $E(g(Y)|Y = y) = \sum_Y \{g(Y)P(Y \neq y|Y = y) + g(y)P(Y = y|Y = y)\} = g(y)$ since $P(Y = y|Y = y) = 1$ and $P(Y \neq y|Y = y) = 0$.

Answers to Problem Set 3

[1] By Chebyshev's two-tailed inequality bound, $P(|X - \mu| \geq a) \leq \frac{\sigma^2}{a^2}$. By Cantelli's inequality or the one-tailed bound, $P(X - \mu \geq a) \leq \frac{\sigma^2}{a^2+\sigma^2}$. Thus also, $P(-[X - \mu] \leq -a) \leq \frac{\sigma^2}{a^2+\sigma^2}$. Combining the last two inequalities, therefore $P(|X - \mu| \geq a) \leq \frac{2\sigma^2}{a^2+\sigma^2}$.
Hence $P(|X - \mu| \geq a) \leq \min\left(\frac{\sigma^2}{a^2}, \frac{2\sigma^2}{a^2+\sigma^2}\right)$.

[2] This is just a set of numbers within (0,1] where the first n number of binary digits are specified. The probability of observing this set of numbers is $\frac{1}{2} + \frac{1}{2^2} + \cdots + \frac{1}{2^n} = 1 - (1/2)^n$.

[3] This is similar to Question 6 in Problem set 1, except the solution is a little bit more elegant in that we start of with one parameter a, not two, to solve, thus providing some economy. $E(aX + Y)^2 = a^2E(X^2) +$

$2aE(XY) + E(Y^2) \geq 0$ for any a. Then, choose $a = -\frac{\sqrt{E(Y^2)}}{\sqrt{E(X^2)}}$ to derive the correlation bound or Cauchy-Schwarz inequality.

[4] Risk-free rate of return of bond is $1/B - 1$. The bond's return rate is $1/0.9 - 1 = 11\%$. Under risk neutrality, the bond is preferred.

[5] Suppose \$1 is invested in a stock that either goes up to \$u or down to \$d, or \$1 can be put in a risk-free deposit to yield \$r all by the next period. If $u > d \geq r$, then one can borrow infinitely at the risk-free rate and buy the risky asset, making either $u - r > 0$ or $d - r \geq 0$ per dollar of risk-free loan, thus making risk-free arbitrage profit. If $r \geq u > d$, then one can shortsell infinitely the risky asset and lend at the risk-free rate, making either $r - u \geq 0$ or $r - d > 0$ per dollar of shortsale, thus making risk-free arbitrage profit. In order there should be no arbitrage, then we must have $u > r > d$.

[6] $C[u^2d] = [0.4(1.5) + 0.6(0.5)]/1.015 = 0.887$. If share price is 2 and $K = 1$, exercise profit $= 1$, so exercise if it is American call. The American call price should be 1. $C[ud]$=0.536, the European call price at node [UD]. We assumed the probabilities p, q are the risk-neutral probabilities.

[7] (a) $N(d_2)$. Thus, $N(d_2)$ in the Black–Scholes call option pricing model is the risk-neutral probability of maturing in-the-money for the call.

$$P(S_\tau \geq K) = P(\ln S_\tau - \ln S_0 \geq \ln K - \ln S_0) = P\left(\ln \frac{S_\tau}{S_0} \geq \ln \frac{K}{S_0}\right)$$

Since $\ln \frac{S_\tau}{S_0} \sim N\left((r - 1/2\sigma^2)\tau, \sigma^2\tau\right)$, then

$$\begin{aligned}
&P\left(\ln \frac{S_\tau}{S_0} \geq \ln \frac{K}{S_0}\right) \\
&= P\left(\frac{\ln(S_\tau/S_0) - (r - 1/2\sigma^2)\tau}{\sigma\sqrt{\tau}} \geq \frac{\ln(K/S_0) - (r - 1/2\sigma^2)\tau}{\sigma\sqrt{\tau}}\right) \\
&= \int_{\frac{\ln(K/S_0)-(r-1/2\sigma^2)\tau}{\sigma\sqrt{\tau}}}^{\infty} \phi(z)dz \\
&= \int_{-\infty}^{\frac{\ln(S_0/K)+(r-1/2\sigma^2)\tau}{\sigma\sqrt{\tau}}} \phi(z)dz \\
&= N(d_2)
\end{aligned}$$

where $d_2 = \frac{\ln(S_0/K)+(r-1/2\sigma^2)\tau}{\sigma\sqrt{\tau}}$, and we have made use of the property of the standard normal PDF, i.e., it is symmetrical about the mean zero.

(b) From (a), we see that event $(S_\tau \geq K)$ is identical to

$$\left(\frac{\ln(S_\tau/S_0) - (r - 1/2\sigma^2)\tau}{\sigma\sqrt{\tau}} \geq \frac{\ln(K/S_0) - (r - 1/2\sigma^2)\tau}{\sigma\sqrt{\tau}}\right).$$

Moreover, $\frac{\ln(\tilde{S}_\tau/S_0) - (r-1/2\sigma^2)\tau}{\sigma\sqrt{\tau}}$ is equivalent to a standard normal RV $\tilde{Z}$. Or

$$\tilde{S}_\tau \equiv S_0 \exp\left((r - 1/2\sigma^2)\tau + \sigma\sqrt{\tau}\,\tilde{Z}\right).$$

Then

$$\begin{aligned}
&E(S_\tau|S_\tau \geq K) \\
&\quad = E\left(S_0 \exp\left((r - 1/2\sigma^2)\tau + \sigma\sqrt{\tau}\tilde{Z}\right)\Big| Z \right. \\
&\qquad \left. \geq \frac{\ln(K/S_0) - (r - 1/2\sigma^2)\tau}{\sigma\sqrt{\tau}}\right) \\
&\quad = \left(\int_{\frac{\ln(K/S_0)-(r-1/2\sigma^2)\tau}{\sigma\sqrt{\tau}}}^{\infty} S_0 \exp((r - 1/2\sigma^2)\tau \right. \\
&\qquad \left. +\sigma\sqrt{\tau}z)\phi(z)dz\right)/P(S_\tau \geq K) \\
&\quad = S_0 \exp\left((r - 1/2\sigma^2)\tau + \frac{1}{2}\sigma^2\tau\right) \\
&\qquad \times \left[\int_{\frac{\ln(K/S_0)-(r-1/2\sigma^2)\tau}{\sigma\sqrt{\tau}}}^{\infty} \exp\left(-\frac{1}{2}\sigma^2\tau + \sigma\sqrt{\tau}z - \frac{1}{2}z^2\right)\right. \\
&\qquad \left. \times \frac{1}{\sqrt{2\pi}}dz\right]/P(S_\tau \geq K) \\
&\quad = S_0 \exp\left((r - 1/2\sigma^2)\tau + \frac{1}{2}\sigma^2\tau\right) \\
&\qquad \times \left[\int_{\frac{\ln(K/S_0)-(r-1/2\sigma^2)\tau}{\sigma\sqrt{\tau}}}^{\infty} \exp\left(-\frac{1}{2}\left(\sigma\sqrt{\tau} - z\right)^2\right)\right. \\
&\qquad \left. \times \frac{1}{\sqrt{2\pi}}dz\right]/P(S_\tau \geq K)
\end{aligned}$$

(doing a change of variable $y = \sigma\sqrt{\tau} - z$)

$$= S_0 \exp(r\tau)$$

$$\times \left[\int_{-\infty}^{\frac{\ln(S_0/K)+(r+1/2\sigma^2)\tau}{\sigma\sqrt{\tau}}} \exp(-\frac{1}{2}y^2)\frac{1}{\sqrt{2\pi}}dy \right] / P(S_\tau \geq K)$$

$$= S_0 e^{r\tau} N(d_1)/N(d_2)$$

where $d_1 = \frac{\ln(S_0/K)+(r+1/2\sigma^2)\tau}{\sigma\sqrt{\tau}}$

(c)

$$\begin{aligned}
E[C_\tau] &= E[\max(S_\tau - K, 0)] \\
&= E[\max(S_\tau - K, 0)|S_\tau \geq K]\ P(S_\tau \geq K) \\
&\quad + E[\max(S_\tau - K, 0)|S_\tau < K]\ P(S_\tau < K) \\
&= E(S_\tau - K|S_\tau \geq K)\ P(S_\tau \geq K) \\
&\quad + E(0|S_\tau < K)\ P(S_\tau < K) \\
&= E(S_\tau|S_\tau \geq K)\ P(S_\tau \geq K) \\
&\quad - K\ E(1|S_\tau \geq K)\ P(S_\tau \geq K) \\
&= E(S_\tau|S_\tau \geq K)\ P(S_\tau \geq K) \\
&\quad - K\ P(S_\tau \geq K).
\end{aligned}$$

Note that the last term is due to the fact that $E(1|S_\tau \geq K) = 1$ since it is integrating over the conditional PDF $f(S_\tau|S_\tau \geq K)$ which itself is a proper PDF that integrates to 1. Hence

$$\begin{aligned}
C_0 &= e^{-r\tau} E[C_\tau] \\
&= e^{-r\tau}\Big(E[S_\tau|S_\tau \geq K]\ P[S_\tau \geq K] - K\ P[S_\tau \geq K]\Big) \\
&= e^{-r\tau}\Big(E[S_\tau|S_\tau \geq K]\ N(d_2) - K\ P[S_\tau \geq K]\Big) \\
&\quad \text{(using results from (b) and (a))} \\
&= e^{-r\tau}\left(S_0\ e^{r\tau}\ N(d_1) - K\ N(d_2)\right) \\
&= S_0 N(d_1) - K e^{-r\tau}\ N(d_2).
\end{aligned}$$

[8] $0.251

Answers to Problem Set 4

[1] $U' > 0$. So, $U' = a - 2bW > 0 \Rightarrow W < a/2b$. If we include having display of risk aversion or concave utility, i.e., $U'' < 0$, then $b > 0$ and also $a > 0$ since $W > 0$.

[2] $A \succ B$, so

$$U(100K) > 0.1U(1M) + 0.89U(100K) + 0.01U(0)$$
$$\Rightarrow 0.11U(100K) > 0.1U(1M) + 0.01U(0).$$

Thus, $0.11U(100K) + 0.89U(0) > 0.1U(1M) + 0.9U(0)$. But $C \succ D$ is contradictory. This is called the Allais paradox.

[3] Note e in V does not affect the FOC as in Eqs. (4.4) and (4.5) since it is a constant $e > 0$. e satisfies the IC conditions. The solutions are similarly $Q_B = Q_G = K$.

[4] For any X, X', and any $q \in (0, 1)$, strict concavity of $U \Rightarrow U(qX^T P_j + [1-q]X'^T P_j) > qU(X^T P_j) + (1-q)U(X'^T P_j)$. Since this holds for every P_j, then

$$\sum_j \pi_j U(qX^T P_j + [1-q]X'^T P_j) > \sum_j \pi_j \left(qU(X^T P_j) + (1-q)U(X'^T P_j)\right).$$

Or

$$\sum_j \pi_j U([qX^T + [1-q]X'^T]P_j) > q\sum_j \pi_j U(X^T P_j) + (1-q)\sum_j \pi_j U(X'^T P_j)$$

LHS is $V(qX + [1-q]X')$, and RHS is $qV(X) + (1-q)V(X')$. Hence, V is strictly concave.

[5] We shall illustrate with discrete probability. If $\tilde{Y} = \tilde{X} + \tilde{\epsilon}$, then for every joint occurrence of $(x_i \in X, e_j \in \epsilon)$, where $e_j \leq 0$, there exists $y_{ij} \in Y$ s.t. $y_{ij} = x_i + e_j$. Suppose there are N sample points $\{x_i\}_{i=1,2,\ldots,N}$ and M sample points $\{e_i\}_{i=1,2,\ldots,M}$, then there are NM sample points (not necessarily all distinct) $\{y_{ij}\}_{i\in[1,N];j\in[1,M]}$. We can without loss of generality, sequence the outcomes $x_N > x_{N-1} > x_{N-2} > \cdots > x_2 > x_1$.

The table below shows the joint event (x_i, e_j). We assume there is some i where $e_i < 0$, i.e., $\epsilon \leq 0$ (not $\epsilon \leqq 0$).

(X, ϵ)	e_1	e_2	$\cdots\cdots$	e_M	$P(x_i)$
x_1	$y_{11} \leq x_1$	$y_{12} \leq x_1$	$\cdots$	$y_{1M} \leq x_1$	p_1
x_2	$y_{21} \leq x_2$	$y_{22} \leq x_2$	$\cdots$	$y_{2M} \leq x_2$	p_2
x_3	$y_{31} \leq x_3$	$y_{32} \leq x_3$	$\cdots$	$y_{3M} \leq x_3$	p_3
$\vdots$	$\vdots$	$\vdots$	$\vdots$	$\vdots$	$\vdots$
x_N	$y_{N1} \leq x_N$	$y_{N2} \leq x_N$	$\cdots$	$y_{NM} \leq x_N$	p_N

Key point to note is that any event $Y = y_{ij} \leq x_i$ has a probability $P(Y = y_{ij} \leq x_i)$ or simply written as $P(y_{ij} \leq x_i)$. This probability is interpreted as $Y = y_{ij}$ and that $y_{ij} \leq x_i$. Thus, $P(Y = y_{ij} \leq x_i) = P(\{Y = y_{ij}\} \bigcap \{y_{ij} \leq x_i\}) = P(y_{ij})$ that is exactly equal to $P(x_i, e_j)$, the joint probability of x_i and e_j.
Then, taking the first row

$$P(Y < x_1) = \sum_{j=1}^{M} P(y_{1j} \leq x_1) = \sum_{j=1}^{M} P(y_{1j}) = p_1 = P(X < x_2).$$

Taking the first two rows

$$P(Y < x_2) = \sum_{i=1}^{2} \sum_{j=1}^{M} P(y_{ij} \leq x_2) = p_1 + p_2 = P(X < x_3).$$

In general, taking the first k rows, for any $1 \leq k \leq N - 1$

$$\begin{aligned} P(Y < x_k) &= \sum_{i=1}^{k} \sum_{j=1}^{M} P(y_{ij} \leq x_k) \\ &= p_1 + p_2 + \cdots + p_k = P(X < x_{k+1}). \end{aligned}$$

Then, it can be seen that since $P(X < x_1) = 0$ and $P(X \leq x_N) = 1$,

$$\begin{aligned} P(X \leq x_1) &< P(X < x_2) = P(Y < x_1) \\ P(X \leq x_2) &< P(X < x_3) = P(Y < x_2) \\ P(X \leq x_3) &< P(X < x_4) = P(Y < x_3) \\ &\cdots \quad \cdots \\ P(X \leq x_{N-1}) &< P(X < x_N) = P(Y < x_{N-1}) \\ &\text{and } P(X \leq x_N) = P(Y < x_N). \end{aligned}$$

Thus, for any x within the support (range of the values taken by the RV except the last value x_N) of the distribution of X, $P_X(X \leq x) < P_Y(Y < x)$, which by definition means X first-order stochastically dominates (FSD) Y.

For the converse, if X FSD Y, then by definition, $P_X(X \leq x) < P_Y(Y < x)$ for any x in the support of the distributions. Since we can always define a RV ϵ such that its realization e is the difference between any values taken by $Y - X$, then $y = x + e$, and we can write $\tilde{Y} = \tilde{X} + \tilde{\epsilon}$. Thus, we can again use the table before. In the table, the first row consists of all joint events. For all j, $y_{1j} = x_1 + e_j \leq x_1$, in order that $P(Y < x_1) = p_1 > P(X < x_1) = 0$ as per the FSD relationship. Similarly, the second row consists, for all j, $y_{2j} = x_2 + e_j \leq x_2$, in order that $P(Y < x_2) = p_1 + p_2 > P(X < x_2) = p_1$ as per the FSD relationship. Recall that not all e_j's $= 0\ \forall j$. Is it possible that some small number of events in rows $c \leq i$ may deviate and not follow $y_{cj} = x_c + e_j < x_c$? e.g., $y_{cj} = x_c + e_j > x_c$ instead? And still satisfy $P(Y < x_i) > P(X < x_i)$.

We claim this is not possible as such deviation can result in $P(Y < x_c) < P(X \leq x_c)$ if there are many joint events (x_c, e_j) with zero probability.

Hence, for all $x_1 < x_2 < \cdots < x_N$, FSD implies $y_{ij} = x_i + e_j \leq x_i$, and thus $e_j \leq 0$ for all j but where $e_j < 0$ for some j. Thus, $\tilde{Y} = \tilde{X} + \tilde{\epsilon}$, with $\epsilon \leq 0$. Iff relationship is equivalence.

Answers to Problem Set 5

[1] The Arrow–Debreu U-certificate pays state-contingent unit dollar, \$1, supposing at a current $t = 0$, its price is $\$a_0^u$. It can be sold in Utopia at $a_0^u y$ utopi with a state-U payoff of y utopi or else zero. Hence $\pi_0^u = a_0^u y$, and $X^u = y$. D-security price in Utopia is $a_0^d y$ utopi with state-D payoff of y or else zero.

[2] An arbitrage portfolio is $\langle x_1 - y_1, x_2 - y_2, \ldots, x_M - y_M \rangle$.

[3]

$$(0)(0.5) + (1)(0.3) + (-1)(0.07) + (5)(0.03) = 0.38.$$

$$\text{Risk-free bond price } = 0.5 + 0.3 + 0.07 + 0.03 = 0.9.$$

$$\text{Risk-free rate is } \frac{1}{0.9} - 1 = \frac{1}{9} \text{ or } 11.11\%$$

[4] $\exists Q^{-1} \ni QQ^{-1} = I_S$. Each kth column of Q^{-1} represents a portfolio and is a $S \times 1$ vector $(y_1, y_2, \ldots, y_S)^T$ where y_j is the number of

units of the jth security held in the portfolio. This portfolio replicates the kth-state security with payoff \$1 only in the kth state and zero otherwise. In the second case with Q_M payoff matrix, its rank is $S < M$. Hence, there is $M - S$ number of redundant securities, in the sense they can be replicated by any S number of linearly independent portfolios.

[5] The put's payoff vector is $(1, 0, 0)$. This is the payoff of state-1 security, which therefore has the same price P. Form a portfolio comprising 2 stocks, 3 short calls, and 2 short puts. This has payoffs $(0, 1, 0)$ and is thus the same payoff as a state 2 security. Thus, the price of state-2 security is $2S - 3C - 2P$. Form a portfolio comprising 2 calls, 1 put, and 1 short stock. This has payoffs $(0, 0, 1)$ and is thus the same payoff as a state-3 security. Thus, the price of state-3 security is $2C + P - S$.

[6] The risk-free zero-coupon discount bond with price \$ $1/(1 + R)$ will pay \$1 at the end of period. A portfolio of one state-D security and one dual-state security produces payoff vector \$ $(1, 1, 1)$. The latter is the payoff of the discount bond costing \$ $1/(1 + R)$. Hence, the price of the dual state security is \$ $1/(1 + R) - c_d$.

[7] The situation with a stock and a bond has the advantage of reaching any of the 3 states. Thus if state D, e.g., is a very bad state that will cause ruin for the investor who will certainly like to insure against it and be paid when D occurs, then the situation of stock and bond can allow partial insurance, and is certainly preferred to a situation when only state U and state M securities are traded. In the latter, state D can never be insured. However, if e.g., state M and state D securities are available, and state U has low probability and is not seriously damaging to the investor if it occurs, then being able to invest in state payoffs under M or D may be better than the stock and bond whereby the allocation to state M and D cannot be exact since any combination of the payoffs of the stock and bond may not necessarily produce a scalar multiple of vector payoff (0,1,1).

Answers to Problem Set 6

[1] No, if there is arbitrage opportunity, his/her wealth could be increased infinitely so that he/she will not find a finite maximum to his/her expected utility.

[2] Let $m = E(W_1)$ and expand a Taylor series about m:

$$\begin{aligned}
E[U(W_1)] &= E\Big[U(m) + (W_1 - m)U'(m) + \frac{1}{2!}(W_1 - m)^2 U''(m) \\
&\quad + \frac{1}{3!}(W_1 - m)^3 U'''(m) + \cdots\Big] \\
&= U(m) + \frac{U''(m)}{2!}E(W_1 - m)^2 \\
&\quad + \frac{U'''(m)}{3!}E(W_1 - m)^3 + \cdots \\
&= f(\mu_2, \mu_3, \mu_4, \ldots)
\end{aligned}$$

where μ_z is the zth central moment of normal RV W_1. Since such moments are of the form $\mu_z = 0$ if $z > 1$ is odd, and $\mu_z = \frac{z!}{(z/2)!}\frac{\sigma^2}{2^{z/2}}$ for z even and with $\mu_2 = \sigma^2$, then clearly $f(\mu_2(m,\sigma^2), \mu_3(m,\sigma^2), \mu_4(m,\sigma^2), \ldots)$ is a function of only m and σ^2.

[3] $dV = \frac{\partial V}{\partial x} \cdot dx$ where $\cdot$ is element by element multiplication. Thus, dV is $N \times 1$. Total second derivative (which is negative if V is maximized at the point where $dV = 0$) is

$$\begin{aligned}
d^2V = d(dV) &= d\left(\frac{\partial V}{\partial x} \cdot dx\right) \\
&= \frac{\partial}{\partial x}\left(\frac{\partial V}{\partial x} \cdot dx\right)dx \\
&= dx^T\left(\frac{\partial^2 V}{\partial x^2}\right)dx \\
&= dx^T\left(2\frac{\partial V}{\partial \sigma_P^2}\Sigma\right)dx < 0
\end{aligned}$$

as Σ is a strictly positive definite matrix since it is a covariance matrix, and $\frac{\partial V}{\partial \sigma_P^2} < 0$.

[4] $\max_x EU([1 + r_f + x(r_M - r_f)]W_0) \Rightarrow E[U'(W_1)(r_M - r_f)] = 0$. Hence, $E(r_M) - r_f = -\frac{\text{cov}(U', r_M)}{EU'} > 0$ since $EU' > 0$ and $\text{cov}(U', r_M) < 0$ because as r_M increases, U' should decrease due to concave $U'' < 0$. $r_M - r_f$ can be < 0, only its expectation is strictly positive. Yes, for $\beta_i < 0, E(r_i) < r_f$.

[5] We solve for state prices c_1, c_2 using 2 price equations: $6 = [c_1 * 9 + c_2 * 6]$ and $6/7 = c_1 + c_2$. Thus, the state prices are $c_1 = 2/7$ and $c_2 = 4/7$.

No, by the theorem, since state prices exist. Price of European put = $2/7^* \max(7\text{-}9,0)+4/7^* \max(7-6,0) = 4/7$. Hence, it is better to exercise rightaway at $7-6=\$1$ than wait. American put price is therefore \$1.

[6] The FOCs are for every j,

$$\lambda = \sum_{i=1}^{S} \theta_i q_{ij}/W_1(\omega_i)$$

and Eqs. (6.12) to (6.15).

[7] Using the APT, $\lambda_0 = 0.10$. $0.13 = 0.10 + 3\lambda_1 + 2\lambda_2, 0.15 = 0.10 + 2\lambda_1 + 4\lambda_2$, so solving, $\lambda_1 = 0.0025, \lambda_2 = 0.01125$.

Answers to Problem Set 7

[1] First, note that $E(X_{n+1}|\mathcal{F}_n) = 1/2 - 1/2 = 0$. As an aside, recall that $E(X_n|\mathcal{F}_n) = X_n$.
Now, $S_{n+1} = S_n + X_{n+1}$, so $E(S_{n+1}|\mathcal{F}_n) = E(S_n|\mathcal{F}_n) + E(X_{n+1}|\mathcal{F}_n) = S_n + 0$. Hence, S_{n+1} is a martingale.
Now for any function $f(\cdot), E(f(S_{n+1})|\mathcal{F}_n) = E(f(S_n + X_{n+1})|\mathcal{F}_n) = g(S_n)$ for a function $g(\cdot)$. The last statement is true because for any given S_n, the LHS is mapped to a number that only depends on S_n. X_{n+1} is i.i.d. and is integrated out in the expectation operator. Hence, S_{n+1} is a Markov process.

[2] Let $\text{var}X_t = \text{var}Y_t = \sigma^2$ since $X_t \sim Y_t$. Let constant or deterministic m_t be the mean of X_t. $Z_1 \triangleq E(X_{t+1}|\mathcal{F}_t) - E(X_t|\mathcal{F}_t) = m_t - X_t$, and $Z_2 \triangleq E(Y_{t+1}|\mathcal{F}_t) - E(Y_t|\mathcal{F}_t) = Y_{t+1} - Y_t$. Then, $\frac{\text{var}Z_1}{\text{var}Z_2} = \frac{\sigma^2}{2\sigma^2} = 1/2$. In additon, $E(X_{t+1}Y_{t+1}|\mathcal{F}_t) = Y_{t+1}E(X_{t+1}|\mathcal{F}_t) = Y_{t+1}m_t$ which exists since m_t and Y_{t+1} exist.

[3] $E(M_{t+1}|M_t, M_{t-1}, \ldots) = M_t$ and $E(M_{t+2}|M_{t+1}, M_t, \ldots) = M_{t+1}$ since M is a martingale. By iterated expectations

$$\begin{aligned} E[E(M_{t+2}|M_{t+1}, M_t, \ldots)|M_t, M_{t-1}, \ldots] &= E(M_{t+1}|M_t, M_{t-1}, \ldots) \\ &= M_t \end{aligned}$$

Similarly, we can show $E(M_s|M_t, M_{t-1}, \ldots) = M_t$ for every $s > t$.

[4] $E(S_3|S_2 = \$6 \text{ at UD}) = 0.5(10+8) = \$9$.
$E(S_3|S_2 = \$6 \text{ at DU}) = 0.5(9+7) = \$8$. If S_n is Markov, then $E(S_3|S_2 = 6) = g(S_2) = g(6)$ for a certain function g. But since $g(6)$ cannot be both \$9 and also \$8, therefore, in general we cannot find g to satisfy the Markov condition that $E(S_3|S_2) = g(S_2)$. Hence, S_n is not a Markov process.

[5] Yes, in this case when $R = 1$, S_n is Markov as well as a martingale. For C_n, no, it is not a martingale. This is because for this type of option, its future expected value is path dependent, not just dependent on the last underlying stock price.

[6] $Y_t \triangleq E(X_n|\mathcal{F}_t), n \geq t$. Thus

$$E(Y_{t+1}|\mathcal{F}_t) = E(E(X_n|\mathcal{F}_{t+1})|\mathcal{F}_t) = E(X_n|\mathcal{F}_t) \triangleq Y_t.$$

Hence, Y_t is a martingale. This conditional expectation as a martingale is called the "Tower property".

[7]

$$\begin{aligned} E[Z_{t+1}|\mathcal{F}_t] &= E[N_{t+1} - q(t+1)|\mathcal{F}_t] \\ &= E[N_{t+1} - N_t + N_t|\mathcal{F}_t] - q(t+1) \\ &= E[N_{t+1} - N_t|\mathcal{F}_t] + (E[N_t|\mathcal{F}_t] - qt) - q \\ &= N_t - qt \equiv Z_t. \end{aligned}$$

Hence, Z_{t+1} is a martingale.

[8] If $C_T^*(S_T^*)$ is the terminal value of a contingent claim in a market with no arbitrage opportunity, by Theorem 7.3, $\exists$ an equivalent martingale probability measure Q s.t. $E_Q(C_T^*|\mathcal{F}_0) = C_0^*$. Hence, starting with C_0^*, we can always construct a self-financing trading strategy s.t. $X_T \cdot S_T^* = C_T^*$. By Lemma 7.2, all such generating strategies for C_T^* are the same, i.e., it is unique.

[9] Expected payoff is $\frac{1}{2} \times 1 + \frac{1}{4} \times 1 + \frac{1}{8} \times 1 + \cdots = \$1$.

[10] No. The whole idea in this last section is that the state contingent claims are attainable by self-financing trading strategies involving trading only in the stock and the bond and where discounted stock price process is a martingale. In general, continuous time markets need not be completed by an infinitely large number of tradable state-contingent claims in the market; it can be completed by a finite number of securities trading in the market on a dynamic basis with continuous rebalancing of portfolio on a self-financing strategy. In such a situation, the finite number of securities are said to dynamically span the continuous-time market.

[11] $E_t(X_{t+1}S^*_{t+1}) = X_{t+1}E_t(S^*_{t+1}) = X_{t+1}S^*_t$. By Eq. (7.2), therefore $E_t(X_{t+1}S^*_{t+1}) = X_tS^*_t$. Taking iterated expectation

$$\begin{aligned} E_{t-1}(E_t(X_{t+1}S^*_{t+1})) &= E_{t-1}(X_{t+1}S^*_{t+1}) \\ &= E_{t-1}(X_tS^*_t) = X_tE_{t-1}(S^*_t) \\ &= X_tS^*_{t-1} = X_{t-1}S^*_{t-1}. \end{aligned}$$

Hence, continuing with using iterated expectations and self-financing as in Eq. (7.2), we obtain $E_0(X_{t+1}S^*_{t+1}) = X_1S^*_0$. Hence, $X_{t+1}S^*_{t+1}$ is also a martingale w.r.t. the same measure.

Answers to Problem Set 8

[1] $\theta = 5.04\%$; 2.52%. Yes, for CARA negative exponential utility has increasing relative risk aversion with increase in wealth.

[2] (a) From Eq. (8.3), in a single-period world, ignore subscript t. Remove superscript k since all are representative agents in the economy.

(b)

$$p_j = \sum_s^S \phi_s x_s = \sum_s^S \pi_s \left[\frac{U'(C_s)}{U'(C_0)} x_s\right].$$

(c) Divide equation in (b) by p_j on LHS and RHS to obtain

$$1 = E\left[\frac{U'(\tilde{C})}{U'(C_0)}(1+\tilde{r}_j)\right],$$

where

$$1+\tilde{r}_j = \frac{x_s}{p_j}.$$

Hence

$$1 = \text{cov}\left(\frac{U'(\tilde{C})}{U'(C_0)}, \tilde{r}_j\right) + \frac{E(1+r_j)}{1+r_f}.$$

So, $\forall j$

$$E(\tilde{r}_j - r_f) = -(1+r_f)\text{cov}\left(\frac{U'(\tilde{M})}{U'(C_0)}, \tilde{r}_j\right)$$

where M is the end-of-period remaining wealth to be used as consumption $\tilde{C}$.

Then

$$E(\tilde{r}_j - r_f) = \frac{\text{cov}(U'(\tilde{M}), \tilde{r}_j)}{\text{cov}(U'(\tilde{M}), \tilde{r}_m)} E(\tilde{r}_m - r_f).$$

Assuming $\tilde{r}_j, \tilde{r}_m$ are bivariate normal under CAPM, then applying the hint,

$$E(\tilde{r}_j - r_f) = \frac{E(U''(\tilde{r}_m)) \, \text{cov}(r_j, r_m)}{E(U''(\tilde{r}_m)) \text{cov}(r_m, r_m)} E(\tilde{r}_m - r_f)$$

Hence the Sharpe-Lintner CAPM.

[3]

$$E(\tilde{r}_i - r_f) = \frac{\text{cov}\big(\tilde{r}_i, -1/\tilde{R}_M\big)}{E\big(1/\tilde{R}_M\big)}$$

RHS becomes

$$-\frac{\text{cov}\big(\tilde{r}_i, 1/\tilde{R}_M\big)}{E\big(1/\tilde{R}_M\big)} = \frac{E\big(1/R_M^2\big)}{E\big(1/R_M\big)} \text{cov}(r_i, r_M)$$

Hence

$$E(\tilde{r}_i - r_f) = \frac{E\big(1/R_M^2\big)}{E\big(1/R_M\big)} \text{cov}(r_i, r_M)$$

And

$$E(\tilde{r}_M - r_f) = \frac{E\big(1/R_M^2\big)}{E\big(1/R_M\big)} \text{var}(r_M)$$

Dividing one by the other, we obtain the CAPM.

[4] For the power utility, put $d = \gamma^{-\gamma}, A = 0, B = \gamma^{-1}$. For the log utility case, put $d = 1, A = 0, B = 1$, i.e., $\gamma = 1$ in the power case.

[5] FOCs are: $1/C_0 = \lambda, 1/(2C_s) - \lambda\phi_s = 0$ for $s = u, d, 1/(4C_{ss}) - \lambda\phi_{ss} = 0$ for $ss = uu, ud, du, dd$. And budget constraint $C_0 + \sum_s \phi_s C_s + \sum_{ss} \phi_{ss} C_{ss} = W_0 = 100$. Then, $C_0 = 1/3W_0 = \$33.33$. $C_u = W_0/(6\phi_u) = 100/(0.66) = 151.51$. $C_d = W_0/(6\phi_d) = 100/(4.8) = 20.83$.

[6] (a) λ is the Lagrange multiplier, and *de facto*, the individual chooses to consume C_{ts} at t if state s by buying C_{ts} units of the t, s state-contingent

claims at $\$\phi_{ts}$ each. The FOCs are, $\forall t, s$:

$$\gamma \pi_{ts} \delta^t C_{ts}^{\gamma-1} - \lambda \phi_{ts} = 0 \Rightarrow \phi_{ts} = \frac{\gamma \pi_{ts} \delta^t C_{ts}^{\gamma-1}}{\lambda}$$

$$\gamma C_0^{\gamma-1} - \lambda = 0 \Rightarrow \lambda = \gamma C_0^{\gamma-1}$$

$$W_0 = C_0 + \sum_t \sum_s \phi_{ts} C_{ts}.$$

Therefore, $\phi_{ts} = \frac{\gamma \pi_{ts} \delta^t C_{ts}^{\gamma-1}}{\gamma c_0^{\gamma-1}}$.

(b) From Part (a) and the results of Section 8.4,

$$\forall t, s : C_{ts} = C_0 \left(\frac{\phi_{ts}}{\delta^t \pi_{ts}} \right)^{1/(\gamma-1)}$$

Solve for C_0 and C_{ts} explicitly in terms of W_0.

Let $\left(\frac{\phi_{ts}}{\delta^t \pi_{ts}} \right)^{1/(\gamma-1)} = e_{ts}$ for $t = 1, 2, \ldots, T$; $s = 1, 2, \ldots, S$, then

$$\begin{pmatrix} A_{(T\times S)\times 1} & -I_{(T\times S)\times(T\times S)} \\ 1_{1\times 1} & B_{1\times(T\times S)} \end{pmatrix} \begin{pmatrix} C_0 \\ \vdots \\ C_{ts} \\ \vdots \\ C_{TS} \end{pmatrix}_{(T\times S+1)\times 1} = \begin{pmatrix} 0 \\ \vdots \\ 0 \\ \vdots \\ W_0 \end{pmatrix}$$

where

$$A = \begin{pmatrix} e_{11} \\ e_{12} \\ \vdots \\ e_{1S} \\ e_{21} \\ e_{22} \\ \vdots \\ e_{2S} \\ e_{31} \\ \vdots \\ e_{TS} \end{pmatrix}$$

and

$$B = (\phi_{11}, \phi_{12}, \ldots, \phi_{ts}, \ldots, \phi_{TS}).$$

We can notate the above in the form $\Sigma C = W_0 E$, so $C = W_0[\Sigma^{-1}E]$ where $E = (0, 0, \ldots, 0, 1)^T$.

[7] In a single-period model (or "two-period time-points"),

$$M_1 = \delta U'([W_0 - C_0](1 + r_1^M))/U'(C_0)$$

is a function of r_1^M. For jointly normal r_1^j and r_1^M

$$\begin{aligned} \text{cov}_0(r_1^j, M_1) &= E_0[\delta(W_0 - C_0)U''([W_0 - C_0](1 + r_1^M))/U'(C_0)] \\ &\quad \times \text{cov}_0(r_1^j, r_1^M) \\ &= B\text{cov}_0(r_1^j, r_1^M) \end{aligned}$$

where $B = E_0[\delta(W_0 - C_0)U''([W_0 - C_0](1 + r_1^M))/U'(C_0)]$ is a constant. Then, $\text{cov}_0(r_1^j, M_1 - Br_1^M) = 0$. Thus, for any stochastic M_1, the term $M_1 - Br_1^M$ must be equal to a constant A to attain the zero covariance. Hence, $M_1 = A + Br_1^M$.

Answers to Problem Set 9

[1] (a) For $s > t > 0, W_s - W_t$ and $W_t - W_0$ are independent, being BM. Therefore, $E[W_s - W_t|W_t, W_0] = E[W_s - W_t] = 0$. But $E[W_s - W_t|W_t] = E[W_s|W_t] - W_t$. Hence, $E[W_s|W_t] - W_t = 0$ or $E[W_s|W_t] = W_t$ which proves BM W_s is a martingale. (b) To prove W_s is Markov, we require that for any measurable function f, there exists a measurable function g such that

$$E[f(W_s)|W_t] = g(W_t).$$

Let LHS be $E[f(W_s - W_t + W_t)|W_t] = g(x) = E[f(W_s - W_t + x)]$ where we treated the given W_t as if it is a constant x. Now, $W_s - W_t \sim N(0, s - t)$. So

$$g(x) = \frac{1}{\sqrt{2\pi(s-t)}} \int_{-\infty}^{\infty} f(u + x) e^{-\frac{u^2}{2(s-t)}} du.$$

Hence, a function $g(W_t)$ exists on the RHS when we put $x = W_t$.

[2]

$$\begin{pmatrix} E[W_{t_1}^2] & E[W_{t_1}W_{t_2}] & \cdots & E[W_{t_1}W_{t_n}] \\ E[W_{t_2}W_{t_1}] & E[W_{t_2}^2] & \cdots & E[W_{t_2}W_{t_n}] \\ \vdots & \vdots & & \vdots \\ E[W_{t_n}W_{t_1}] & E[W_{t_n}W_{t_2}] & \cdots & E[W_{t_n}^2] \end{pmatrix}$$

$$= \begin{pmatrix} t_1 & t_1 & \cdots & t_1 \\ t_1 & t_2 & \cdots & t_2 \\ \vdots & \vdots & & \vdots \\ t_1 & t_2 & \cdots & t_n \end{pmatrix}$$

[3] For $\triangle = T/n$

$$\begin{aligned} \int_0^T |dW_t|dt &= \lim_{n\to\infty} \sum_{j=0}^{n-1} \left|W_{(j+1)\triangle} - W_{j\triangle}\right|\triangle \\ &\leq \lim_{n\to\infty} \max_{0\leq j\leq n} \left|W_{(j+1)\triangle} - W_{j\triangle}\right| \left(\sum_{j=0}^{n-1} \triangle\right) \\ &= \lim_{n\to\infty} \max_{0\leq j\leq n} \left|W_{(j+1)\triangle} - W_{j\triangle}\right|T \\ &= 0 \end{aligned}$$

since W_t is continuous, and $\lim_{n\to\infty}\left[W_{(k+1)\triangle} - W_{k\triangle}\right] = 0$.

$$\begin{aligned} \int_0^T (dt)^2 &= \lim_{n\to\infty} \sum_{j=0}^{n-1} \left(t_{(j+1)\triangle} - t_{j\triangle}\right)^2 \\ &\leq \lim_{n\to\infty} \max_{0\leq j\leq n} \left(t_{(j+1)\triangle} - t_{j\triangle}\right) \left(\sum_{j=0}^{n-1} t_{(j+1)\triangle} - t_{j\triangle}\right) \\ &= \lim_{n\to\infty} \max_{0\leq j\leq n} \left(t_{(j+1)\triangle} - t_{j\triangle}\right)T \\ &= 0 \end{aligned}$$

[4] $W_b \sim N(0,b), W_a \sim N(0,a)$. $E(W_b - W_a) = 0, \mathrm{var}(W_b - W_a) = \mathrm{var}(W_b)$ $+ \mathrm{var}(W_a) - 2\mathrm{cov}(W_a, W_b) = b + a - 2a = b - a$. Thus, $W_b - W_a \sim N(0, b-a)$.

[5] Length of path of $W_t^{1/n}$ over interval $1/n$ is $\left|\frac{X_{1/n}}{\sqrt{n}}\right| = \frac{1}{\sqrt{n}}$. Since $W_t^{1/n}$ is sum of tn i.i.d. $\frac{X_{1/n}}{\sqrt{n}}$, over interval $[0,t]$, its total variation is $t\frac{n}{\sqrt{n}} = t\sqrt{n} \to \infty$ as $n \to \infty$.

[6]
$$\begin{aligned} dXY &= XdY + YdX + \frac{1}{2}\left(\frac{\partial^2 XY}{\partial X^2}\right)(dX)^2 \\ &\quad + \frac{1}{2}\left(\frac{\partial^2 XY}{\partial Y^2}\right)(dY)^2 + \frac{\partial^2 XY}{\partial X \partial Y}(dX)(dY) \\ &= XdY + YdX + dXdY. \end{aligned}$$

This equals

$$\begin{aligned} &XY(edt + fdW) + YX(adt + bdW) + bfXYdt \\ &= XY((a + e + bf)dt + (b + f)dW). \end{aligned}$$

[7] $d(1/Y) = -(1/Y^2)dY + 1/2[2/Y^3](dY)^2 = \frac{1}{Y}([-\mu + \sigma^2]dt - \sigma dW)$.

[8]
$$\begin{aligned} \int_{-\infty}^{\infty} e^{i\lambda z} u_{zz} dz &= \int_{-\infty}^{\infty} e^{i\lambda z} du_z \\ &= [e^{i\lambda z} u_z]_{-\infty}^{\infty} - \int_{-\infty}^{\infty} u_z de^{i\lambda z} \\ &= [e^{i\lambda z} u_z]_{-\infty}^{\infty} - i\lambda \int_{-\infty}^{\infty} u_z e^{i\lambda z} dz \\ &= [e^{i\lambda z} u_z]_{-\infty}^{\infty} - i\lambda \int_{-\infty}^{\infty} e^{i\lambda z} du \\ &= [e^{i\lambda z} u_z]_{-\infty}^{\infty} - [i\lambda e^{i\lambda z} u]_{-\infty}^{\infty} + i\lambda \int_{-\infty}^{\infty} u de^{i\lambda z} \\ &= [e^{i\lambda z} u_z]_{-\infty}^{\infty} - [i\lambda e^{i\lambda z} u]_{-\infty}^{\infty} + (i\lambda)^2 \hat{u}(\lambda) \\ &= -\lambda^2 \hat{u}(\lambda). \end{aligned}$$

Looking at the boundary conditions, u and u_z approach zero as z approaches $-\infty$ or ∞.

Answers to Problem Set 10

[1] We prove by contradiction. Suppose $u > d > e^{(r-\delta)h} > 0$. This $\Rightarrow$ $ue^{\delta h} > de^{\delta h} > e^{rh} > 0$. We can make pure arbitrage money at $t = 0$

by borrowing \$1 and paying \$ e^{rh} at $t = h$, investing in the dividend-yielding stock at $t = 0$ and receiving cum-dividend stock value at least as large as \$ $de^{\delta h}$ which is more than the borrowing cost.

Suppose $e^{(r-\delta)h} > u > d > 0$. This $\Rightarrow e^{rh} > ue^{\delta h} > de^{\delta h} > 0$. We can make pure arbitrage money at $t = 0$ by short-selling dividend-yielding stock at $t = 0$ and paying cum-dividend value at $t = h$ of at most \$ $ue^{\delta h}$, lending at risk-free rate at $t = 0$ and receiving principal and interest payback of \$ e^{rh} at $t = h$, which is more than the short-selling buyback cost.

[2] $c_S - p_S = 1$ and $c_{SS} - p_{SS} = 0$ from the put-call parity. European put delta is $p_S = c_S - 1 = N(d_1) - 1 = -[1 - N(d_1)] = -N(-d_1) < 0$. Delta of 5 ATM puts is $-5N\left(-\frac{r}{\sigma\sqrt{T}} - \frac{1}{2}\sigma\sqrt{T}\right)$.

[3] $\triangle_1^U = \frac{C_{uu} - C_{ud}}{uS_0 e^{\delta h}(u-d)}$ and $B_1^U = \frac{uC_{ud} - dC_{uu}}{R(u-d)}$. Initial portfolio cost $C_0 = \triangle_0 S_0 + B_0 = \frac{p^2 C_{uu} + 2p(1-p)C_{ud} + (1-p)^2 C_{dd}}{R^2}$.

[4] $\frac{f^{uu} - f^{ud}}{S^{uu} - S^{ud}}$.

[5] Let X be the number of shares, Y be the number of units of Call 2 and Z be the number of units of Call 3 in the delta-gamma-vega self-financing neutral hedge. Then

$$0.7 + X + 0.5Y + 0.6Z = 0$$

$$0.065 + 0.047Y + 0.053Z = 0$$

$$5.3 + 6.5Y + 4.8Z = 0$$

$$C_1 + 10X + YC_2 + ZC_3 + B = 0.$$

No, it is not necessary that the delta-gamma-vega hedge will perform better than the delta-gamma hedge. It will perform better generally when there is a large movement in the stock's volatility.

[6]

$$dM_t = \begin{cases} 1 - \lambda dt & \text{with probability } \lambda dt \\ -\lambda dt & \text{with probability } 1 - \lambda dt \end{cases}$$

$$E(M_t) = 0.$$

[7] Form the following portfolio (one can also short this portfolio for similar results): A short position in an American put, a long position in an American call, a short position in a stock, and a long position in a risk-free bond of value K (lending). The contingent payoffs of the portfolio are shown below. Either the put could be exercised early at at $t <$

T, or if not, there are two possibilities at maturity, either $S_T \leq K$ or $S_T > K$. The $ amounts represent gains (losses) to the portfolio holder.

Portfolio cost	Early exercise at t	$S_T \leq K$	$S_T > K$
$-P_0$	$-(K - S_t)$	$-(K - S_T)$	0
C_0	$C_t > 0$	0	$S_T - K$
$-S_0$	$-S_t$	$-S_T$	$-S_T$
K	Ke^{rt}	Ke^{rT}	Ke^{rT}
$-P_0 + C_0 - S_0 + K$	+	+	+

If there is an early exercise of the put by the buyer, then the portfolio holder should hold on to the call which still has positive value $C_t > 0$, but close the stock short position and also receive back the loan. It is seen that under all contingencies, the payoffs on the portfolio is always strictly positive. Thus, to prevent arbitrage, the portfolio cost at the start must be strictly positive. Hence, $-P_0 + C_0 - S_0 + K > 0$, or $C - P > S - K$.

Now, form the second portfolio (one can also short this portfolio for similar results): A short position in an American call, a long position in an American put, a long position in a stock, and a short position in a risk-free bond of value $Ke^{-r\tau}$ (borrowing). The contingent payoffs of the portfolio are shown below. Either the call could be exercised early at $t < T$, or if not, there are two possibilities at maturity, either $S_T \leq K$ or $S_T > K$. The $ amounts represent gains (losses) to the portfolio holder.

Portfolio cost	Early exercise at t	$S_T \leq K$	$S_T > K$
$-C_0$	$-(S_t - K)$	0	$-(S_T - K)$
P_0	$P_t > 0$	$K - S_T$	0
S_0	S_t	S_T	S_T
$-Ke^{-rT}$	$-Ke^{-r(T-t)}$	$-K$	$-K$
$-C_0 + P_0 + S_0$ $-Ke^{-r\tau}$	+	0	0

If there is an early exercise of the call by the buyer, then the portfolio holder should hold on to the put which still has positive value $P_t > 0$, but sell the stock and also pay back the loan. It is seen that under all contingencies, the payoffs on the portfolio is always positive. Thus, to prevent arbitrage, the portfolio cost at the start must be positive. Hence, $-C_0 + P_0 + S_0 - Ke^{-r\tau} \geq 0$, or $S - Ke^{-r\tau} \geq C - P$.

[8] This is not a perfect hedge and suffers from 3 key transaction costs that may or may not be substantial, but are nevertheless costly. First is the actual stock exchange transaction fees paid for buying and selling of the stock, and the repeated costs if the stock price crosses the strike price line many times. Second, for one crossing to $K+\epsilon$, $\epsilon > 0$, and a purchase, there is a slippage cost of ϵ. For two crossings, buy at $K+\epsilon$ and then sell at $K-\epsilon$, there is a slippage cost of 2ϵ. In general, for $2n$ crossings, there is a slippage cost of $2n\epsilon > 0$. The slippage occurs because it is not possible to buy or sell at exactly K since the signal to buy has to occur only when price exceeds K, and the signal to sell has to occur only when price is below K. Third, there is interest compounding cost. For example, if the last purchase is at $t < T$, where T is the maturity time, and the continuously compounded risk-free rate is r, then the interest cost is $\geq (K+\epsilon)e^{r(T-t)} - 1$. The last term excludes interest cost on the earlier slippages. Only in the case where the stock price starts below K and remains below K till T, then there is zero transaction costs. When price exceeds K and the call ends in the money at $S_T > K$, the gain from this stop-loss strategy is presumably $S_T - K$ less all the transaction costs mentioned. This would cancel out hopefully most of the loss in the otherwise short call position. However, there is a fourth cost, which is the trading risk that the market price jumps up so fast at $t \leq T$ that a placed limit order to buy at $K+\epsilon$ was not effectuated, being superceded by higher-price buy orders at that time. In such a case, if the stock price remains above K, and the eventual scramble to buy occurs at price $M < S_T$, then the stop-loss strategy only gains $M - K < S_T - K$, and there is a further loss of $M - K$ in addition to all the 3 transaction costs.

Answers to Problem Set 11

[1] Male: $Q(50) = 8,966,618/9,022,649 = 99.379\%$.
Female: $Q(50) = 9,219,130/9,262,013 = 99.537\%$.
Survival for 49 years, Male: $1 - P(50,49) = 10,757/9,022,649 = 0.119\%$.
Female: $1 - P(50,49) = 30,698/9,262,013 = 0.331\%$.

[2] $Q(52,13) = 7329740/8841435 = 0.829022$.
Price $= \$100,000 \times 0.829022/(1.03)^{13} = \$56,452$.

[3] $P\sum_{n=1}^{5}(1/1.03)^n Q(56,n) = 20,000\sum_{t=1}^{5}(1/1.03)^{5+t}Q(61,t)$. Under a non-flat interest rate term structure, we use $P\sum_{n=1}^{5}(1/[1+0.03+0.001n])^n Q(56,n) = 20,000\sum_{t=1}^{5}(1/[1+0.035+0.001t])^{5+t}Q(61,t)$.

[4] Yes, an MC as it is a Markov process with countable state space.

	$-N+1$	$-N+2$	$-N+3$	$-N+4$	$-N+5$	$\cdots$	$N-2$	$N-1$	N
$-N$	2/3	0	0	0	0	$\cdots$	0	0	0
$-N+1$	0	2/3	0	0	0	$\cdots$	0	0	0
$-N+2$	1/3	0	2/3	0	0	$\cdots$	0	0	0
$-N+3$	0	1/3	0	2/3	0	$\cdots$	0	0	0
$\vdots$	$\vdots$	$\vdots$	$\vdots$	$\vdots$	$\ddots$	$\vdots$	$\vdots$		
$N-2$	0	0	0	0	0	$\cdots$	0	2/3	0
$N-1$	0	0	0	0	0	$\cdots$	1/3	0	2/3
N	0	0	0	0	0	$\cdots$	0	1/3	0

[5] All row and column variables are multiples of S_0.

	u^{-N}	u^{-N+1}	u^{-N+2}	u^{-N+3}	u^{-N+4}	$\cdots$	u^{N-2}	u^{N-1}	u^N
u^{-N}	1	0	0	0	0	$\cdots$	0	0	0
u^{-N+1}	1-p	0	p	0	0	$\cdots$	0	0	0
u^{-N+2}	0	1-p	0	p	0	$\cdots$	0	0	0
u^{-N+3}	0	0	1-p	0	p	$\cdots$	0	0	0
$\vdots$	$\vdots$	$\vdots$	$\vdots$	$\vdots$	$\vdots$	$\ddots$	$\vdots$	$\vdots$	$\vdots$
u^{N-2}	0	0	0	0	0	$\cdots$	0	p	0
u^{N-1}	0	0	0	0	0	$\cdots$	1-p	0	p
u^N	0	0	0	0	0	$\cdots$	0	0	1

[6] Subset $\{D\}$. No, not an irreducible MC. MC is transient, not recurrent, because D is an absorbtion state.

[7] Let the invariant distribution be $(p \quad 1-p)^T$. Then

$$(p \quad 1-p)\begin{pmatrix} 1-a & a \\ b & 1-b \end{pmatrix} = (p \quad 1-p)$$

or the two equations to be solved are: $p = b + p(1-a-b) \to p = \frac{b}{a+b}$, and $1-p = a + (1-p)(1-b-a) \to 1-p = \frac{b}{a+b}$. So, the invariant distribution is $\left(\frac{b}{a+b}, \frac{a}{a+b}\right)^T$.

[8]

$$\begin{aligned} P(T_y = 2|X_0 = z) &= \sum_{k \neq y} P(y|k)P(k|z) \\ &= P^2(y|z) - P(y|y)P(y|z) \\ &< P^2(y|z) \end{aligned}$$

[9] For each rating z, first find the T-year credit migration MC equal to $(M)^T$ where M is the one-year transition matrix. (We have to assume time-homogeneity if we are only given the 1-year credit migration transition matrix.) The z-row last column of $(M)^T$ gives the probability of default starting at z after T years, which is $1 - \Pi_z(0,T)$, where $\Pi_z(0,T)$ is the probability of survival up to T years. Then, apply Eq. (11.8).
A 2-year treasury or risk-free interest yield is 2% p.a. If $\Pi_A(0,2)$ is given as 0.95, and $\delta_A = 0.1$, then from Eq. (11.8), the credit risk premium is $C_A(2) = \frac{1}{2}\ln\left[\frac{1}{0.1+0.9\times0.95}\right] = 2.3\%$. The 2-year A-rated discount bond thus yields $2\% + 2.3\% = 4.3\%$. Hence, the price of the 2-year A-rated discount bond is $\$100/(1+4.3\%)^2 = \$91.92$ or you may use $\$100e^{-2(0.043)} = \$91.76$. (In practice, the market uses the first formula, while continuous compounding is usually mathematically more tractable in theoretical literature.)

[10] Unconditional probability of default at time $n+1$ is conditional probability $\times$ probability of survival up to n, or $\Pi_z(0,n) - \Pi_z(0,n+1)$. Unconditional probability of default up to time $n+1$ is $1 - \Pi_z(0,n+1)$.

Answers to Problem Set 12

[1] Let $y_n = \left(1+\frac{r}{n}\right)^{nt} \cdot \ln y_n = \frac{t\ln(1+r/n)}{1/n}$. Differentiating the numerator and denominator w.r.t. n

$$\begin{aligned}\lim_{n\to\infty}\ln y_n &= t\left[\lim_{n\to\infty}\frac{(1/[1+r/n])(-r/n^2)}{-1/n^2}\right]\\ &= t\left[\lim_{n\to\infty} r(1/[1+r/n])\right]\\ &= rt\end{aligned}$$

Hence, $\lim_{n\to\infty} y_n = e^{rt}$.

[2] From definition, LHS is

$$\frac{B(t,s)}{B(t,u)} \times \frac{B(t,u)}{B(t,w)} = \frac{B(t,s)}{B(t,w)} \equiv \text{RHS}$$

[3] (a) $B^0(0,1) = 0.9804, B^0(0,2) = 0.9630, B^0(0,3) = 0.9460$. (b) On p.a. basis: $r(0,1) = 2.00\%$, $r(0,2) = 1.90\%$, $r(0,3) = 1.87\%$. (c) $F(0,1,2) = B(0,1)/B(0,2) = 0.9804/0.963 = 1.0181$, and $F(0,2,3) = B(0,2)/B(0,3) = 0.963/0.946 = 1.0180$.

[4] Payoffs at $t = 2$ are as follows. State uu : $0.9843 - 0.98 = 0.0043$; ud : $0.9804 - 0.98 = 0.0004$; du : $0.9814 - 0.98 = 0.0014$; dd : 0. $C_1^u = (0.75[0.0043] + 0.25[0.0004])/1.017 = 0.00327$. $C_1^d = (0.75[0.0014] + 0.25[0])/1.022 = 0.00103$. Hence, $C_0 = (0.75[0.00327] + 0.25[0.00103])/1.02 = 0.00266$. He or she gains in the call option when spot interest rate falls and the call gets into the money. Thus, he or she is reducing risk exposure to falling interest rate.

[5] The zeros at $t = 0$ are $B^0(0,1) = 0.9804, B^0(0,2) = 0.963, B^0(0,3) = 0.946$. Hence, PV of coupon cash-inflow at $t = 1$ is $B^0(0,1) \times \$5 = \$4.902$. PV of coupon cash-inflow at $t = 2$ is $B^0(0,2) \times \$5 = \$4.815$. PV of coupon cash-inflow cum principal redemption at $t = 3$ is $B^0(0,3) \times \$105 = \$99.33$. Hence, price of coupon bond at $t = 0$ is

$$\$99.33 + 4.815 + 4.902 = \$109.05.$$

[6] Using Eq. (12.20), treating r_t as a RV, then

$$\begin{aligned}\text{cov}(r_s, r_v) &= \text{cov}(r_t, r_t)e^{-a(v+s)+2at} \\ &\quad + \sigma^2 e^{-a(v+s)} \int_t^{v \wedge s} e^{2au} du \\ &= e^{-a(v+s-2t)}\text{var}(r_t) + \frac{\sigma^2}{2a} e^{-a(v+s)} \left[e^{2a(v \wedge s)} - e^{2at}\right].\end{aligned}$$

[7] $\frac{dC}{C} = \frac{[C_t + C_X \mu X + \frac{1}{2} C_{XX} \sigma^2 X^2]}{C} dt + \frac{C_X \sigma X}{C} dW^P$ under the physical P-measure where $C_t, C_X, C_{XX}, \mu, \sigma$ are possibly functions of $X(t)$ and t, and thus adapted to $\mathcal{F}_t$. Then, in a similar way that we transform $\frac{d(X/M)}{(X/M)} = (\mu(t,X) - r)dt + \sigma(t,X)dW^P$ to $\frac{d(X/M)}{(X/M)} = \sigma(t,X)dW^Q$ via Girsanov theorem, $C(X,t)/M$ can become a martingale. Then, using the fundamental asset pricing theorem, its current price given terminal boundary conditions, can be found by taking expectations in the Q-measure.

[8] The value of the payer swap is

$$G \times \left\{ [P(0,0.5) - P(0,2.5)] - 0.5K \left[\sum_{j=1}^{4} B(0, 0.5j + 0.5) \right] \right\}.$$

INDEX